Design Center
Stuttgart

des Landesgewerbeamts
Baden-Württemberg

FOCUS
BALANCE

Design Center Stuttgart
des Landesgewerbeamts Baden-Württemberg

INTERNATIONALER DESIGNPREIS
BADEN-WÜRTTEMBERG 2003
UND MIA SEEGER PREIS 2003

BADEN-WÜRTTEMBERG
INTERNATIONAL DESIGN AWARD 2003
AND MIA SEEGER PRIZE 2003

FOCUS BALANCE

ARNOLDSCHE

FOCUS

INHALT

CONTENTS

DR. WALTER DÖRING
Wirtschaftsminister und Stellvertretender Ministerpräsident Baden-Württembergs
Baden-Württemberg Minister of Economic Affairs and Technology, and Deputy Minister President

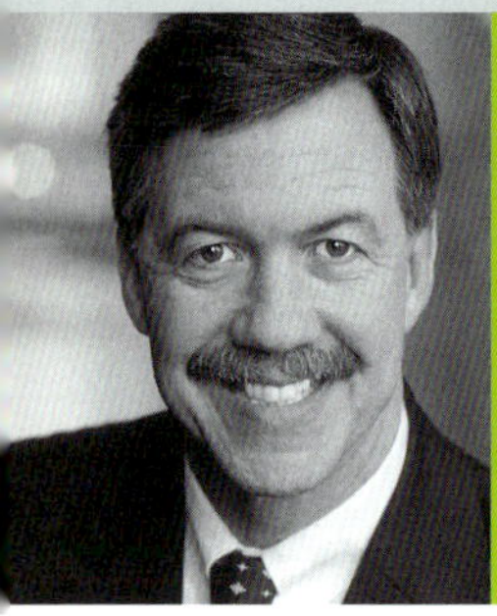

BALANCE _ WIRTSCHAFT . ECONOMY

Work-Life-Balance

Neben steigenden fachlichen Qualifikationen erwarten Unternehmen heute von ihren Mitarbeitern auch sogenannte weiche Faktoren wie Fähigkeit zur Eigenverantwortung, zum selbstständigen Handeln und zum unternehmerischen Denken. Gruppenarbeit, flexible Arbeitszeiten, „Vertrauensarbeitszeit" oder Telearbeit sind Ausdruck der neuen Arbeitswelt. Damit einher gehen Risiken wie Selbstüberforderung bis zur Selbstausbeutung der Mitarbeiter. Folgen davon können Burn-out-Syndrom und körperliche und psychische Probleme sein. Ursache ist die fehlende „Work-Life-Balance". Sie ist Basis für ein erfülltes Leben, das eine befriedigende berufliche Karriere umfasst.

Bürgerschaftliches Engagement

Immer mehr Unternehmer fühlen sich als „Bürger" in ihrem gesellschaftlichen Umfeld und engagieren sich selbst in sozialen und kulturellen Projekten: „Corporate Citizenship". Zugleich ermutigen und fördern sie bürgerschaftliches Engagement ihrer Mitarbeiter. Ich begrüße es, dass immer mehr Unternehmen diesem Leitbild folgen. Dies hat wiederum positive Auswirkungen auf die Motivation und das Leistungsvermögen der Mitarbeiter, die Erfahrungen vielfältiger Art aus ihrem bürgerschaftlichen Engagement gewinnen, wenn sie auf diesem Wege ihre „Work-Life-Balance" mit Leben erfüllen.

Work-Life Balance

Apart from ever better professional qualifications, today's companies also expect their employees to display so-called soft factors such as the ability to assume responsibility, to act independently, and to think entrepreneurially. Group work, flexible working hours, "trust time" or teleworking are the expression of this new world of work. This involves risks for employees, from demanding too much of themselves to exploiting themselves. The results of this may be burn-out, and physical or mental problems. The reason for this is the lack of a "work-life balance". This is the basis for a fulfilled life, which includes a satisfactory professional career.

Citizenship

More and more businesspeople are seeing themselves as "citizens" in their social environment, becoming involved in social and cultural projects: what is known as "corporate citizenship". At the same time, they encourage and promote good citizenship on the part of their employees. For me, the fact that more and more companies are following this example is something to be welcomed. In its turn, this has positive effects on the motivation and performance of employees, who gather a wealth of experience from their social involvement if they use it as a way of filling their "work-life balance" with life.

DR. FRIEDRICH BULLINGER
Präsident des Landesgewerbeamts Baden-Württemberg
President of the Baden-Württemberg Office for the Promotion of Trade and Industry

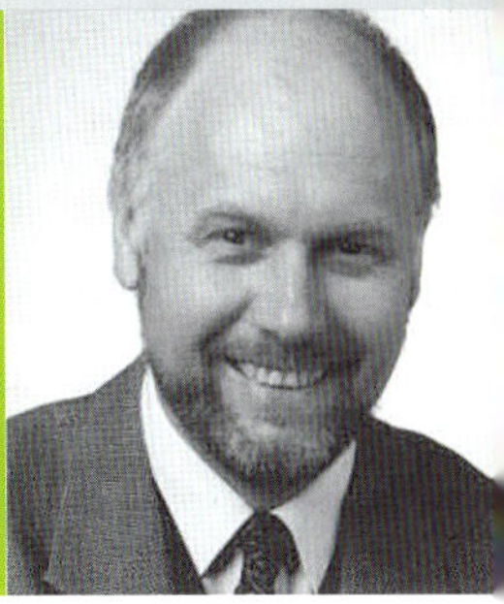

BALANCE _ WETTBEWERB . COMPETITION

Thema und Markt

Psychisches und physisches Gleichgewicht zu erlangen und zu bewahren rückt zunehmend in den Brennpunkt des öffentlichen Interesses. Viele Wirtschaftszweige spüren das und profitieren schon heute davon: beispielsweise die Gesundheitsbranche, der Freizeitmarkt oder der Tourismus. Parallel zu dieser Entwicklung steigt die Nachfrage nach gut gestalteten Produkten, die zur Balance beitragen – hier ist der Markt gefordert! Und schließlich harmoniert „Balance" hervorragend mit dem aktuellen Jahresthema des Landesgewerbeamts Baden-Württemberg „Von Kopf bis Fuß: Gesundheit als Wirtschaftsfaktor".

Wettbewerb

Mit „Focus Balance" haben wir einen sehr aktuellen Bezug zu einem gesellschafts- und wirtschaftsrelevanten Thema herstellen können. Außerdem liegt der einzigartige Reiz des Wettbewerbs darin, Trends aufzugreifen, in diesem Sinne besonders gut gestaltete Produkte zu prämieren und der Öffentlichkeit zu zeigen. Da aber bislang das physische und psychische Wohlbefinden, das Produkte bewirken können, bei der Gestaltung eine eher untergeordnete Rolle einnahm, können wir mit dem diesjährigen Wettbewerb einen Meilenstein setzen. Außerdem ist die nicht-kommerzielle Ausrichtung ideal für kleine und mittlere Unternehmen, und so können wir auch Produktinnovationen präsentieren, die es woanders nicht zu sehen gibt.

Topic and market

Attaining and preserving mental and physical balance is a subject that is increasingly at the centre of public interest. Many branches of industry have noticed this and are already profiting from it today: for example the healthcare industry, the leisure market, tourism. Running parallel to this development, there is an increasing demand for well designed products that contribute to balance – this is the challenge for the market. And finally, "Balance" harmonizes excellently with the current annual theme of the Baden-Württemberg Office for the Promotion of Trade and Industry – "From head to toe: health as an economic factor".

The competition

With "Focus Balance", we have created a very topical link to a socially and economically relevant subject. Moreover, the unique attraction of a competition such as this is to pinpoint trends, to single out especially well designed products that belong to those trends, and to present them to the public. In doing so, we are creating a milestone, because the personal well-being that products can bring about has so far only occupied a fairly subordinate role in design. Moreover, the competition's non-commercial character is ideal for small and medium-sized companies, and in this way we are also able to present innovative products that cannot be seen elsewhere.

HENNING HORN
Leiter des Design Center Stuttgart des Landesgewerbeamts Baden-Württemberg
Head of the Design Center Stuttgart at the Baden-Württemberg
Office for the Promotion of Trade and Industry

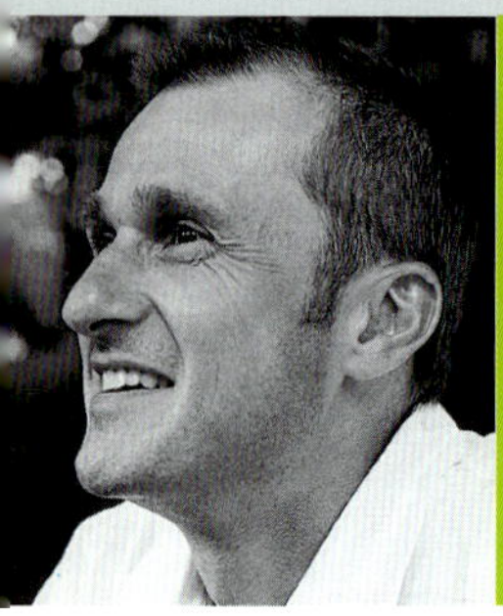

BALANCE _ DESIGN

Partnerschaft

Der Wettbewerb hat einen großen Schritt nach vorn gemacht: Mit dem Magazin FOCUS haben wir einen wichtigen Partner und Multiplikator gewonnen. Künftig fokussieren wir gemeinsam gesellschaftlich relevante Themen aus dem Blickwinkel des Design. Damit sind wir noch näher am Puls der Zeit. Und immer mehr Menschen werden erkennen, wie professionelle Gestaltung ihren Alltag bereichern und erleichtern kann in allen Lebenslagen, beruflich und privat. Auch mit dem neuen Partner bleiben wir unserem hohen Anspruch treu: Der Internationale Designpreis Baden-Württemberg ist fair und seriös. Mit dem „Focus Balance" verfolgt das Design Center Stuttgart kein eigenes wirtschaftliches Interesse. Produkte, an deren Entwicklung unsere Juroren beteiligt waren, sind vom Wettbewerb ausgeschlossen. So bleiben wir unabhängig – und so soll es auch bleiben.

Design für „Balance"

„Balance", das bedeutet Produkte zu entwerfen, die nicht nur innovativ sind in ihrer Funktionalität. „Balance" heißt auch, mit Handicaps umzugehen und Dinge zu gestalten für Menschen, die mit Einschränkungen leben müssen. Mit dem „Focus Balance" sind erstmals im Wettbewerb Produkte gefragt, die auf Geist und Seele wirken, die Stimmungen berücksichtigen, Atmosphäre schaffen und die vor allen Dingen Wohlbefinden fördern. Wer hier Gefühlsduselei oder gar Esoterisches erwartet, irrt gewaltig. LED-beleuchtete Objekte, deren farbiges Licht emotionalen Einfluss ausübt, oder wegweisende Sitzgelegenheiten: Neben Produkten für den Freizeit- und Wellness-Bereich spielen Hightechlösungen am Arbeitsplatz eine wichtige Rolle.

BALANCE _ DESIGN

Partnership

The competition has taken a great step forward: in the magazine FOCUS, we have gained an important partner and multiplier. In future, we will be focussing together on socially relevant topics from a design perspective. Our finger will be even closer to the pulse of things. And more and more people will recognize how professional design can enrich and simplify their everyday lives, be it at work or in the home. Even with our new partner, we will remain true to the high standards we have set ourselves: the Baden-Württemberg International Design Award is fair and respectable. In organizing "Focus Balance", the Design Center Stuttgart is not pursuing any economic interests of its own. Products in whose development our judges were involved are excluded from competing. In this way, we remain independent – and that is how it should stay.

Design for "balance"

"Balance" means designing products that are not just innovative in the way they work. "Balance" also means coping with handicaps and designing objects for people who have to live with constraints. For the first time in the history of the competition, "Focus Balance" calls for products that appeal to the intellect and the soul, that consider moods, create atmosphere and, above all, promote well-being. But anyone expecting to find mawkishness or even esoteric products here is barking up the wrong tree. Objects illuminated with LEDs whose coloured light influences the emotions, or pioneering seating design: apart from products for the leisure and wellness area, high-tech solutions at the workplace play an important role.

UNIV.-PROF. DR. HORST W. OPASCHOWSKI
Der Autor lehrt an der Universität Hamburg und leitet das BAT Freizeit-Forschungs-Institut.
Er ist Verfasser des Buches „Was uns zusammenhält. Krise und Zukunft der westlichen Wertewelt".

BALANCE _ LEBEN IM GLEICHGEWICHT

Ich tue, was ich will.

Anfang der neunziger Jahre interviewten zwei italienische Psychologen italienische Bauern in den hochgelegenen Bergtälern der Alpen, die von der industriellen Revolution weitgehend verschont geblieben sind. In ihren Interviews kam zum Ausdruck, dass die Bauern ihre Arbeit nicht von ihrer Freizeit unterscheiden konnten. Bei den Interviewern entstand ein doppelter Eindruck: Die Bauern arbeiteten sechzehn Stunden am Tag oder sie arbeiteten überhaupt nicht. Sie melkten Kühe, mähten Wiesen, erzählten ihren Enkeln Geschichten, spielten Akkordeon für Freunde. Und auf die Frage, was sie denn gern tun würden, wenn sie mehr Zeit zur Verfügung hätten, kam die Antwort: Kühe melken, Wiesen mähen, Geschichten erzählen, Akkordeon spielen ... Für ihr ganzes Leben galt und gilt eigentlich nur ein Grundsatz: „Ich tue, was ich will" (Massimini u.a. 1991). Das Leben, auch das Arbeitsleben, bot und bietet ständig und gleichermaßen Herausforderungen dafür.

Mehr fließende Übergänge zwischen Berufs- und Privatleben schaffen.

Politik und Wirtschaft sollten sich rechtzeitig auf den sich ankündigenden Wertewandel in Richtung auf eine neue **Gleichgewichtsethik** einstellen und mehr fließende Übergänge zwischen Berufs- und Privatleben schaffen. Insbesondere die junge Generation befindet sich derzeit auf dem Wege zu einer neuen **Lebensbalance**. Leistung und Lebensgenuss sind für sie keine Gegensätze mehr. Ganz anders, als es in den siebziger bis neunziger Jahren befürchtet und diagnostiziert worden war, hat sich die Einstellung der jungen Generation zu Arbeit und Leistung entwickelt: Die befürchtete Leistungsverweigerung fand und findet nicht statt. Im Zeitvergleich der letzten Jahre ist beispielsweise erkennbar, dass **Leistung und Lebensgenuss immer gleichgewichtiger** beurteilt werden. Also: kein Lebensgenuss ohne Leistung. Umgekehrt gilt aber auch: Lebensgenuss lenkt nicht mehr automatisch von Leistung ab. Und wer sein Leben nicht genießen kann, wird auf Dauer auch nicht leistungsfähig sein.

Arbeit ohne Lust und Freizeit ohne Leistung kann der Mensch auf Dauer nicht ertragen.

Der Mensch kann nicht untätig in seinen eigenen vier Wänden verweilen. Er braucht eine Aufgabe. Die Passivität und Untätigkeit des Menschen ist offensichtlich nicht im Plan der Schöpfung vorgesehen. Der Mensch ist eher als gefährdetes Wesen geschaffen, das um sein Überleben kämpfen muss wie andere Lebewesen auch. Aus der Sicht der Evolutionsbiologie ist der Mensch geradezu auf Anstrengung programmiert (von Cube 1997), auf den ganzen Einsatz seiner Kräfte. Hingegen führt Lust ohne Anstrengung zu Langeweile oder gar Selbstzerstörung. Daraus folgt: Arbeit ohne Lust und Freizeit ohne Leistung kann der Mensch auf Dauer nicht ertragen.

Veränderung der Lebensprioritäten: Die Interessen richten sich auf das ganze Leben.

Die demographische Entwicklung wird in den nächsten Jahrzehnten zu einer Veränderung der Lebensprioritäten führen. Die Interessen richten sich dann auf das ganze Leben: Die Aufspaltung in die Teilbereiche „Arbeit" und „Freizeit" wird weitgehend entbehrlich. Dieser Wandel geht mit einem **Bedeutungszuwachs von Gesundheit, Natur und Religion** einher. Diese drei Bereiche erweisen sich als konstante Lebensqualitäten bis ins hohe Alter.

Der Trend nach dem Trend in der Zeit nach 2010 wird wohl der **Sinnorientierung** gehören. **Von der Flucht in die Sinne zur Suche nach dem Sinn:** Diese Bewegung wird aber nur bedingt kommerzialisierbar und konsumierbar sein. Die Sinnorientierung wird zur wichtigsten Ressource der Zukunft und zu einer großen Herausforderung der Wirtschaft in Wohlstandsländern werden. Mit jedem neuen Konsumangebot muss zugleich die Sinnfrage „Wofür das alles?" beantwortet werden.

Die Sinnorientierung wird zur wichtigsten Ressource der Zukunft und zu einer großen Herausforderung der Wirtschaft werden.

Zukunftsmärkte werden immer auch **Sinnmärkte** sein – bezogen auf Gesundheit und Natur, Kultur und Religion. Letztlich geht es um Authentizität und Lebensqualität. **Wertebotschaften** statt Werbebotschaften heißt dann die Forderung der Verbraucher, die sich auch als eine **Generation von Sinnsuchern** verstehen. Von Konsumverzicht will diese wenig wissen, dafür umso mehr von der Werthaltigkeit des Konsums: Das Selbst„wert"gefühl soll gestärkt werden, ohne dabei den Sozial„wert" mit Prestigecharakter aus den Augen zu verlieren.

Das Gleichgewicht von materiellem Wohlstand und sozialem Wohlbefinden wiederfinden.

Die Menschen wollen das Gleichgewicht von materiellem Wohlstand und sozialem Wohlbefinden wiederfinden. Gesucht wird eine **Balance von Leistungs-, Genuss- und Sozialorientierung.** Die Menschen legen besonderen Wert darauf, im Leben etwas zu leisten und das Leben zu genießen. Bemerkenswert hoch ist aber auch der Wunsch in der Bevölkerung ausgeprägt, anderen helfen zu wollen.

Die Bevölkerung erwartet von der Zukunft nicht das ganz große Glück. Es sind eher die kleinen Glücksmomente des Lebens in einer entspannten und störungsfreien Atmosphäre: Stimmung, Harmonie, Geborgenheit. Und so altmodisch es auf den ersten Blick erscheinen mag: Auf die Frage, was „im neuen Jahrtausend wichtig und wertvoll" ist, nennen die Bürger im 21. Jahrhundert „soziale Verantwortung" (51%) genauso wie „Freiheit" (50%). Beides sind in Zukunft **Garanten** dafür, dass man unbeschwert leben und sich freuen kann. Wenn diese Balance als Lebensprinzip für den Einzelnen und als Basis für das soziale Zusammenleben gelingt, dann kann das 21. Jahrhundert eine Ära der Verantwortung und eine Ära der Freiheit sein.

Wer soziales Wohlbefinden (und nicht nur materiellen Wohlstand) erreichen will, sollte – neben den biblischen Geboten natürlich – **die folgenden zehn Gebote für ein ausbalanciertes Leben** im 21. Jahrhundert beherzigen:

1 Bleibe nicht dauernd dran; schalte doch mal ab.

2 Jage nicht ständig schnelllebigen Trends hinterher.

3 Kaufe nur das, was du wirklich willst, und mache dein persönliches Wohlergehen zum wichtigsten Kaufkriterium.

4 Versuche nicht, permanent deinen Lebensstandard zu verbessern oder ihn gar mit Lebensqualität zu verwechseln.

5 Entdecke die Hängematte wieder. Lerne wieder, „eine Sache zu einer Zeit“ zu tun. Lieber einmal etwas verpassen als immer dabei sein.

6 Genieße nach Maß, damit du länger genießen kannst.

7 Mache nicht alle deine Träume wahr; heb dir noch unerfüllte Wünsche auf.

8 Du allein kannst es – aber du kannst es nicht allein; hilf anderen, damit auch dir geholfen wird.

9 Tu nichts auf Kosten anderer oder zu Lasten nachwachsender Generationen; sorge nachhaltig dafür, dass das Leben kommender Generationen lebenswert bleibt.

10 Verdiene dir deine Lebensqualität – durch Arbeit oder gute Werke: Es gibt nichts Gutes, außer man tut es.

GRUNDLAGENLITERATUR

Cube, F. von:
Fordern statt Verwöhnen,
9. Auflage, München 1997

Massimini, F. u. a.:
Flow und biokulturelle Evolution.
In: M. u. I. Csikszentmihalyi (Hrsg.):
Die außergewöhnliche Erfahrung im Alltag,
Stuttgart 1991, S. 77–102

Opaschowski, H. W.:
Deutschland 2010.
Wie wir morgen arbeiten und leben.
Voraussagen der Wissenschaft
zur Zukunft unserer Gesellschaft,
2. neubearbeitete Auflage, Hamburg 2001

PROF. DR. HORST W. OPASCHOWSKI
The author lectures at the University of Hamburg and is head of the BAT Freizeit-Forschungs-Institut [Leisure Research Institute]. He is the author of "Was uns zusammenhält. Krise und Zukunft der westlichen Wertewelt" [What keeps us together. Crisis and future of the Western world of values].

BALANCE – LIFE ON AN EVEN KEEL

At the beginning of the 1990s, two Italian psychologists interviewed Italian farmers in the high valleys of the Alps who had remained largely untouched by the Industrial Revolution. In their interviews, it became apparent that the farmers were unable to distinguish their work from their leisure time. For the interviewers themselves, the impression that arose was a dual one: the farmers worked sixteen hours a day, or they did not work at all. They milked cows, mowed meadows, told their grandchildren stories, played the accordion for friends. And when asked what they would like to do if they had more time at their disposal, their answer was: milk cows, mow meadows, tell stories, play the accordion ... In fact, only one principle has only ever been and is still valid for their whole life: "I do what I like" (Massimini et al. 1991). Life, and that includes working life, has always offered and continues to offer challenges to this.

I do what I like.

Politics and business life should adjust in good time to the imminent change in values towards a new **ethics of balance** and create more transitions between working life and private life that are characterized by flux. The young generation in particular is currently moving towards a **new life-work balance.** For them, working hard and enjoying life are no longer polar opposites. The young generation's attitude to work and achievement has developed quite differently than was feared and diagnosed in the period between the 1970s and 1990s: the dreaded rejection of hard work has not happened and is not happening. Looking back over recent years, for example, it is clear that **working hard and enjoying life** are being regarded as **ever more equal in weight.** In other words, you can't enjoy life without working hard. But the converse is also true. The enjoyment of life no longer automatically distracts people from work. And someone who is incapable of enjoying life will in the long run be incapable of working hard.

Create more transitions between working life and private life that are characterized by flux.

People cannot sit around idly inside their own four walls. They need something to do. Clearly, it is not part of creation's plan for people to be passive and idle. Rather, man was created as an endangered creature that has to fight for its survival just as other creatures do. From the point of view of evolutionary biology, man is virtually programmed to exert himself (von Cube 1997), and to do so with all the energy at his disposal. By contrast, pleasure without exertion leads to boredom or indeed self-destruction. The bottom line, in the long run, is that people cannot bear work without pleasure and leisure without hard work.

People cannot bear work without pleasure and leisure without hard work.

A change in existential priorities:
Interest will focus on the whole of a person's life.

In the next few decades, demographic developments will lead to a change in existential priorities. Interest will then focus on the whole of a person's life: we will be more or less able to dispense with a division into the sub-categories of "work" and "leisure". This change is going hand in hand with an **increase in the significance of health, the environment and religion.** These three areas are proving to be constant qualities of life right into old age.

In the time after 2010, the trend to follow the trend will probably be one of **meaning focus. From the flight into sensuality to the search for meaning:** however, this movement will only partly be consumable and open to commercialization. Meaning focus will become the most important resource of the future and a great challenge for the economy of affluent countries. Every product offered for consumption will also have to answer the question as to its meaning: "What is the point of all this?"

Meaning focus will become the most important resource of the future and a great challenge for the economy.

The markets of the future will always also be **markets of meaning** – in relation to health and the environment, culture and religion. In the end, the issue is one of authenticity and quality of life. **Value messages** instead of advertising messages will then be what consumers demand, and they will also see themselves as a **generation in search of meaning.** They will be less interested in doing without consumption, but all the more interested in consuming objects with inherent value: they will be interested in strengthening their feeling of personal value, but without losing sight of the social value that goes with prestige.

People want to rediscover the balance between material affluence and social well-being. What they are looking for is a **balanced focus on hard work, enjoyment and the social sphere.** People are especially concerned to achieve something in their lives and to enjoy life. However, there is also a surprisingly pronounced wish in society to help others.

Rediscover the balance between material affluence and social well-being.

People do not expect the future to bring them blissful happiness. Rather, what they hope for are those fleeting moments of happiness in a relaxed, harmonious atmosphere: mood, harmony, security. And old-fashioned as it may seem at first glance, the citizens of the 21st century, when asked what they thought was "important and valuable in the new millennium", state both "social responsibility" (51%) and "liberty" (50%). In the future, both of these are **guarantees** that we can live without care and take pleasure in life. If this balance manages to assert itself as an existential principle for the individual and as a basis for social coexistence, then the 21st century may be an era of responsibility and an era of liberty.

Anyone wishing to achieve social well-being (and not just material affluence) should – apart from the Biblical Ten Commandments, naturally – take to heart the following **ten commandments for a balanced life** in the 21st century:

1. Don't stick at it all the time; switch off now and then.
2. Stop constantly pursuing fleeting trends.
3. Only buy what you really want, and make personal well-being your principal criterion for buying.
4. Stop permanently trying to improve your standard of living, or indeed confusing your standard of living with quality of life.
5. Rediscover the hammock. Relearn how to "do one thing at a time". Better to miss out now and then than always be involved.
6. Enjoy in moderation, so that you can enjoy longer.
7. Don't make all your dreams come true; keep a few unfulfilled dreams on one side.
8. You alone can do it – but you can't do it alone; help others as you would be helped too.
9. Don't do anything at other people's expense or to the disadvantage of future generations; take steps to ensure that the life of future generations remains worth living.
10. Earn your quality of life – by work or good works: nothing good ever came of waiting for something good to happen.

REFERENCES

Cube, F. von:
Fordern statt Verwöhnen,
9th edition, Munich 1997

Massimini, F. et al.:
Flow und biokulturelle Evolution.
In: M. and I. Csikszentmihalyi (eds):
Die außergewöhnliche Erfahrung im Alltag,
Stuttgart 1991, pp. 77–102

Opaschowski, H. W.:
Deutschland 2010.
Wie wir morgen arbeiten und leben.
Voraussagen der Wissenschaft
zur Zukunft unserer Gesellschaft,
2nd revised edition, Hamburg 2001

FOCUS BALANCE _ DER WETTBEWERB

Wenn der Kopf vom Arbeiten am Bildschirm oder nach langen Besprechungen raucht – mal eben ins firmeneigene Fitnessstudio gehen, um dort zu strampeln und zu schwitzen, bis der Kopf wieder frei ist. Oder spezielle Seminare, die Entspannung und Gesundheit etwa von Führungskräften zum Inhalt haben – denn laut der Bundesvereinigung der deutschen Arbeitgeberverbände sollen gerade gesundheitliche Probleme von Führungskräften die Wirtschaft pro Jahr durchschnittlich mit fünf Milliarden Euro belasten. Dies sind Beispiele für Unternehmen, die sich um Fitness, Wohlbefinden und Gesundheit ihrer Mitarbeiter kümmern. Die Folgen sind größere Motivation, effizienteres Arbeiten und ein spürbar sinkender Krankenstand.

Hier verschwimmt die strikte Trennung von Arbeit und Freizeit sowie Privatbereich immer mehr. Gesundheit, Umweltbedingungen, physisches und psychisches Wohlbefinden: Das Gleichgewicht des Menschen steht zunehmend im Zentrum des allgemeinen Interesses. Aber auch die Frage der Vereinbarkeit von beruflicher und privater Lebenswelt ist inzwischen ins Zentrum gesellschaftspolitischer Diskussionen vorgerückt. Dabei gewinnt gerade das Schlagwort „Work-Life-Balance" zunehmend an Bedeutung. Es zählt nicht mehr ein einseitiges Arbeitsethos sondern ein gesundes Gleichgewicht zwischen Arbeit und Privatleben beziehungsweise Freizeit. So werden in immer mehr Unternehmen Mitarbeiter und Chefs bereits nach ihrer „Work-Life-Balance" beurteilt. Will man auf Dauer wettbewerbsfähig bleiben, ist Balance ein wichtiger Faktor.

Die Gesundheitsbranche, der Freizeitbereich, Wellness und Fitness nehmen so eine immer größere Bedeutung ein. Dabei kommt es zu einer steigenden Nachfrage nach guten Produkten in diesen Bereichen. Die Designer sind hier gefragt, Produkte zu entwerfen, die die Balance bei Menschen fördern. Zunehmend tauchen Produkte auf, die Stimmungen ansprechen, Atmosphären schaffen und das Wohlbefinden fördern, zum Beispiel mit LED beleuchtete Produkte, deren farbiges Licht man seinen Stimmungen anpassen kann. Produkte, die nicht nur funktional und innovativ sind, sondern auf Seele und Geist wirken. Ein weiterer Aspekt ist die zunehmende ältere Bevölkerungsgruppe, die einen wachsenden Markt für speziell für sie gestaltete Produkte bildet. Beispiele hierfür sind Produktentwürfe für gesundes und sicheres Wohnen.

So wurde dieses Jahr der Wettbewerb unter das Thema „Focus Balance" gestellt. Der Internationale Designpreis Baden-Württemberg wird jährlich als Staatspreis für zukunftsweisende und herausragende Gestaltungsleistungen an Hersteller und Designer verliehen. Seit geraumer Zeit wird er unter einem

Thema ausgeschrieben, das jährlich wechselt – eine besondere Herausforderung für die Einreicher. In der Vergangenheit waren das beispielsweise „Arbeitswelten", „Mobilität" oder „Lebensart". Gut gestaltete Produkte erhalten den Focus in Silber, die besten davon – pro Kategorie jedoch nur ein Produkt – den Focus in Gold.

Seit 2002 wird auch der Mia Seeger Preis für Nachwuchsdesigner unter dem gleichen thematischen Dach ausgeschrieben. Hiervon profitieren alle Beteiligten. Die Junior-Designer zeigen interessante konzeptionelle Ideen und Visionen, die Profis des Internationalen Designpreises präsentieren Serienprodukte mit Weltklasseniveau – eine gelungene Ergänzung.

Beim Internationalen Designpreis Baden-Württemberg 2003 konnten Firmen und Designer für den Wettbewerb „Focus Balance" industrielle Serienprodukte einreichen, die nicht länger als zwei Jahre auf dem deutschen Markt sind. Zugelassen waren aber auch Prototypen, deren Serienreife gewährleistet war, jedoch keine Unikate. Eine weitere Bedingung war, dass Produkte der Juroren nicht am Wettbewerb teilnehmen konnten; denn beim Internationalen Designpreis Baden-Württemberg wird größter Wert auf Neutralität der Juroren gelegt.

Außerdem sollten die eingereichten Produkte sich in eine der folgenden 12 Kategorien einordnen lassen:

- **Medizin, Rehabilitation, Geriatrie:** Gehhilfen, Bandagen, medizinische Geräte
- **Fitness, Wellness, Sport:** Sportartikel, Saunazubehör
- **Freizeit, Outdoor, Spiel:** Reise-, Camping-, Gartenbedarf, Spielzeug
- **Wohnen:** Einrichtungen und Accessoires
- **Küche, Haushalt:** Kücheneinrichtungen, Haushaltsutensilien, elektrische Küchengeräte
- **Bad, Sanitär:** Badeinrichtungen, Duschköpfe, Fliesen
- **Objekt-, Büro- und Arbeitplatzausstattung:** Einrichtungen und Accessoires
- **Arbeitssicherheit, Produktion:** Schutzeinrichtungen, Anlagen, Werkzeuge
- **Textilien, Bekleidung:** Teppiche, Gardinen, Stoffe, Kleidung
- **Beleuchtung:** Lampen, Leuchten
- **Kommunikation, Information, Foto, Optik:** elektronische Produkte, Media-Produkte, Kameras, Brillen
- **Mobilität:** Individualfahrzeuge, öffentliche Verkehrsmittel

Insgesamt wurden 440 Produkte von 229 Einreichern aus der ganzen Welt angemeldet. Dabei stammen die meisten ausländischen Hersteller aus Italien, der Schweiz, Österreich, Finnland und den USA. Aber auch Länder wie die Niederlande, Schweden, Spanien und Japan waren vertreten.

Die Mehrzahl der ausländischen Designer kam aus Italien, Österreich, der Schweiz, den USA, Schweden und Finnland. Es waren jedoch auch Produkte von Designern aus Frankreich, Großbritannien, den Niederlanden, Dänemark, Indien, Japan und Neuseeland dabei.

Zwei Tage begutachtete und prüfte eine internationale Expertenjury in Stuttgart die 440 Produkte. Dabei ging sie nach folgenden Kriterien vor:

- Gestaltungsqualität
- Innovationsgehalt
- Ergonomie
- Verwendung neuer Werkstoffe
- ökologischer Aspekt
- Gebrauchsqualität
- Benutzerführung, Produktgrafik beziehungsweise -typographie
- gestalterische Qualität der Produktausstattung: Bedienungs-, Montageanleitung, Verkaufsverpackung
- Wertigkeit, Anmutung und emotionaler Gehalt

Aus den eingereichten Produkten wählte die Jury 68 aus, die sie für auszeichnungswürdig hielt: 9 erhielten den Focus in Gold und 59 den Focus in Silber. Alle 68 Produkte sowie die Wettbewerbsergebnisse des Mia Seeger Preises sind im Jahrbuch publiziert und vom 14. September bis zum 7. Dezember 2003 im Haus der Wirtschaft in Stuttgart ausgestellt. Die Produkte, die den Focus in Gold gewonnen haben, werden darüber hinaus im Internet veröffentlicht. Besucher der Ausstellung wählen ihren persönlichen Produktfavoriten und ermitteln damit den Publikumspreis, der am Ende der Ausstellung verliehen wird.

FOCUS BALANCE _ THE COMPETITION

If their heads hurt from working at the computer or after long meetings, employees make their way to the company's own gym to work out until their heads are clear again. Or there are special seminars teaching relaxation techniques and giving health advice to executives, for example. This is not surprising, given that the German Employers' Association estimates that executives' health problems alone cost the economy an average of five billion euro every year. These are examples of companies that care about their employees' fitness, well-being and health. The results are greater motivation, more efficient work and an appreciable drop in the number of days lost due to sickness.

In cases like these, the strict separation of work and leisure/private sphere is becoming increasingly blurred. Health, environmental conditions, physical and mental well-being: more and more, the main focus of general interest is people's balance. Moreover, the question of the compatibility of working and private life has in the meanwhile found its way to the centre of socio-political debate. In this context, increasing importance is being given to the idea of a "work-life balance". A work ethic to the exclusion of everything else is no longer what counts; instead, what is needed is a healthy balance between work and private life or leisure. In more and more companies, employees and bosses are being assessed in terms of their "work-life balance". If remaining competitive in the long term is the aim, balance is an important factor.

As a result, the healthcare industry, the leisure sector, wellness and fitness are becoming ever more important. Demand for good products in these areas is increasing. For designers, this means they have to come up with products that promote people's balance. Increasingly, products are emerging that appeal to moods, create atmosphere and promote well-being; products illuminated by LEDs, for example, whose coloured light can be adjusted to suit current mood. These products are not just functional and innovative, but also appeal to the soul and the intellect. One further aspect is the growing number of elderly people, who represent a burgeoning market for products designed with their specific needs in mind, such as designs for healthy, safe homes.

For this reason, "Focus Balance" was chosen as the theme for this year's competition. The Baden-Württemberg International Design Award is given annually to manufacturers and designers as a government prize for forward-looking and outstanding design work. For some time now, the competition has been centred around a theme which changes from year to year – quite a challenge for entrants. In the past, these themes have included "working environments", "mobility" or "way of life". Well designed products are awarded the Focus in Silver, while the best products – but only one product per category – receive the Focus in Gold.

Since 2002, the Mia Seeger Prize for young design talents has invited entrants to compete under the same theme. All those involved profit from this. The young designers present interesting conceptual ideas and visions, while the professionals entering for the International Design Award present world-class series-manufactured products – and the two sides complement each other perfectly.

For the 2003 Baden-Württemberg International Design Award, companies and designers were allowed to enter industrial series-manufactured products for the "Focus Balance" competition provided the products had not been on the German market for longer than two years. While prototypes whose readiness for series production was guaranteed could also be entered, one-off products could not. One further condition was that the judges' own products were barred from competing; the Baden-Württemberg International Design Award places great importance in its judges' neutrality.

In addition, the products submitted had to fit into one of the following 12 categories:

- **Medicine, rehabilitation, geriatrics:** walking aids, bandages, medical appliances
- **Fitness, wellness, sport:** sports articles, sauna accessories
- **Leisure, outdoor, playtime:** travel, camping, gardening needs, toys
- **Living:** furniture and accessories
- **Kitchen, household:** kitchen furniture, household goods, electrical kitchen appliances
- **Bathrooms, sanitary installations:** bathroom fittings, shower heads, tiles
- **Contract furniture, office and workplace equipment:** furniture and accessories
- **Industrial safety, production:** safety devices, systems, tools
- **Textiles, clothing:** carpets, curtains, fabrics, clothes
- **Lighting:** lamps, luminaires
- **Communication, information, photography, optical equipment:** electronic products, media products, cameras, spectacles
- **Mobility:** private and public transport

229 entrants from all over the world submitted a total of 440 products. Most of the foreign manufacturers were from Italy, Switzerland, Austria, Finland and the USA. However, countries such as the Netherlands, Sweden, Spain and Japan were also represented.

Most of the foreign designers were from Italy, Austria, Switzerland, the USA, Sweden and Finland. But there were also products by designers from France, the UK, the Netherlands, Denmark, India, Japan and New Zealand.

For two days, a panel of international experts met in Stuttgart to assess and examine the 440 products. Their assessment was based on the following criteria:

- design quality
- innovative content
- ergonomics
- use of new materials
- ecological aspect
- quality of use
- user controls, product graphics and typography
- design quality of product peripherals: instructions for use and assembly, sales packaging
- quality, appearance and emotional content

From the products submitted, the judges chose 68 that they felt were worthy of commendation: 9 received the Focus in Gold and 59 the Focus in Silver. All 68 products and the results of the competition for the Mia Seeger Prize are published in the yearbook, and will be on show in the Haus der Wirtschaft in Stuttgart from 14 September to 7 December 2003. The products that have been awarded the Focus in Gold will also be published in the internet. Visitors to the exhibition will choose their personal favourite from the products on show and in this way vote for the winner of the "Popular Award", which is presented when the exhibition closes.

ANDREA SCHOLTZ

BILDER, DIE EINE GESCHICHTE ERZÄHLEN – NEUN PREISTRÄGER DES FOCUS IN GOLD

Volker Schrank, der nicht nur als Fotokünstler sondern auch als Auftragsfotograf arbeitet, hat die neun Produkte fotografiert, die beim Internationalen Designpreis Baden-Württemberg einen Focus in Gold gewonnen haben. Jedes prämierte Objekt stellte er in einen ungewöhnlichen Kontext und setzte jeweils eine Geschichte visuell um. „Ich habe immer versucht, die Produkte in eine Geschichte einzuspannen. Dem Betrachter bleibt es aber überlassen, wie tief die gehen kann“, erzählt Volker Schrank.

Dem Stuttgarter Fotografen, der Visuelle Kommunikation und Foto-Film-Design in Kassel, Dortmund und Irland studiert hat, geht es vor allem darum, mit seinen Fotos Geschichten zu erzählen, die hinter seinen Bildern ablaufen. Er setzt Menschen und Gegenstände in Szene, wie etwa bei seinen Fotoserien „Stars der Wahrheit“, einer preisgekrönten Porträtreihe mit deutschen Nachrichtensprechern, oder „Stars der Unterhaltung“, einer Fotoserie von Radios aus fünf Jahrzehnten.

Was dabei wie zufällig wirkt, ist genau inszeniert. Skurrile Situationen, skurrile Personen oder Randgruppen und skurrile Hintergründe – all das sucht Volker Schrank gezielt aus und baut daraus ein Gesamtbild, wie etwa bei dem Bildmotiv der Bowlingbahn. Dazu sagt er: „Bowling an sich ist für mich persönlich schon skurril genug. Das ist eine Welt, die ich seltsam finde. Ich habe dann ein Model ausgesucht, das dermaßen korpulent ist, dass es im Prinzip schon ein Balanceakt ist, zu bowlen.“ In diesen Kontext hat er das preisgekrönte Produkt „Brille“ gestellt; denn beim Bowlen ist gut peilen und scharf sehen ausschlaggebend.

Die Produkte stehen dabei nicht immer im Vordergrund. Volker Schrank lässt meist den Betrachter das Produkt suchen. Außerdem setzen die meisten Bilder das diesjährige Thema des Internationalen Designpreises Baden-Württemberg „Balance“ um. Dabei ist Volker Schrank wichtig, dass die Interpretation von „Balance“ dem Betrachter überlassen wird: „Ich erzähle einen Teil einer Geschichte, an der jeder selbst weiterspinnen kann.“ Ein anderes Stilelement, das sich durch seine Bilder zieht, sind bewusst konstruierte Räumlichkeiten. Bei einigen Bildern schafft er durch Zentralperspektive und starke Fluchten eine sehr große Tiefe. Bei den anderen dominieren waagrechte Linien und gerade Fronten. Immer komponiert er seine Bilder bewusst mit grafischen Linien und Fluchten und schafft so eine ganz eigene charakteristische Räumlichkeit.

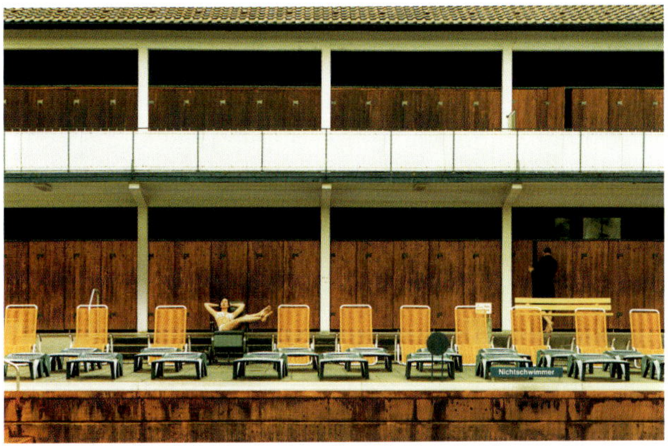

Hier geht meine Balance nicht baden.
I won't lose my balance here.
carryLine: Fahrbare Liege . hospital trolley

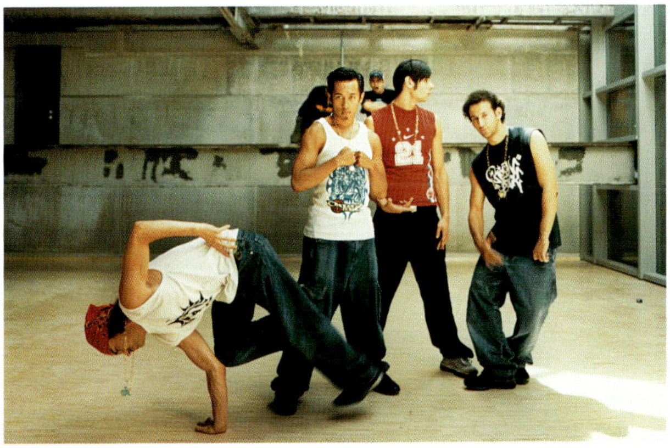

Balanceakt . Balancing act
MAM Preemie: Frühgeborenenschnuller
dummy for premature babies

PICTURES THAT TELL A STORY _ THE NINE FOCUS IN GOLD AWARD WINNERS

Volker Schrank, who does both artistic work of his own and commission work, has photographed the nine products awarded a Focus in Gold in this year's Baden-Württemberg International Design Award. He has placed each award-winning product in an unusual context, telling a story in pictures for each. "I have always tried to spin a story around the products. But it's up to the spectator to decide how far the story goes", says Volker Schrank.

For this Stuttgart-based photographer, who studied visual communication and photographic and film design in Kassel, Dortmund and Ireland, what matters most is to tell the stories behind the pictures he photographs. He stages people and objects, for example in his "Stars of Truth", a prize-winning series of photographic portraits featuring German newscasters, or in his "Stars of Entertainment" series of photographs of radios from five decades.

What appears haphazard at first sight has in fact been precisely staged. Curious situations, oddball people or marginal groups and bizarre backgrounds – Volker Schrank deliberately selects them all, using them to compose a total image, as in the case of the pictorial motif of the bowling alley. He comments: "For me personally, bowling is a strange enough subject as it is. It is a world which I find curious. I then selected a model who is so obese that the act of bowling itself is already a balancing feat". It was into this context that he placed the award-winning spectacles, since taking a good aim and having a good eye are crucial criteria when bowling.

In these pictures, the products are not immediately apparent. Volker Schrank usually makes the spectator search for the product. Moreover, most of the pictures interpret the "balance" theme of this year's Baden-Württemberg International Design Award. What is important for Volker Schrank here is that it is left up to the spectator to interpret "balance": "I tell part of a story, which can be continued at will by the individual spectator". Another style element common to all his pictures is the way the settings have been deliberately constructed. In some pictures, by choosing a central perspective and dramatically receding lines, he creates great depth in his pictures. In other pictures, horizontal lines and straight frontages dominate. He always deliberately uses graphic lines and alignments when composing his pictures, creating a characteristic setting that is very much his own.

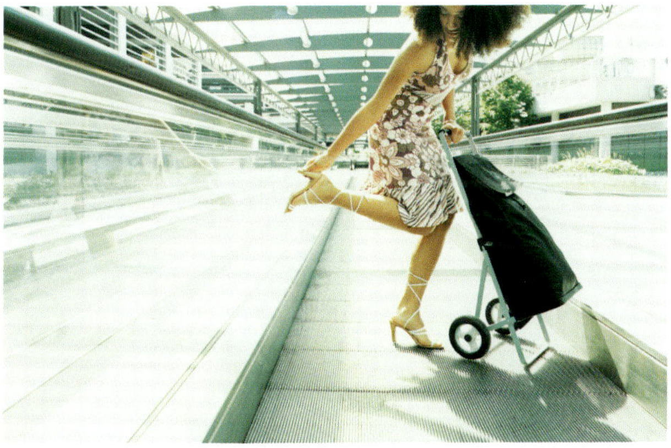

Ganz schön wackelig mit den Schuhen!
It's pretty wobbly on these heels!
Speedy: Marktroller . shopping trolley

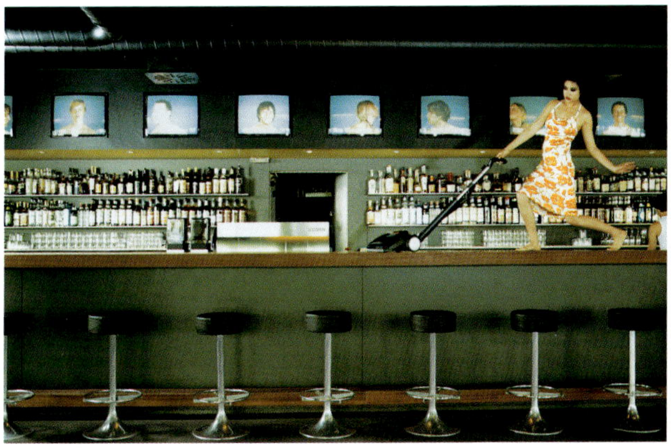

Unglaublich, wo mich mein Chef überall Staub saugen lässt!
I can't believe the places my boss gets me to do the vacuuming!
Art. by Miele: Multifunktionsstaubsauger . multi-functional vacuum cleaner

Ob Captain Kirk standhaft bleibt?
Can Captain Kirk stay upright?
Active Light Wall

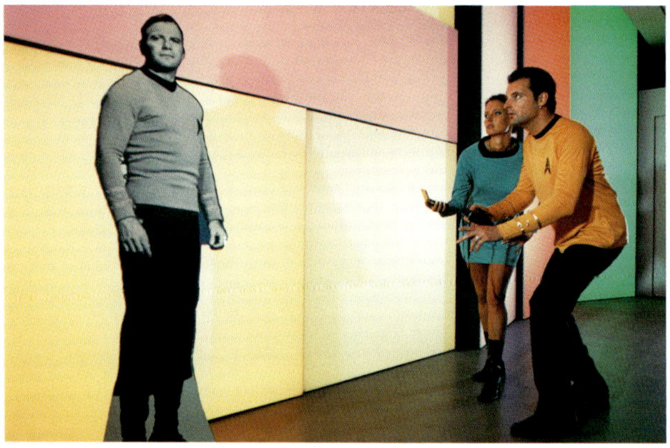

Hoffentlich bewegt er sich nicht ...
Let's hope it doesn't move ...
Solis. Programm 190: Bürostuhl . office chair

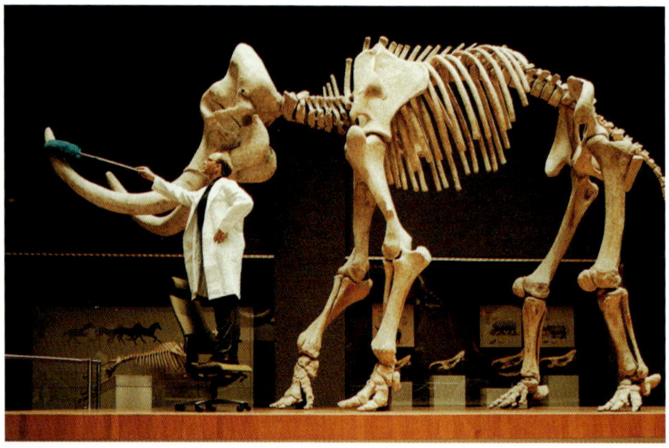

Born to be wild ...
Andersson Flag: Karten-Hänger . card hanger

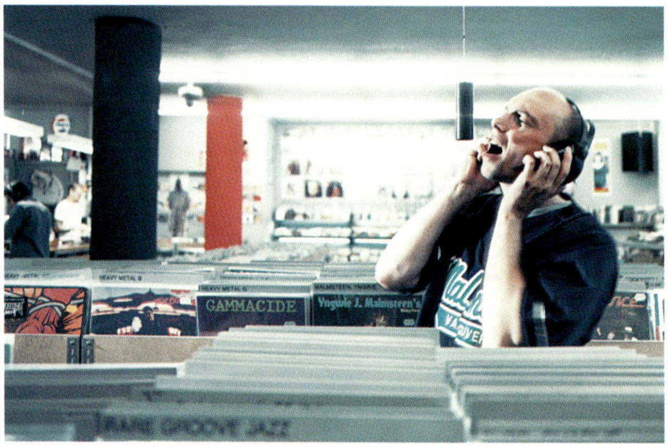

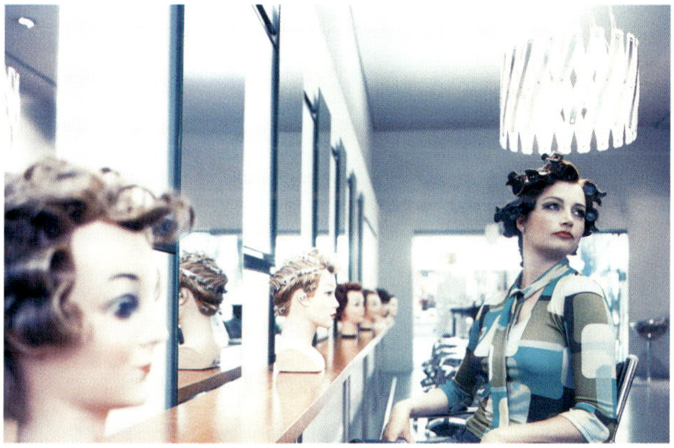

Ganz schön heiß hier drunter.
It ain't half hot under here.
Zoom: Pendelleuchte . pendant lamp

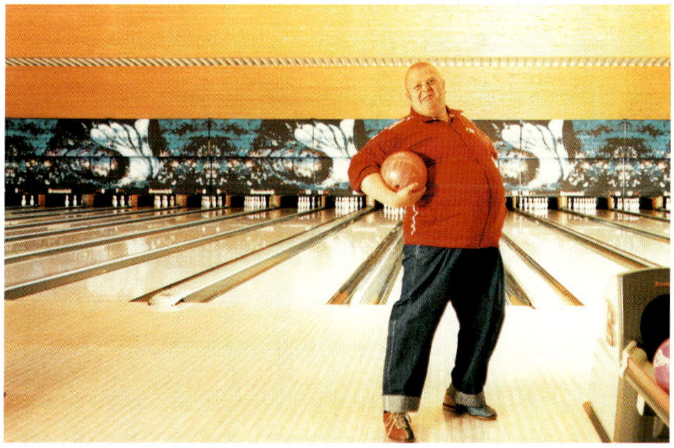

Heute schieb ich mal 'ne ruhige Kugel.
Today I'm going to take things nice 'n' slow.
EBM Balgrip System: Brille . spectacles

MEDIZIN + REHABILITATION

MEDICINE + REHABILITATION

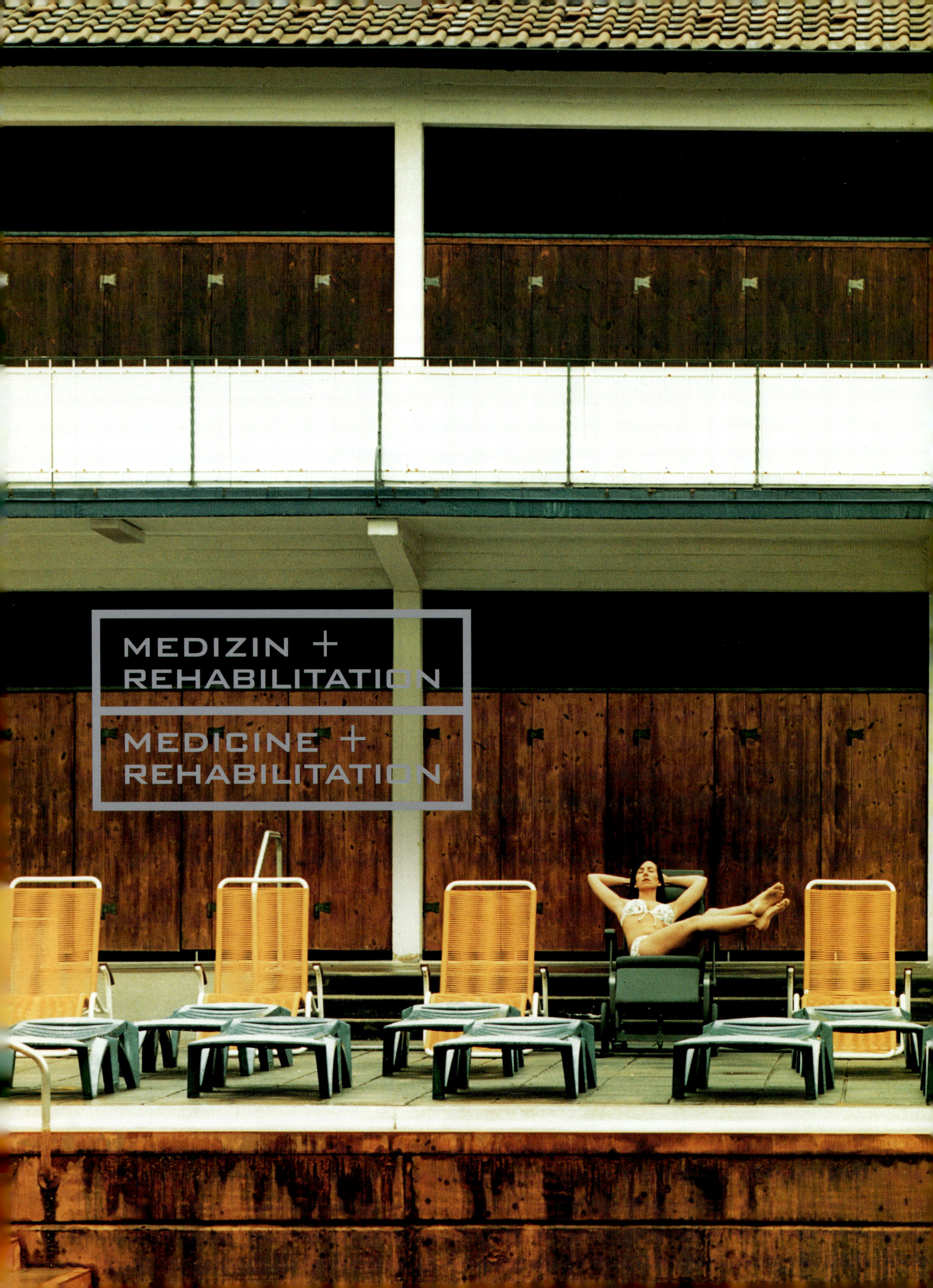

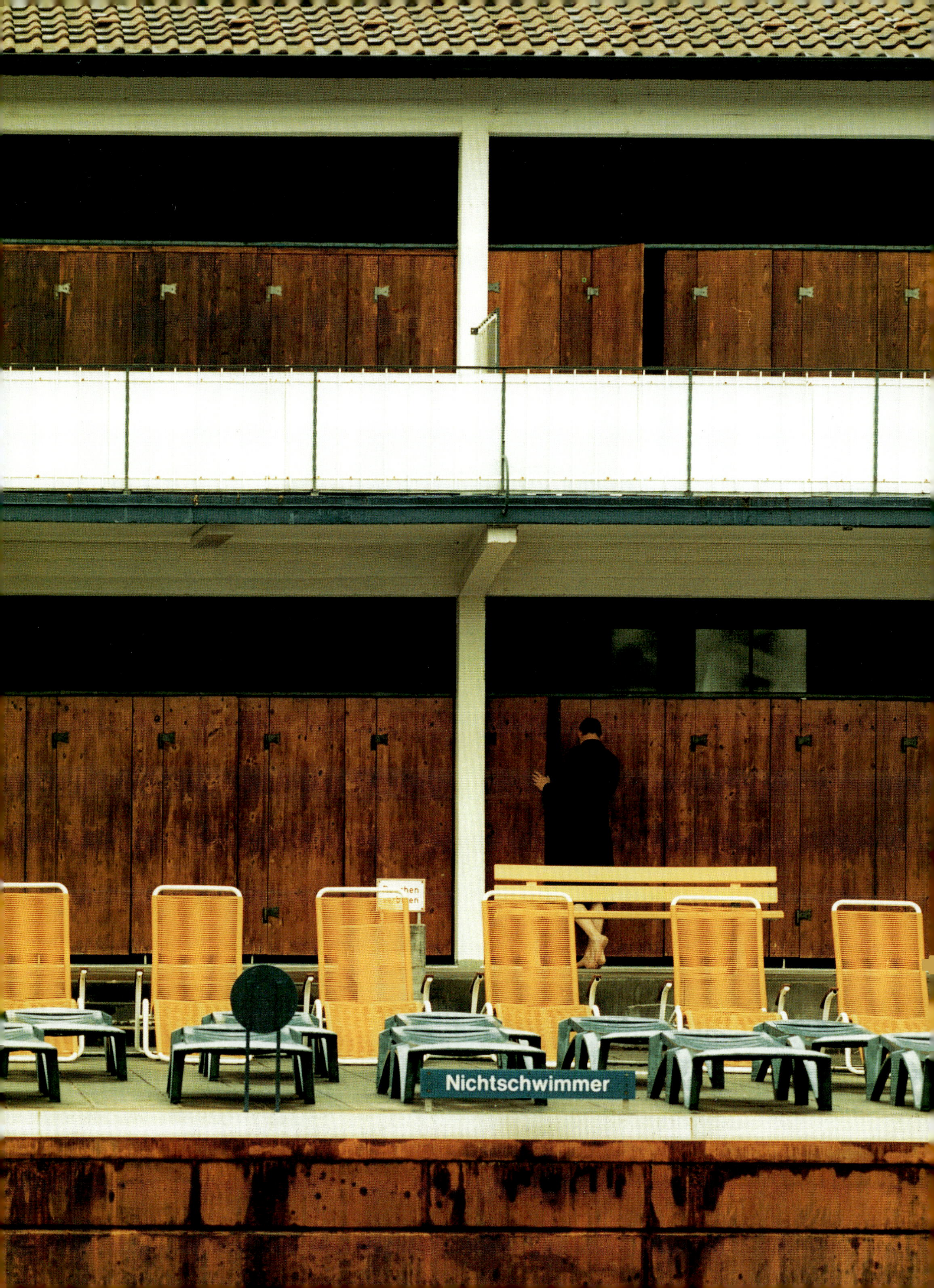
Nichtschwimmer

MEDIZIN + REHABILITATION GOLD

MEDICINE + REHABILITATION GOLD

HERSTELLER . MANUFACTURER
Greiner GmbH
D-74385 Pleidelsheim

VERTRIEB . DISTRIBUTOR
Greiner GmbH
D-74385 Pleidelsheim

DESIGN . DESIGNER
Matthias Ebert
D-70176 Stuttgart

carryLine : Fahrbare Liege . hospital trolley

Die Liege für den Transport von Patienten in Krankenhäusern und Pflegeeinrichtungen zeichnet sich durch sehr gutes Design, hohe Funktionalität und Qualität aus. Die Rückenlehne ist anatomisch geformt und hat einen integrierten Seitenhalt. Sie lässt sich synchron mit dem Beinpolster über eine Sicherheitsgasfeder in die Ruheposition bringen. Die automatische Stufenlagerung der Liegefläche um jeweils 10 Grad bietet einen hervorragenden Liegekomfort und verhindert ein Abrutschen des Patienten.
Dank der Kombination von großen Spurrollen vorne und kleinen Lenkrollen mit Totalfeststeller hinten lässt sich „carryLine" selbst in der Liegestellung leichtgängig und präzise manövrieren. Die leichte Bedienbarkeit setzt sich in dem angenehm gummierten Schiebebügel fort, dessen Form eine Vielzahl von Handstellungen zulässt und somit für kleine und große (Pflege-)Personen bequem zu greifen ist. Die klaren Formen unterstützen die gute Bedienung. Für den seitlichen Einstieg lassen sich die Armauflagen abschwenken.

.

This trolley for transporting patients in hospitals and homes stands out by virtue of its excellent design, high level of functionality and quality. The back is anatomically shaped and has an integrated side support. A pneumatic safety spring allows it to be adjusted to a resting position synchronously with the mattress under the legs. Automatic positioning of the mattress surface in 10° steps provides excellent comfort for the patient lying on it, and prevents him from slipping off.
Thanks to the combination of large safety wheels at the front and small casters fitted with locking brakes at the back, "carryLine" can be manoeuvred easily and precisely even when the patient is in a prone position. This ease of use is also to be found in the pleasantly rubber-coated push bar, whose form allows a wide variety of hand positions and is thus pleasant to grip for tall and short nursing personnel or other persons. Its clear forms contribute to this ease of use. The arm rests can be lowered to allow patients to get on from the side.

Jury . Judges panel

Selten gelingt es, wie bei diesem Produkt, zwischen dem medizinischen und dem Wohnbereich eine Balance zu finden. Wir wünschen uns, dass wir in Zukunft mehr von dieser Zusammenführung zwischen medizinischen Produkten und Wohnbehaglichkeit sehen werden.

.

This product achieves the rare feat of finding a balance between the medical sphere and living room. We would like to see more of this combination of medical products and living comfort in the future.

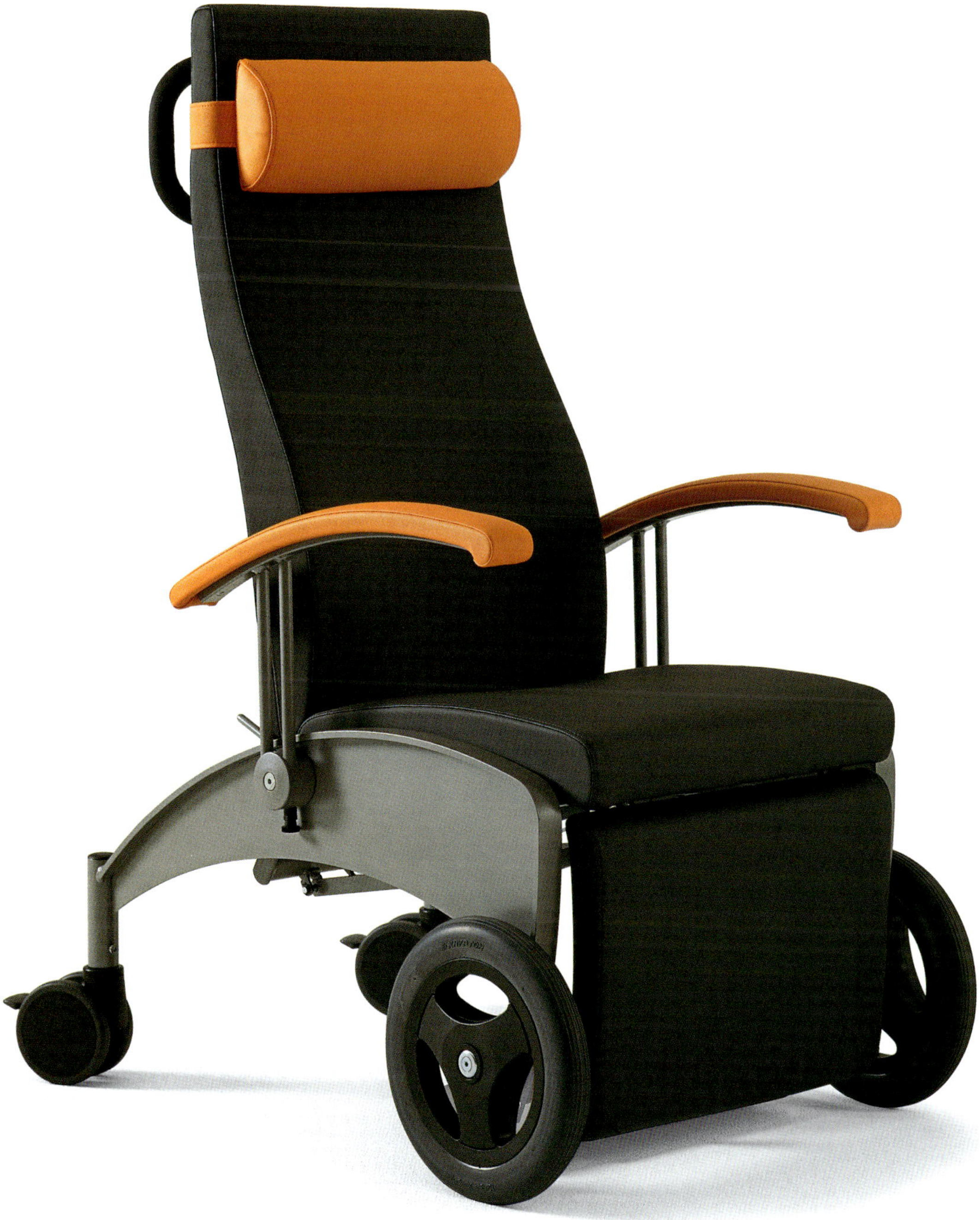

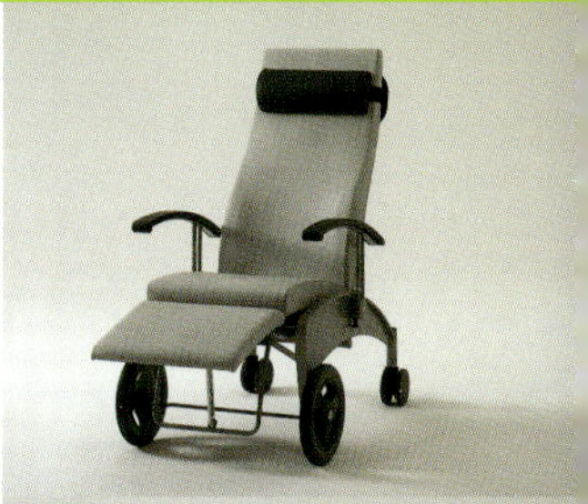

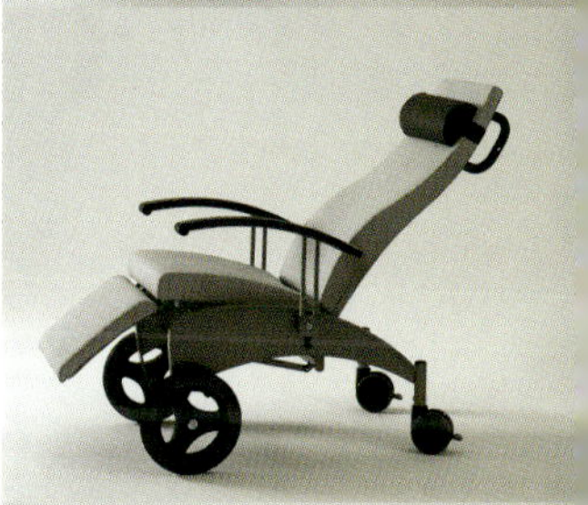

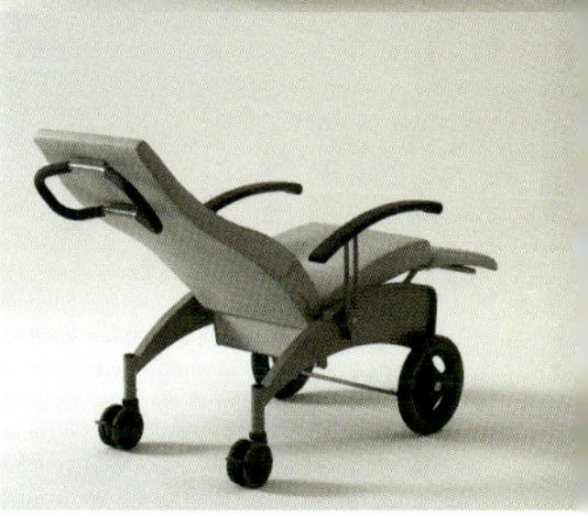

MEDIZIN + REHABILITATION | SILBER

MEDICINE + REHABILITATION | SILVER

HERSTELLER . MANUFACTURER
Beluga Tauchsport GmbH
D-27383 Scheeßel

VERTRIEB . DISTRIBUTOR
Beluga Tauchsport GmbH
D-27383 Scheeßel

DESIGN . DESIGNER
Werksdesign

Sand'sfit : Sandschlange . sand snake

Die Sandschlange wird zum Aufbautraining der Muskulatur und zum Therapieren von Wahrnehmungsdefiziten eingesetzt. Ähnlich einer richtigen Schlange liegt sie gerade oder gekrümmt auf dem Boden. Die Kiesfüllung sorgt für guten Halt, auch wenn kleine Kinderfüße darauf laufen.
Ursprünglich nur zum Balancieren konzipiert, erfreute sich „Sand'sfit" bei vielen Ergo- und Physiotherapeuten schon bald vieler weiterer Einsätze. So bietet das relativ große Gewicht zusammen mit der neuartigen Oberflächenstruktur des Textils vielfältige Möglichkeiten für Hautstimulierung, Lagerung und Bewegungsübungen. Die dichte Innenseite des weichen, elastischen 3D-Textils lässt den Kies nicht durch, die rauhe strukturierte Textiloberfläche gibt taktile Reize:
Sie massiert und sensibilisiert.

.

This sand snake is used for muscle-building exercises and for the therapy of perceptual disorders. Just like a real snake, it can lie straight or curved up on the floor. Its gravel filling allows a firm foothold, even if small children are walking on it. Originally only designed for balancing, "Sand'sfit" has soon found itself being used by many ergotherapists and physiotherapists for many other purposes. Its relative heaviness together with the new surface structure of the fabric cover provides a wealth of possibilities for stimulating the skin, positioning and body exercises. The closely-woven lining of the soft-elastic 3D fabric keeps the gravel in, and the rough, structured surface of the fabric provides tactile stimuli: it massages and sensitizes.

Jury . Judges panel

Ermöglicht eine neue Massage für Kinder. Spielerischer Umgang mit Materialien. Ein empfehlenswertes, auszeichnungswürdiges Produkt.

.

Allows a new form of massage for children. Playful use of materials. A product that can be recommended and deserves to be singled out for commendation.

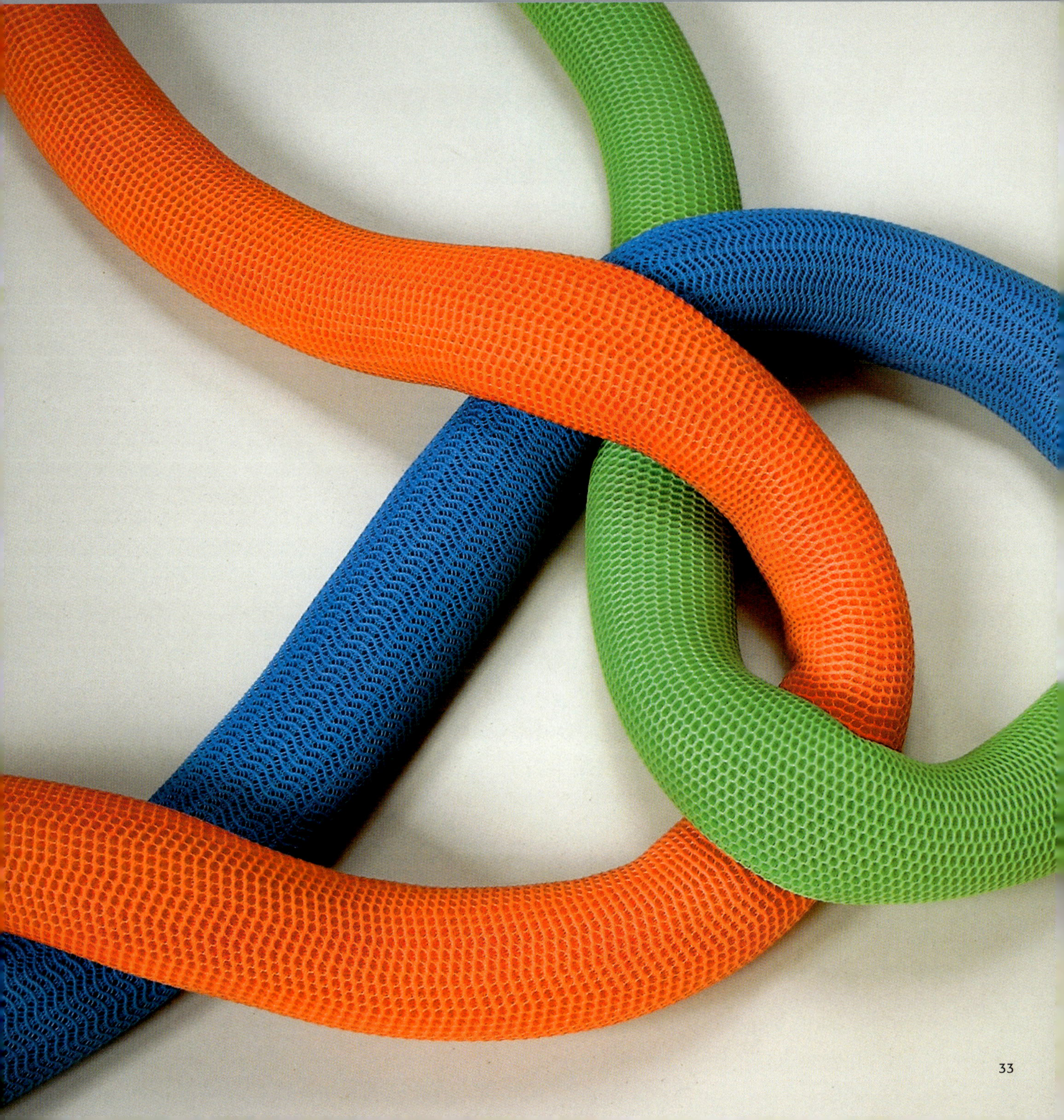

MEDIZIN + REHABILITATION | SILBER

MEDICINE + REHABILITATION | SILVER

HERSTELLER . MANUFACTURER
Condux GmbH
D-70435 Stuttgart

VERTRIEB . DISTRIBUTOR
Condux GmbH
D-70435 Stuttgart

DESIGN . DESIGNER
Reform Design
D-70565 Stuttgart

KS-cfk racing : Krückenski . outrigger

Dieser Krückenski für den professionellen Einsatz revolutioniert den alpinen Behindertensport: sowohl für Monoskifahrer als auch für sogenannte „Einbeiner" in sämtlichen alpinen Skisportarten, wie Slalom, Riesenslalom, Super G und Abfahrt.
Die sehr innovative Ergonomie erlaubt es den Sportlern in der Startphase, sich kraftvoll und in langen Zügen abzustoßen. Möglich ist das durch einen Klappmechanismus. In der Startposition geben die nach unten gerichteten Dorne einen festen Halt. Um in die Fahrposition zu kommen, drückt der Sportler einen im Handgriff integrierten Knopf, der synchron die Kufe, das heißt die Gleitfläche, nach unten bringt und die Unterarmauflage am Basisstab fixiert.
Das Hightechsportgerät vereint sportliche Dynamik mit höchster Produktqualität. Auf aufwändige und anfällige Verstellmechanismen wurde verzichtet. Der Krückenski aus hochwertigsten Materialien, wie Kohlefaser und Aluminium, wird ergonomisch für den jeweiligen Sportler, der genau vermessen wird, nach Maß gefertigt.

.

This professional quality outrigger revolutionizes downhill skiing for the handicapped: whether for monoskiers or for "one-legged skiers", and for all types of downhill skiing, such as slalom, giant slalom, Super G and downhill racing.
The highly innovative ergonomic principles applied here allow the athletes to push off powerfully and with long strokes during the starting phase. This is possible by virtue of a flip up / down mechanism. In the start position, the downward pointing spikes provide a firm grip. To get into a gliding position, the athlete presses a button integrated in the pole handle, which simultaneously lowers the runner (gliding surface) and holds the rest for the lower arm in place on the actual pole.
This high-tech piece of sports equipment combines sporty dynamism and outstanding product quality. Complicated, failure-prone adjusting mechanisms have deliberately been avoided. This outrigger, made of highest quality materials such as carbon fibre and aluminium, is tailor-made ergonomically for the respective athlete, who is measured up precisely for the product.

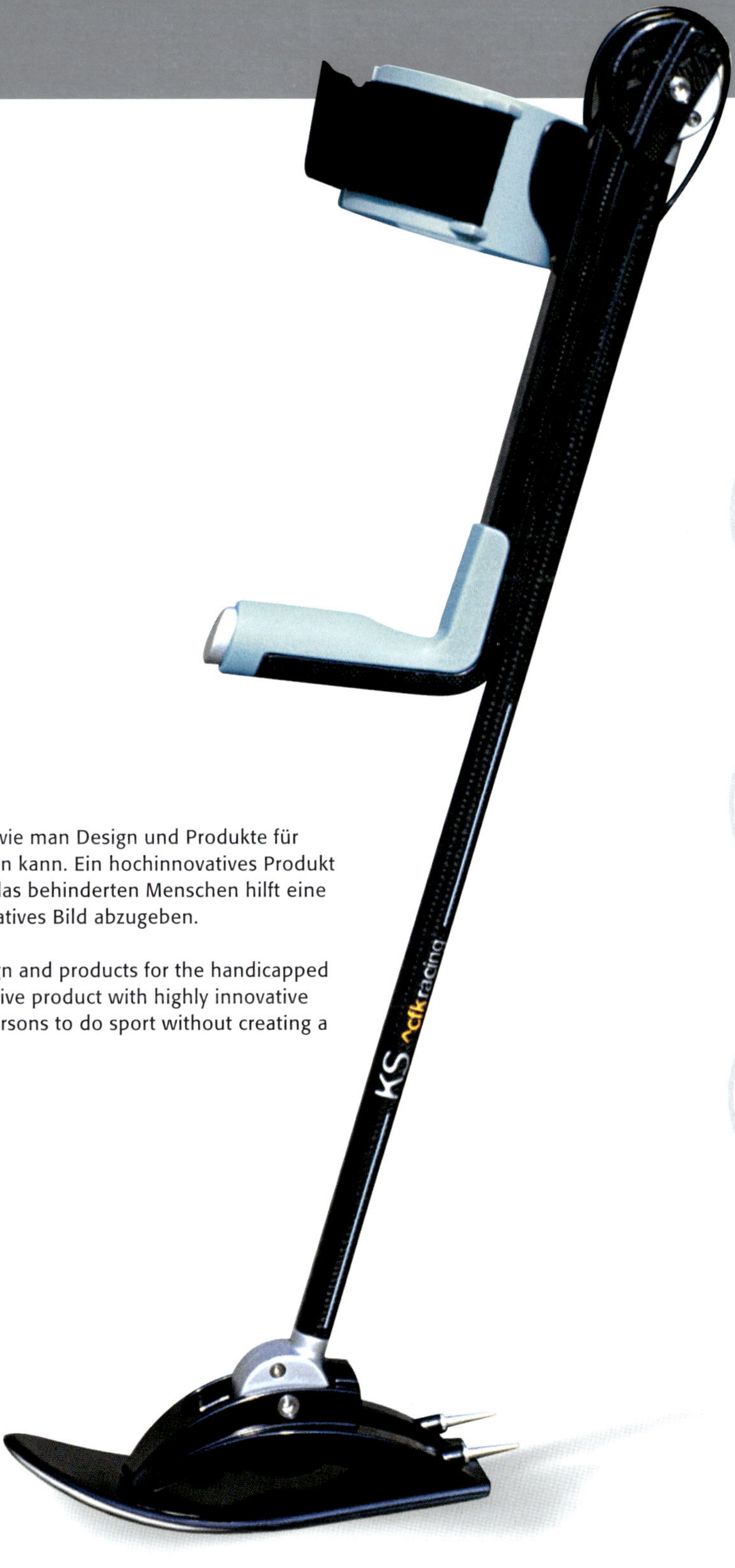

Jury . Judges panel

Ein Produkt, das vorbildlich zeigt, wie man Design und Produkte für Behinderte miteinander verknüpfen kann. Ein hochinnovatives Produkt mit hochinnovativen Materialien, das behinderten Menschen hilft eine Sportart auszuüben, ohne ein negatives Bild abzugeben.

·

An excellent example of how design and products for the handicapped can be combined. A highly innovative product with highly innovative materials, helping handicapped persons to do sport without creating a negative image.

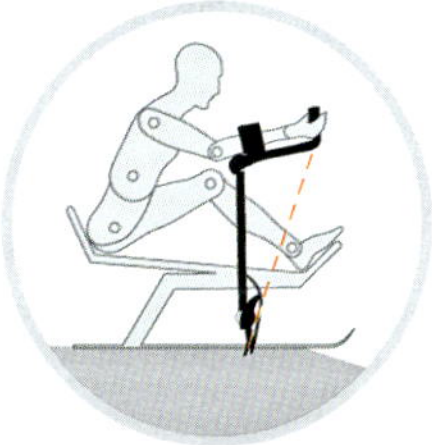

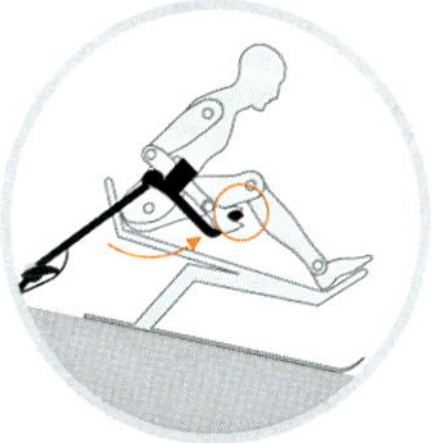

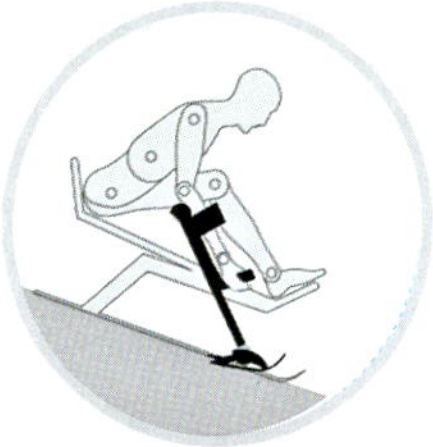

PRODUKTE FÜR KINDER

PRODUCTS FOR CHILDREN

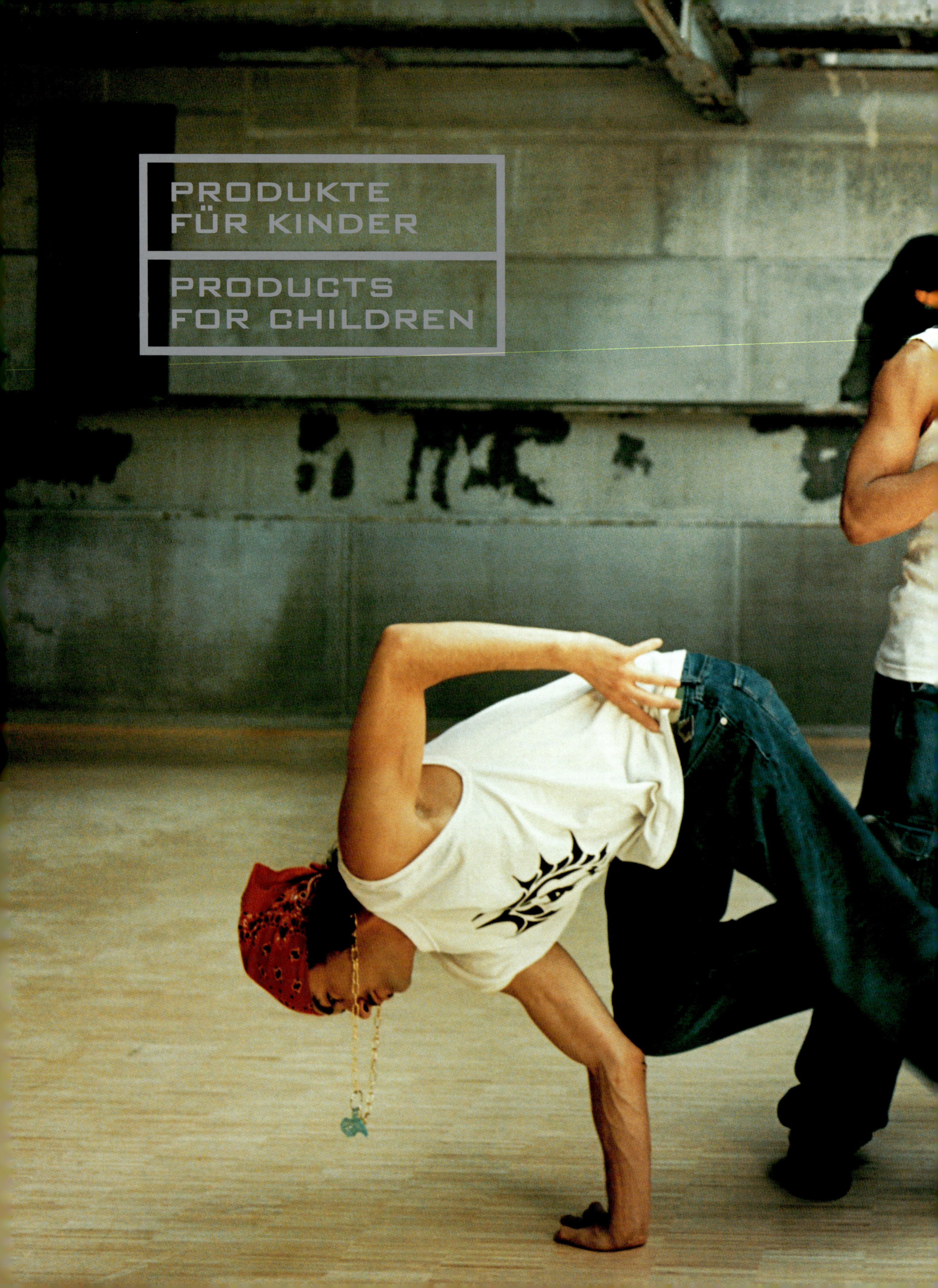

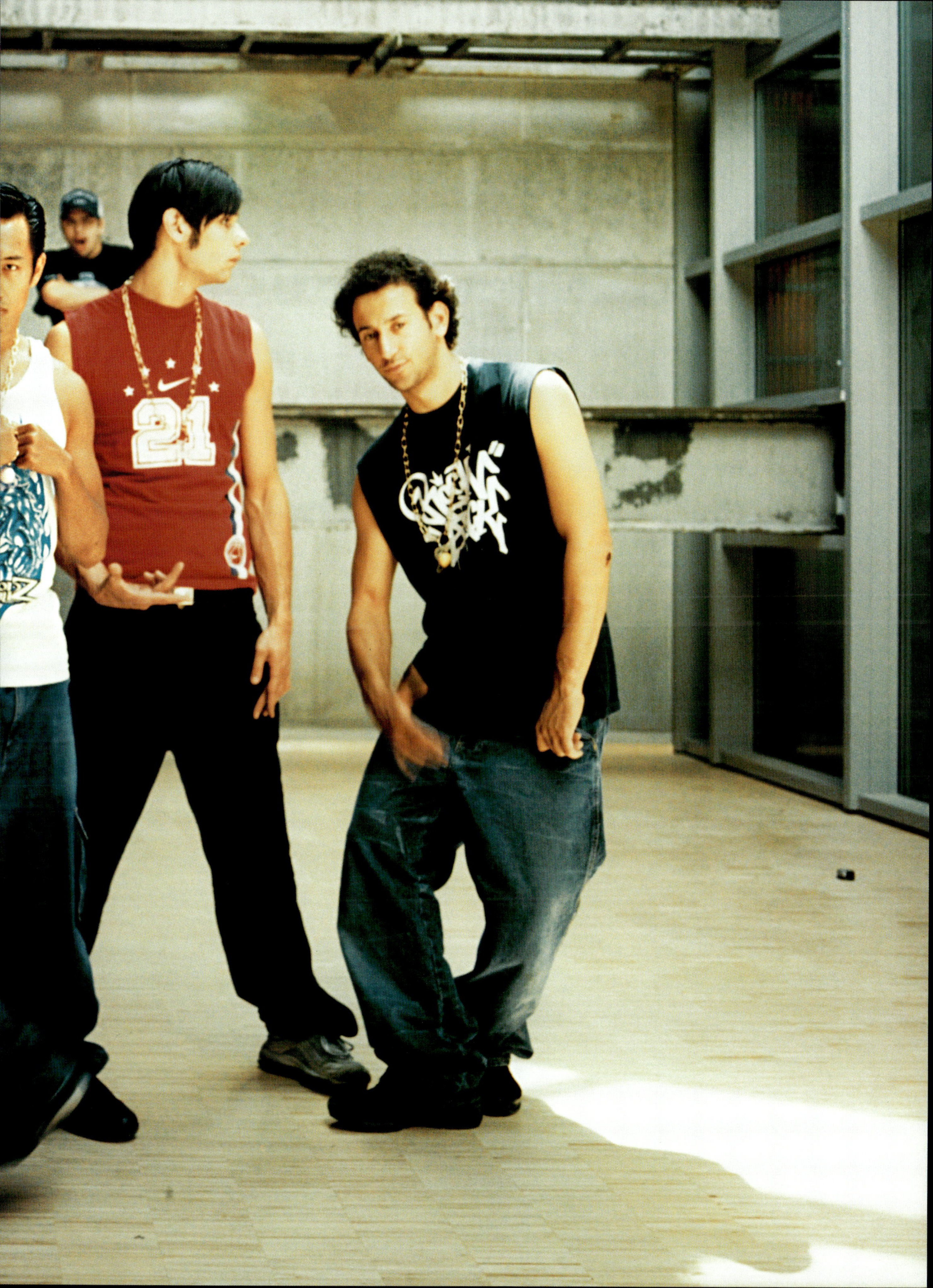
21

HERSTELLER . MANUFACTURER
Bamed AG Ft.
H-9795 Vaskeresztes

VERTRIEB . DISTRIBUTOR
MAM Babyartikel GmbH
D-27383 Scheeßel

DESIGN . DESIGNER
I.D. Pool
A-1010 Wien

MAM Preemie : Frühgeborenenschnuller . dummy for premature babies

„MAM Preemie" wurde speziell für Frühgeborene und Babys mit geringem Geburtsgewicht entwickelt. Die Form der besonders hautfreundlichen und extraweichen Schnuller regt nicht nur die Saugtätigkeit an, sondern sorgt auch für eine gute Position des Kiefers während des Saugens und ermöglicht so ein natürliches Kieferwachstum. Die sterilisierbaren Schnuller unterstützen das Gleichgewicht der Mund- und Gesichtsmuskulatur und fördern dadurch die korrekte Zungentätigkeit. Dabei wird die richtige Nasenatmung angeregt und die Luftwege in Nase und Rachen werden entwickelt. Es gibt drei Schnullergrößen: für die Kleinsten mit einem Gewicht unter 1200 Gramm einen mintfarbenen, für Babys von 1000 bis 1800 Gramm einen weißen Schnuller. Diese beiden Schnuller aus Vollsilikon für den Einsatz in der Klinik können durch Einschneiden des weichen Schildes individuell der Mundform des Babys angepasst und auch bei Mund- und Nasensonden eingesetzt werden. Die dritte Schnullervariante ist gelb, hat eine harte Mundplatte, wird für Babys mit einem Gewicht von 1600 bis 2500 Gramm in der Übergangsphase eingesetzt und kann den Eltern nach Hause mitgegeben werden.

.

"MAM Preemie" has been developed specifically for premature babies and babies whose weight at birth is low. The shape of this dummy, which is especially kind to the skin and extra soft, not only encourages sucking abilities but also ensures that the jaw is well-positioned during sucking, in this way allowing the jaw to grow naturally. The dummies, which can be sterilized, support balance between mouth and face muscles, in this way encouraging correct tongue action. In its turn, this encourages correct breathing through the nose, with the air passages in the nose and throat being developed. The dummy is available in three sizes: there is a mint-coloured dummy for extremely small babies weighing less than 1200 grams, and a white one for babies weighing between 1000 and 1800 grams. Both of these dummies, made completely of silicone for hospital use, can be adjusted to the form of the individual baby's mouth by cutting the soft shield, and can also be used when tubes have been inserted in mouth or nose. The third dummy type is yellow, has a hard shield and is used in the transition phase for babies weighing between 1600 and 2500 grams. Parents can continue giving it to their baby when they return home.

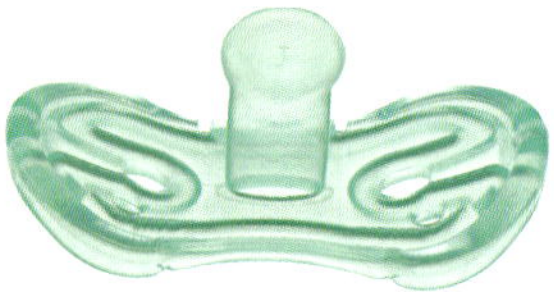

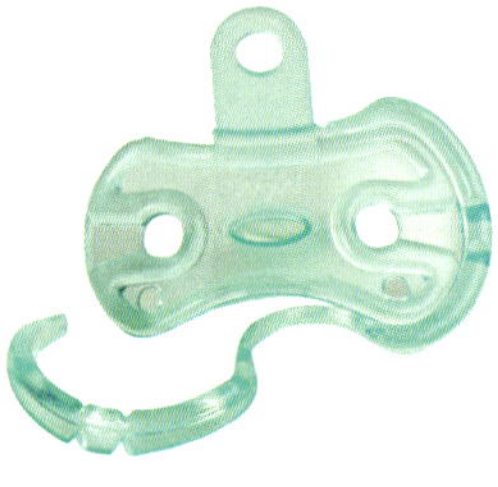

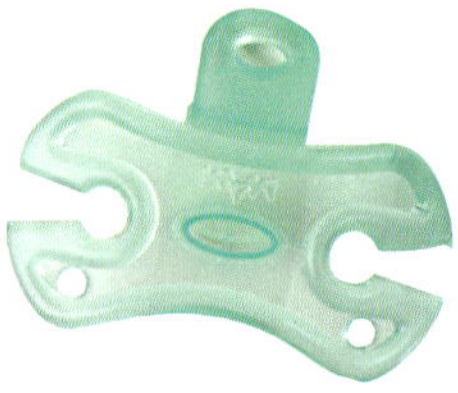

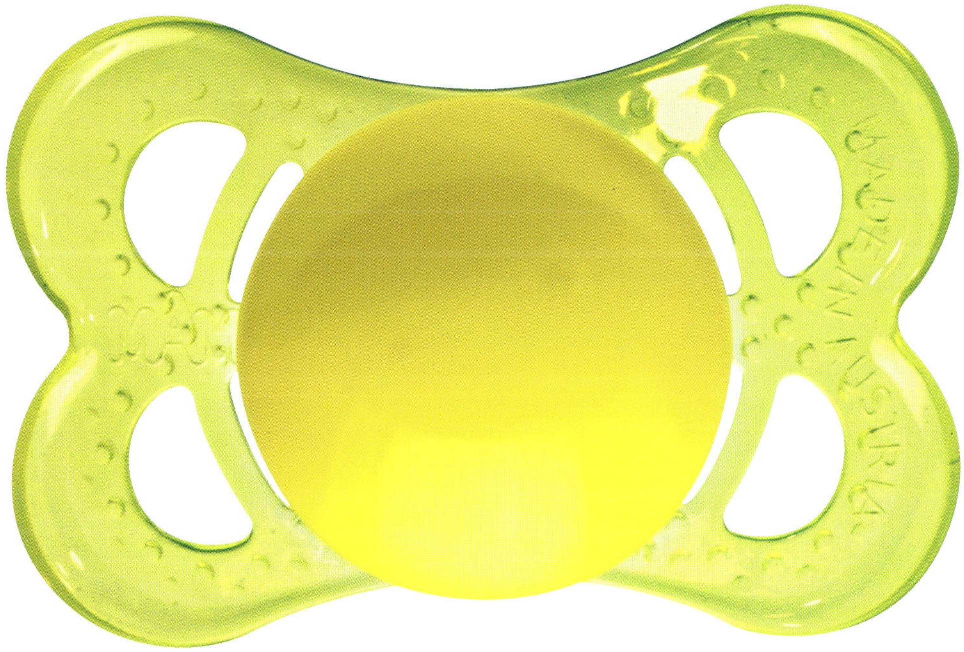

Jury . Judges panel

Ein sehr erfolgreicher Versuch, in einem einzigen Produkt viele Probleme zu lösen. So können mit diesem Schnuller die notwendigen Mund- und Nasenschläuche genutzt und gleichzeitig der Urinstinkt des Saugens angeregt werden. Ein sehr wichtiges und gutes Produkt.

.

A very successful attempt to solve a lot of problems in one single product. For example, any mouth or nose tubes that are needed can be used with this dummy, and at the same time the basic sucking instinct can be stimulated. A very important, good product.

HERSTELLER . MANUFACTURER
Mapa GmbH
D-27404 Zeven

VERTRIEB . DISTRIBUTOR
Mapa GmbH
D-27404 Zeven

DESIGN . DESIGNER
Yellow Circle
D-50676 Köln

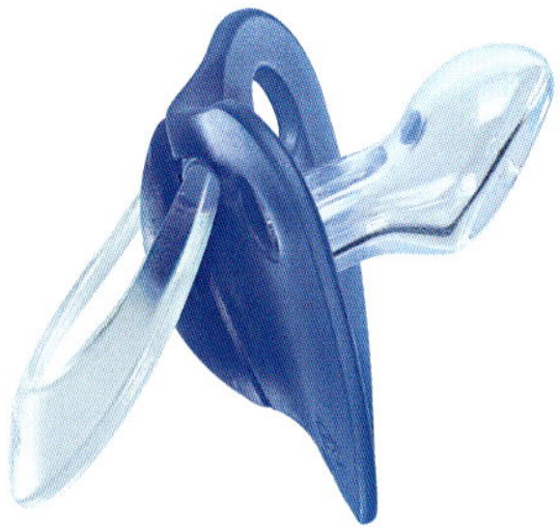

Starlight : NUK Beruhigungssauger . dummy

„Starlight" zeichnet sich nicht nur durch eine klare Formensprache mit fließenden Übergängen sondern auch durch besondere ergonomische Innovationen aus. So wurde ein Zirkulationskanal integriert, der erlaubt, dass Luft zirkulieren und Speichel abfließen kann. Für problemlosen Speichelabfluss aus dem Zirkulationskanal sorgt eine runde Einkerbung in der Mundplatte. Die Lippenauflage ist ergonomisch geformt. Dank des neuen flachen Designs passt sich der Sauger optimal der Gesichtskontur des Babys an. Auch der Greifring fügt sich exakt in die Mundplatte ein, so dass durch die flache Form Druckstellen vermieden werden. Die kiefergerechte Form begünstigt die Entwicklung von Kiefer und Gaumen.
Ein freundliches Produkt mit einer fröhlichen Farbgebung, das ohne die in diesem Bereich übliche Bedruckung mit Motiven auskommt. Eine dazu passende Aufbewahrungsbox sorgt dafür, dass der (Ersatz-)Schnuller immer sauber und hygienisch bleibt, wenn zum Beispiel das Baby mal wieder seinen Schnuller auf den Boden schmeißt.

.

"Starlight" stands out not only because of its clear formal expression (transitions between surfaces flow nicely) but also by virtue of special, innovative ergonomic features. For example, a circulation channel has been integrated, which allows air to circulate and saliva to drain away. A round notch cut into the mouth plate makes it easy for saliva to drain away from the circulation channel. The lip rest is ergonomically shaped. Thanks to its new, flat design, the dummy fits the form of a baby's face perfectly. The ring fits neatly into the mouth plate, and the resulting flat form means that pressure points are avoided. Its shape is kind to the jaw, and fosters the development of jaw and palate.
A friendly, brightly coloured product, which gets by without the ornamentation that is usually printed on such articles. A matching box for safe-keeping ensures that the (stand-by) dummy is always kept clean and hygienic, for example when baby again throws its dummy on the ground.

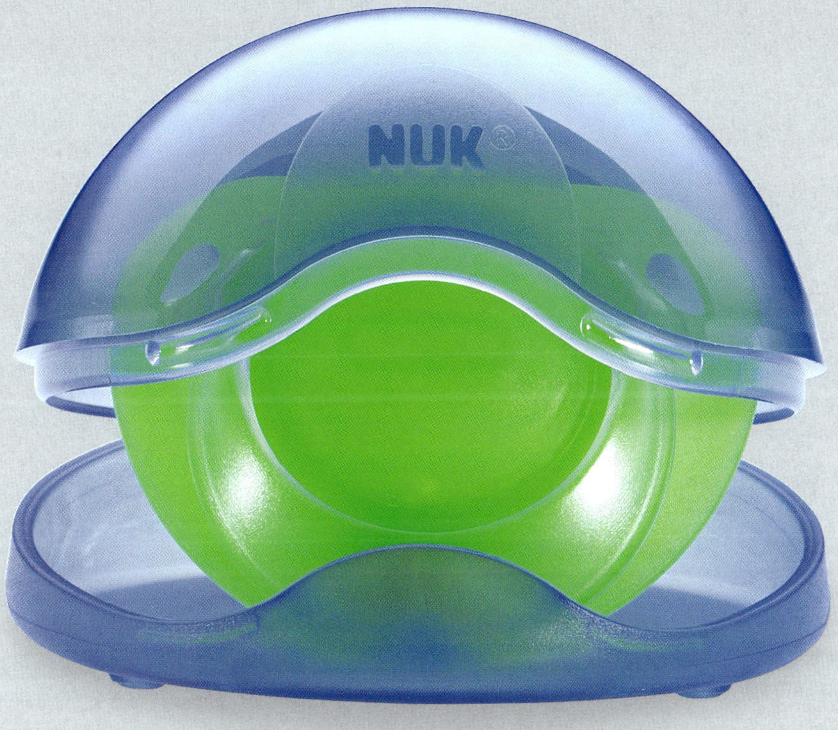

Jury . Judges panel

Ein besonders schönes Beispiel, wie man einer Produktverpackung noch eine andere Funktion gibt, also nicht nur Schutz während des Transports zum Kunden, sondern auch später als Aufbewahrungsbox, die dann weiter mit dem Produkt benutzt wird. Überzeugende, frische Farbgestaltung, gut zum Anfassen, sehr handhabungsfreundlich.

.

A particularly good example of how to give a second function to product packaging. Not only does it protect the product on its way to the customer, but it can also be used subsequently as a box for safe-keeping, and so continue to be used with the product. Convincing, refreshing colour scheme, pleasant to the touch, very user-friendly.

PRODUKTE FÜR KINDER | SILBER
PRODUCTS FOR CHILDREN | SILVER

HERSTELLER . MANUFACTURER
Vaude Sport GmbH & Co. KG
D-88069 Tettnang

VERTRIEB . DISTRIBUTOR
Vaude Sport GmbH & Co. KG
D-88069 Tettnang

DESIGN . DESIGNER
Werksdesign

Butterfly Special : Mitwachsende Kindertrage . extendable baby-carrier

Damit die Familienwanderung ein Genuss wird, muss eine Kindertrage sowohl dem Kind als auch den Eltern hundertprozentig passen. Mit „Butterfly Special" wurde ein mitwachsendes Transportmittel für Kinder im Alter von ein bis drei Jahren entwickelt, das diese Aspekte umgesetzt hat.
Ein verstellbares Torso-Stützelement bietet dem Kind rundum stabilen Halt, ist exakt einstellbar auf jede Kindergröße und sorgt dafür, dass das Kind nicht ungestützt in der Trage hängt. Es wird dagegen ermüdungsfrei und gesundheitsfördernd in aufrechter Sitzposition gehalten, und eine individuell formbare Kopf- und Nackenstütze stützt den Kopf in jeder gewünschten Lage. Die Füße lassen sich bequem auf Fußrasten positionieren. Ein integriertes Klappdach schützt vor Sonne und Regen. Zum bequemen Ein- und Aussteigen lässt sich die rechte Seite der Torsostütze mit einem Griff aufklappen. Ein geräumiger Packsack unter dem Sitz bietet weiteren Komfort zum Verstauen von Proviant und zusätzlicher Kleidung. Das innovative Tergolight-Tragesystem mit dem anatomisch vorgeformten Hüftgurt überträgt die ganze Last optimal.

.

If the family hike is to be enjoyable, then a baby-carrier has to be a perfect fit, both for the child and for the parents. "Butterfly Special" is the result of efforts to develop a papoose that grows with the child from the age of one to three, taking the above considerations into account.
An adjustable torso support provides all-round firm support for the child. It can be adjusted precisely to a child of any size and ensures that, instead of hanging unsupported in the carrier, the child is kept sitting upright, in a healthy position in which it will not tire. An individually adjustable head and neck rest supports its head in any position. Its feet can rest comfortably on footrests. An integrated fold-over canopy protects it from the sun and rain. A single tug is sufficient to fold out the right-hand side of the torso support, which makes getting in and out easy. A roomy storage space under the seat for provisions and extra clothes is an additional convenient feature. Thanks to its innovative Tergolight carrying system, with its anatomically shaped hip belt, the load is spread excellently.

Jury . Judges panel

Ein hochinnovatives Produkt mit einer interessanten Optik, vielseitig verwendbar, wächst mit, schützt das Baby, hohe Anpassungsfähigkeit und akzeptabler Preis. Die Kindertrage kann stehen, das Baby hat eine Kopfstütze, der Träger kann noch zusätzliches Material wie Jacken verstauen.

.

A highly innovative product of interesting appearance, very versatile, grows with the child and protects it, very adjustable, and reasonably priced. The baby-carrier can stand on its own, the baby has a head rest, and other items like anoraks can also be stowed in it.

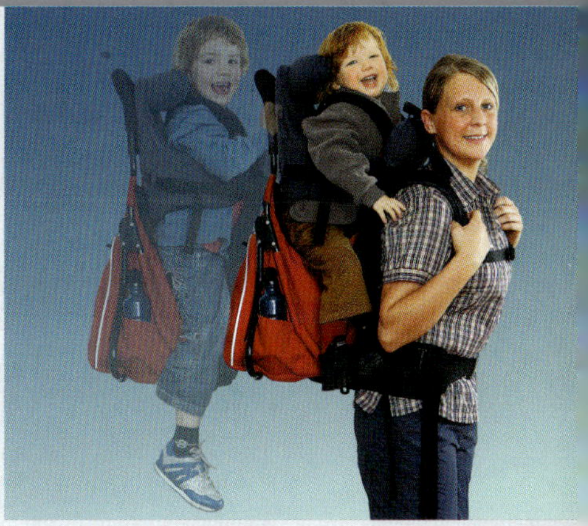

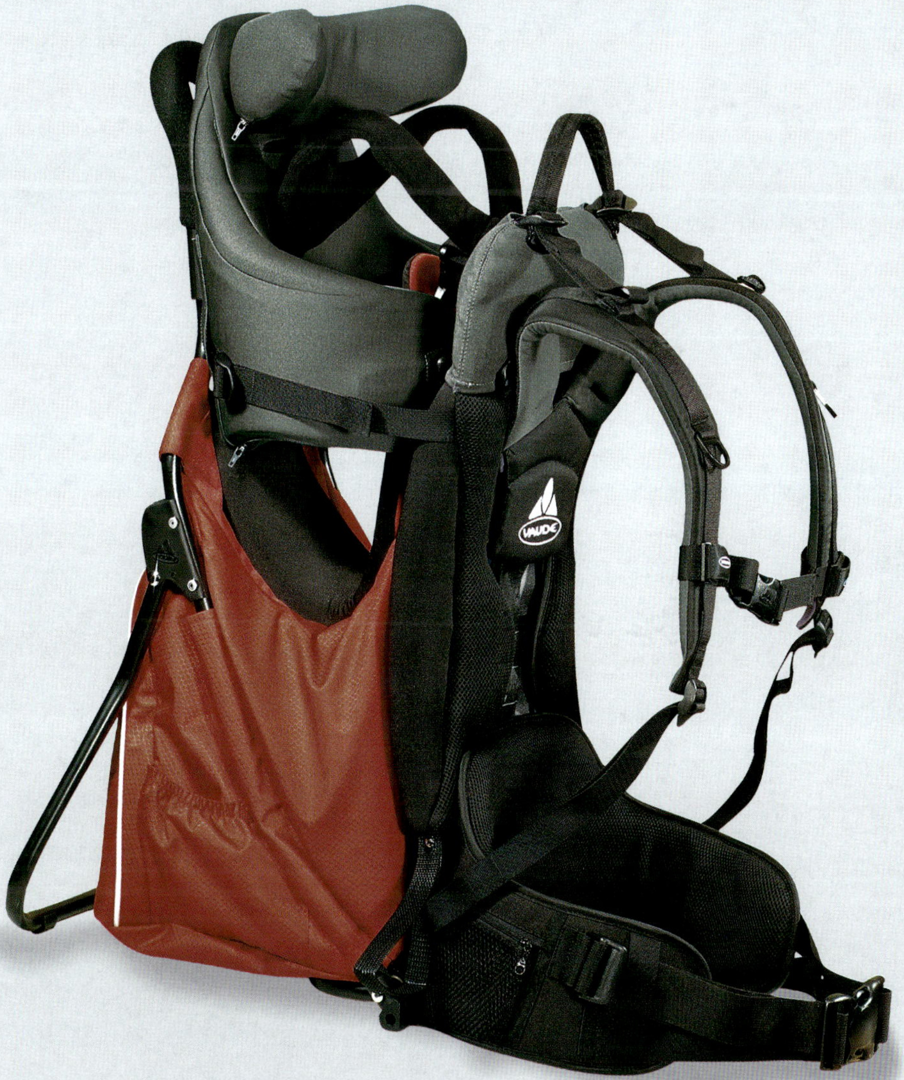
VAUDE

PRODUKTE FÜR KINDER | SILBER
PRODUCTS FOR CHILDREN | SILVER

DESIGN . DESIGNER
Klaus Pfeifer
Architekturbüro a:HP
D-64285 Darmstadt

James : Türente . door duck

Die lustige Türente „James“ sitzt auf der Klinke und verhindert mit ihrem Schnabel, dass die Tür zufällt oder von Kindern unsachgemäß zugezogen wird. Damit sind Kinderfinger vor Quetschungen sicher. Außerdem kann die Tür weder klappern noch durch einen Windzug zugeschlagen werden. „James“ fängt den Stoß ab. Damit verliert auch das Türenzuschlagen durch verärgerte Menschen am Arbeitsplatz und zu Hause seine Wirkung.
Will man die Tür schließen, wird einfach die Türklinke gedrückt, und „James“ nimmt automatisch seinen Schnabel aus dem Türspalt. Die Tür kann somit zugemacht werden, ohne dass vorher der Türstopper – wie bei herkömmlichen Stoppern üblich – aufwändig entfernt werden muss. „James“ kann immer auf seiner Klinke sitzen bleiben. Dank seines flexiblen Halses legt er seinen Schnabel bei geschlossener Tür sanft an die Türzarge.
Der lustige Türstopper wird in einem einzigen Stanzvorgang aus einem unbedenklichen PU-Schaumplattenmaterial hergestellt.

.

The amusing "James" door duck sits on the door handle, its beak preventing the door from slamming and being closed improperly by children. This keeps children's fingers from getting squashed. Moreover, the door can neither bang nor be slammed shut by a sudden draught. "James" absorbs the shock. This also dampens the effect of angry people slamming doors, whether at the workplace or at home.
All that is needed to close the door is to depress the door handle, and "James" will automatically take its beak out of the crack of the door. This means that a door can now be closed without the trouble of having to remove a doorstopper first, as is usual with conventional doorstoppers. "James" can be left in its position on the door handle. Thanks to its flexible neck, its beak rests gently against the door frame when the door is closed.
This amusing doorstopper, made of harmless PU foam laminate, is manufactured in one cutting pass.

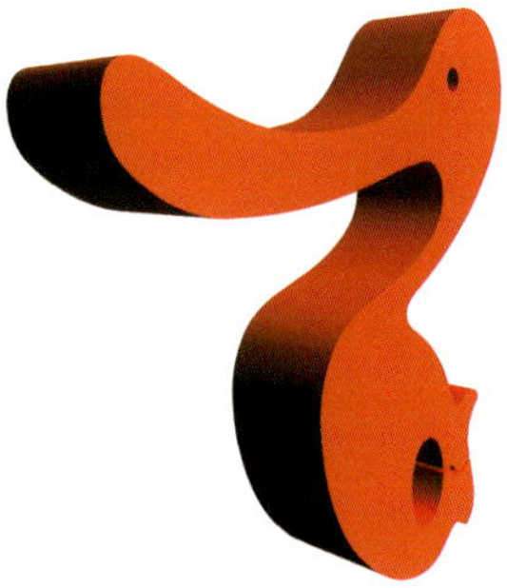

Jury . Judges panel

Lustig ist zu wenig.
Es ist ein intelligentes Produkt, das kindgerecht ist und sehr eindrucksvoll präsentiert, wie man mit einem kleinen Dreh verhindern kann, dass sich die Tür schließt.

.

To call it amusing is not sufficient. It is an intelligent, child-friendly product, demonstrating impressively how a small turn can prevent a door from slamming.

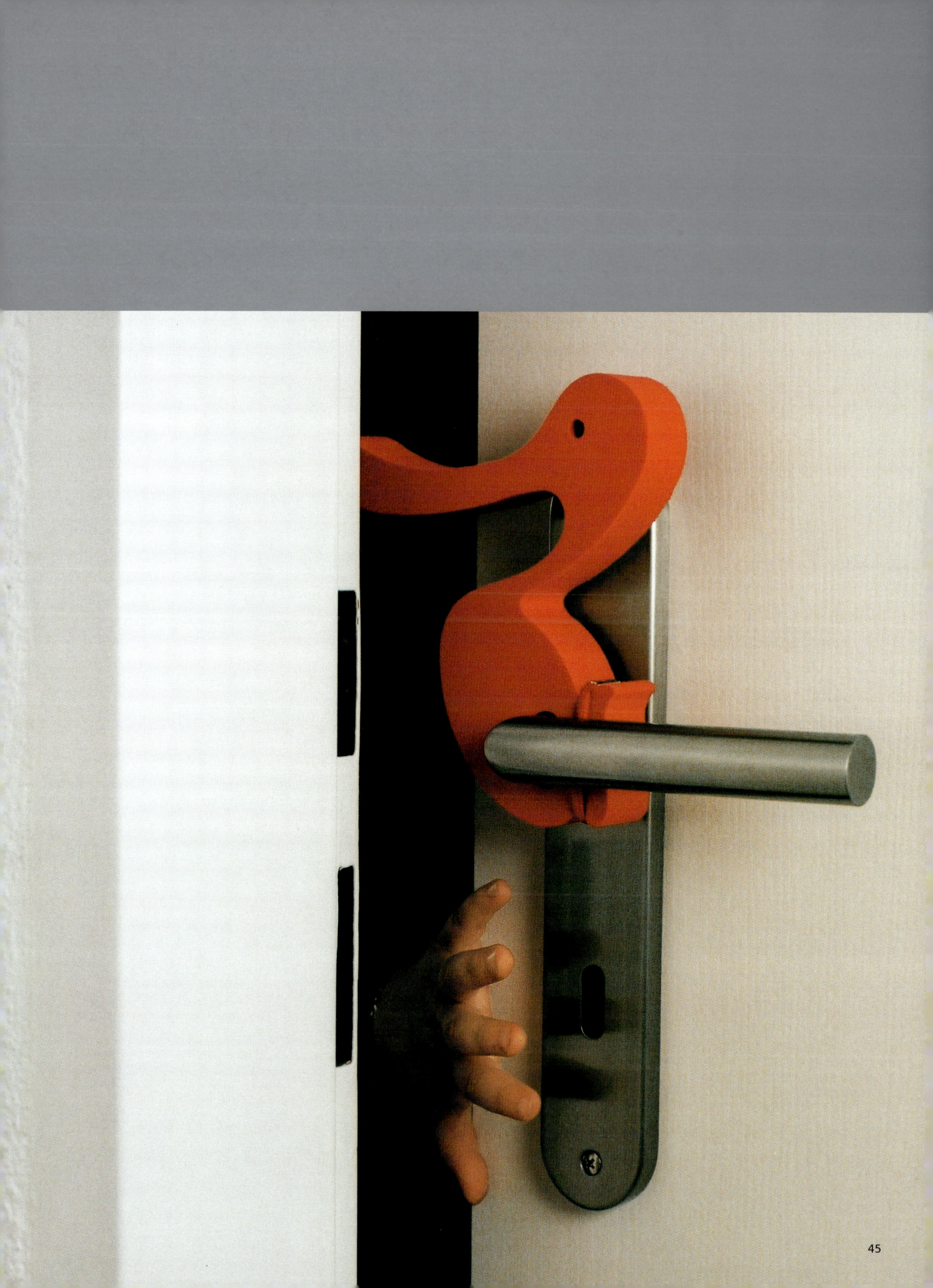

PRODUKTE FÜR KINDER | SILBER

PRODUCTS FOR CHILDREN | SILVER

HERSTELLER . MANUFACTURER
Britax Römer Kindersicherheit GmbH
D-89077 Ulm

VERTRIEB . DISTRIBUTOR
Britax Römer Kindersicherheit GmbH
D-89077 Ulm

DESIGN . DESIGNER
Busse Design Ulm GmbH
D-89275 Nersingen

Römer Baby Safe : Kindersitz . child's seat

Der neue „Römer Baby Safe" bietet für Babys von null bis fünfzehn Monaten noch mehr Sicherheit im Auto. Er hat höhere Seitenwangen und eine tiefere Schale als herkömmliche Modelle. Dadurch bietet er auch im Fall eines Seitenaufpralls größeren Schutz.
Für die ganz Kleinen bis cirka vier Monate gibt es einen Sitzverkleinerer, unter dem zusätzlich ein extra Polster zur Entlastung des empfindlichen Rückens versteckt ist. Der ergonomisch geformte Tragebügel mit seiner dynamisch geschwungenen Form ist dreifach verstellbar und bietet dadurch mehr Raum zum Hineinlegen und Herausnehmen des Babys. Ein Sonnenverdeck, das einfach zu befestigen ist, schützt vor intensiver Sonneneinstrahlung. Der Bezug ist weich gepolstert und lässt sich zum Waschen abnehmen.

.

The new "Römer Baby Safe" provides even more car safety for babies from birth to fifteen months. Compared with conventional models, the side wings are higher and the seat is deeper. This means that it provides greater protection in the case of side impact.
For very small babies up to roughly four months, the size of the seat can be reduced, with an extra cushion to support baby's sensitive back. The ergonomically shaped, dynamically curved carrying handle can be set to three different positions, creating more room when laying the baby down and taking it out again. An easy-to-attach sunshade provides protection against strong sun. The cover is softly padded and can be removed for washing.

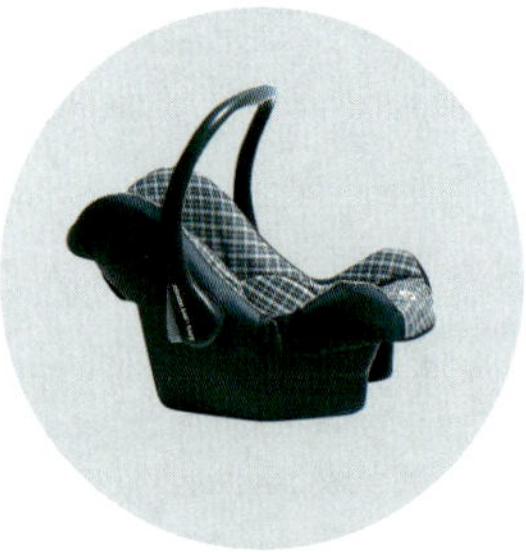

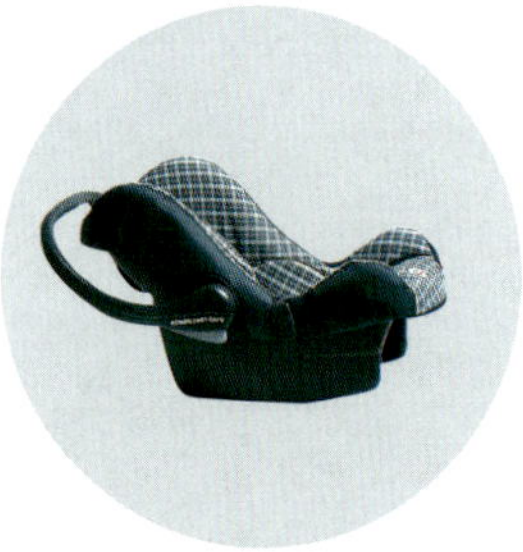

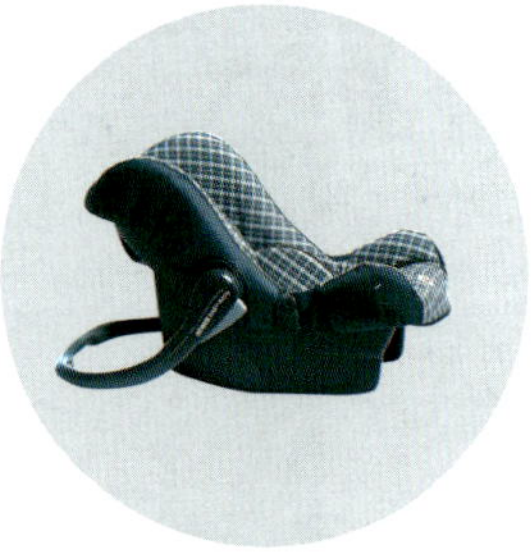

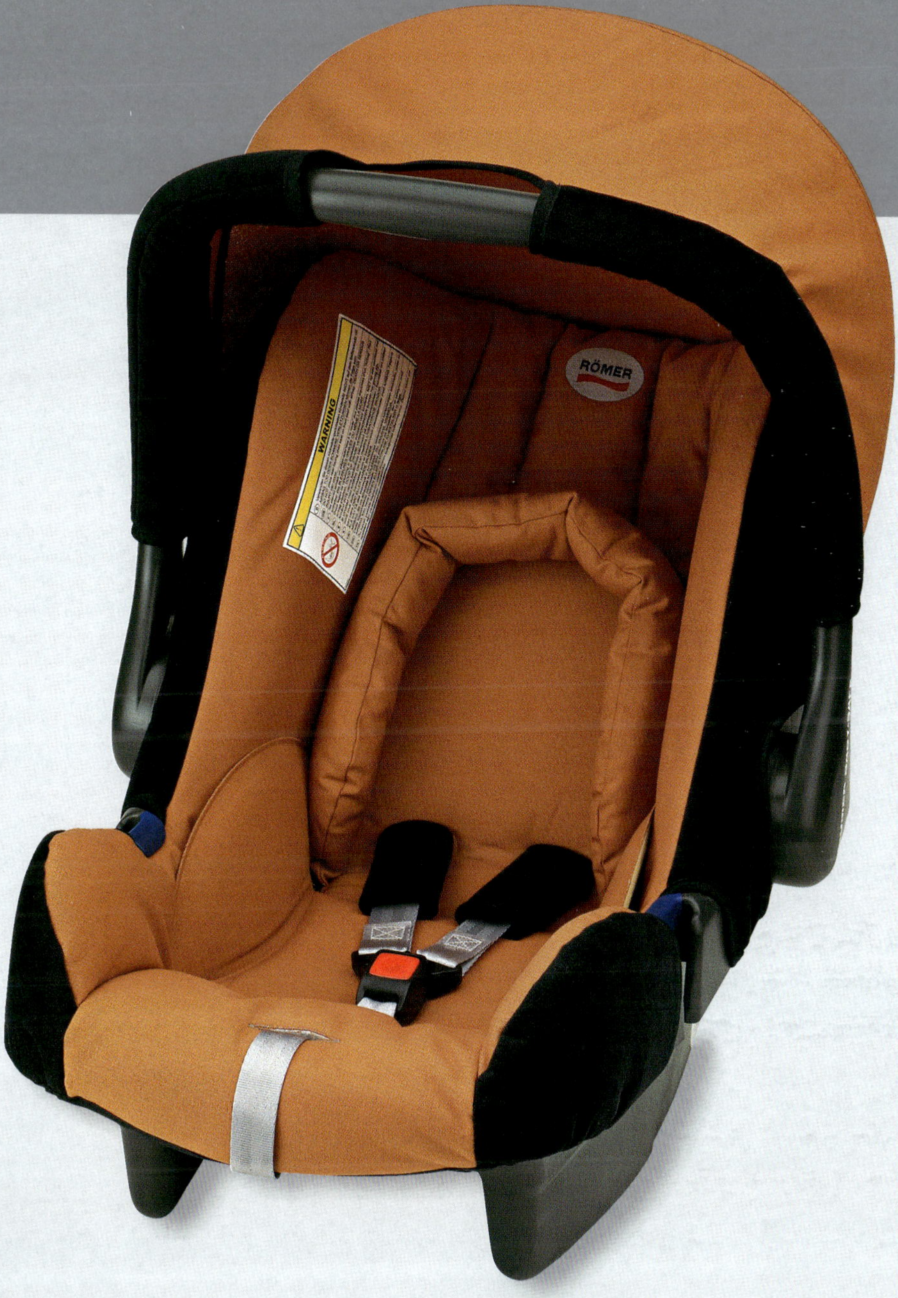

Jury . Judges panel

Es ist sehr schwierig Sicherheit, Komfort und Design gerade beim Kindersitz miteinander zu verbinden. Römer ist hier einen interessanten Weg gegangen. Sehr gute Handhabung, hohe Qualitätsausstrahlung, sehr gute Verbindung von Materialien und Sicherheitsapekt.

In a child's seat, it is especially difficult to combine safety, comfort and design. Römer has taken an interesting approach here. Very easy to use, high-quality appearance, very good combination of materials and safety considerations.

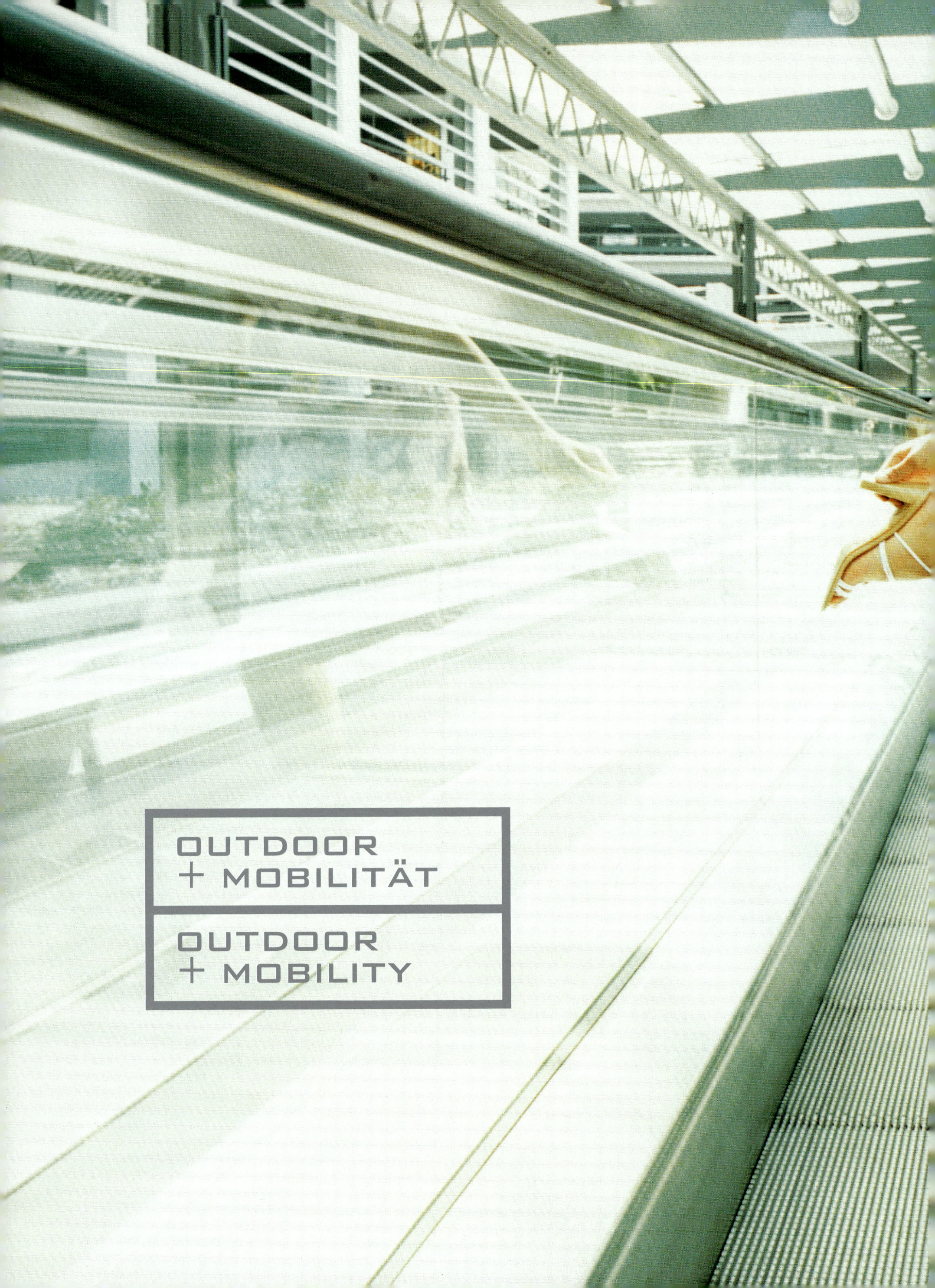

OUTDOOR + MOBILITÄT

OUTDOOR + MOBILITY

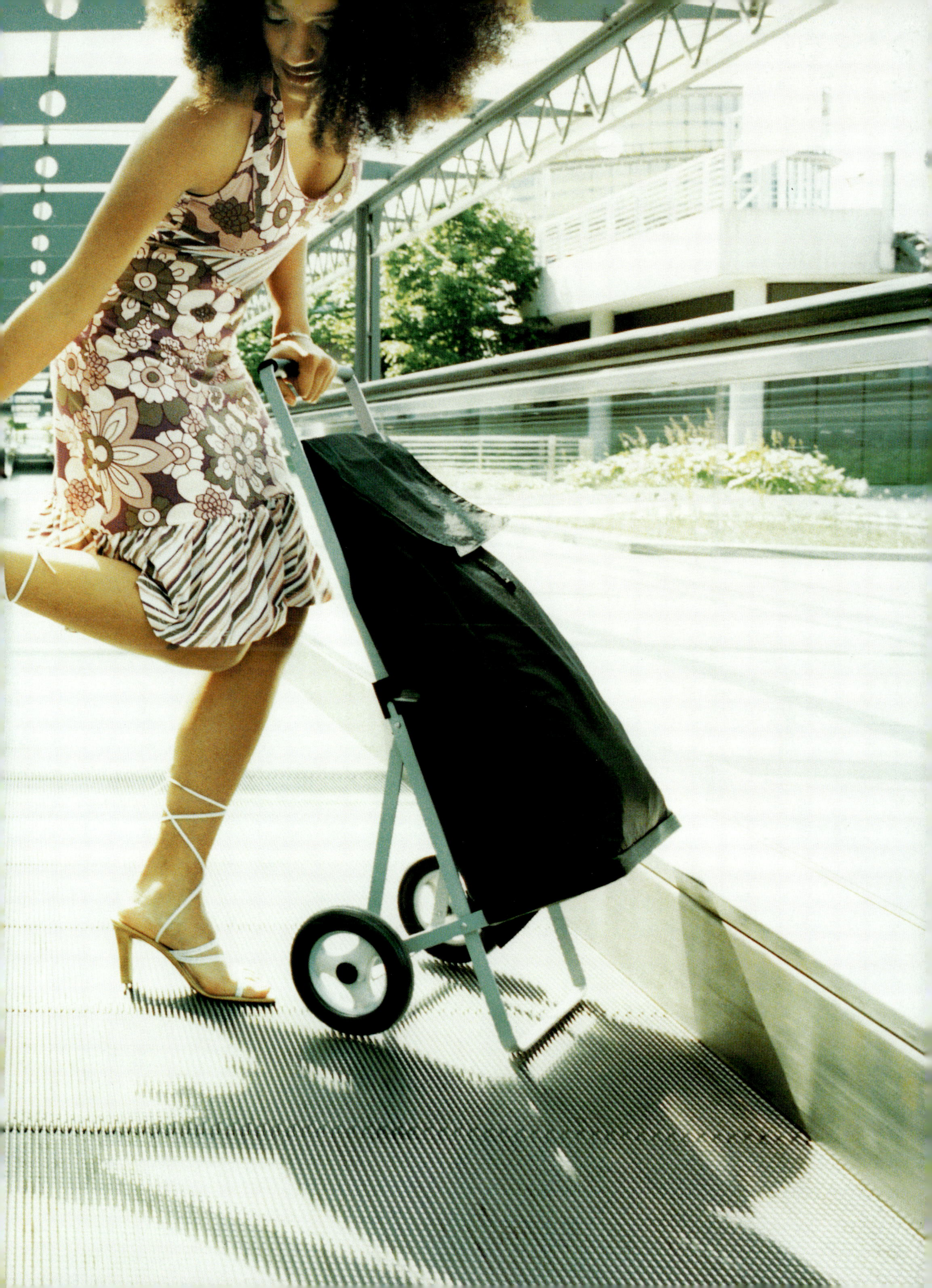

HERSTELLER . MANUFACTURER
Reisenthel Accessoires
D-82178 Puchheim

VERTRIEB . DISTRIBUTOR
Reisenthel Accessoires
D-82178 Puchheim

DESIGN . DESIGNER
Katja Horst
D-80333 München

Speedy : Marktroller . shopping trolley

„Speedy" bringt so manches ins Rollen. „Speedy" ist ein vielseitig einsetzbarer Begleiter für Jung und Alt: zum Einkaufen, auf Messen, für die Reise. Zusammengeklappt hat „Speedy" die Außenmaße für Handgepäck im Flugzeug. Der Marktroller zeichnet sich durch sein extrem leichtes Eigengewicht aus. Er wiegt gerade mal zwei Kilogramm, kann aber bis zu 20 Kilogramm belastet werden. Gestell und Beutel lassen sich auch einzeln nutzen. Ohne Beutel lassen sich mit dem leicht zu bedienenden Marktroller beispielsweise Getränkekästen, Gepäckstücke, Müllsäcke oder andere unhandliche Gegenstände transportieren.
Bei dem Entwurf von „Speedy" wurde viel Wert auf eine kostengünstige Produktion gelegt. Vier U-förmige Bügel, die mit Rundprofilen durch Vernietung verbunden wurden, bilden den Rahmen. Er besteht aus silber oder schwarz lackiertem Aluminium. Der hellgraue Taschenbeutel ist aus leichtem, aber dennoch sehr strapazierfähigem Nylon. Die extragroßen, formschönen Räder im Drei-Speichen-Design lassen sich mit einer Springfeder montieren und demontieren.
.
"Speedy" gets things moving. "Speedy" is a versatile companion for young and old: for shopping, at trade fairs, as a travel companion. When it is folded up, "Speedy's" external dimensions allow it to be taken on board aircraft as hand luggage. The trolley stands out by virtue of its extreme lightness. It weighs just two kilograms but can take loads of up to 20 kilograms. The frame and bag can also be used individually. Without the bag, the easy-to-use trolley can be used to transport things such as bottle crates, pieces of luggage, rubbish bags or other bulky items.
When designing "Speedy", inexpensive production was a priority. The frame is made up of four U-shaped bows, connected with round sections by rivets. It is made of aluminium, painted silver or black. The light grey bag is made of lightweight, yet extremely tough nylon. The elegant wheels, with their three-spoke design, are extra large and can be fitted and taken off by means of a fly spring.

Jury . Judges panel

Ganz simples Produkt aus Halbzeugen gefertigt. Sehr leicht, leicht zu handhaben, leicht zu tragen, leicht zu transportieren. Simples Produkt für jeden. Preiswert, nicht überdesignt. Er räumt mit der Oma-Anmutung auf, die viele dieser Trolleys heute noch haben. Schöne klassische Designsprache.
.
Very simple product made of wrought materials. Very light, easy to handle, easy to carry, easy to transport. A simple product anyone can use. Inexpensive, not over-designed. It does away with the granny image that many of today's trolleys still have. Elegant, classical design language.

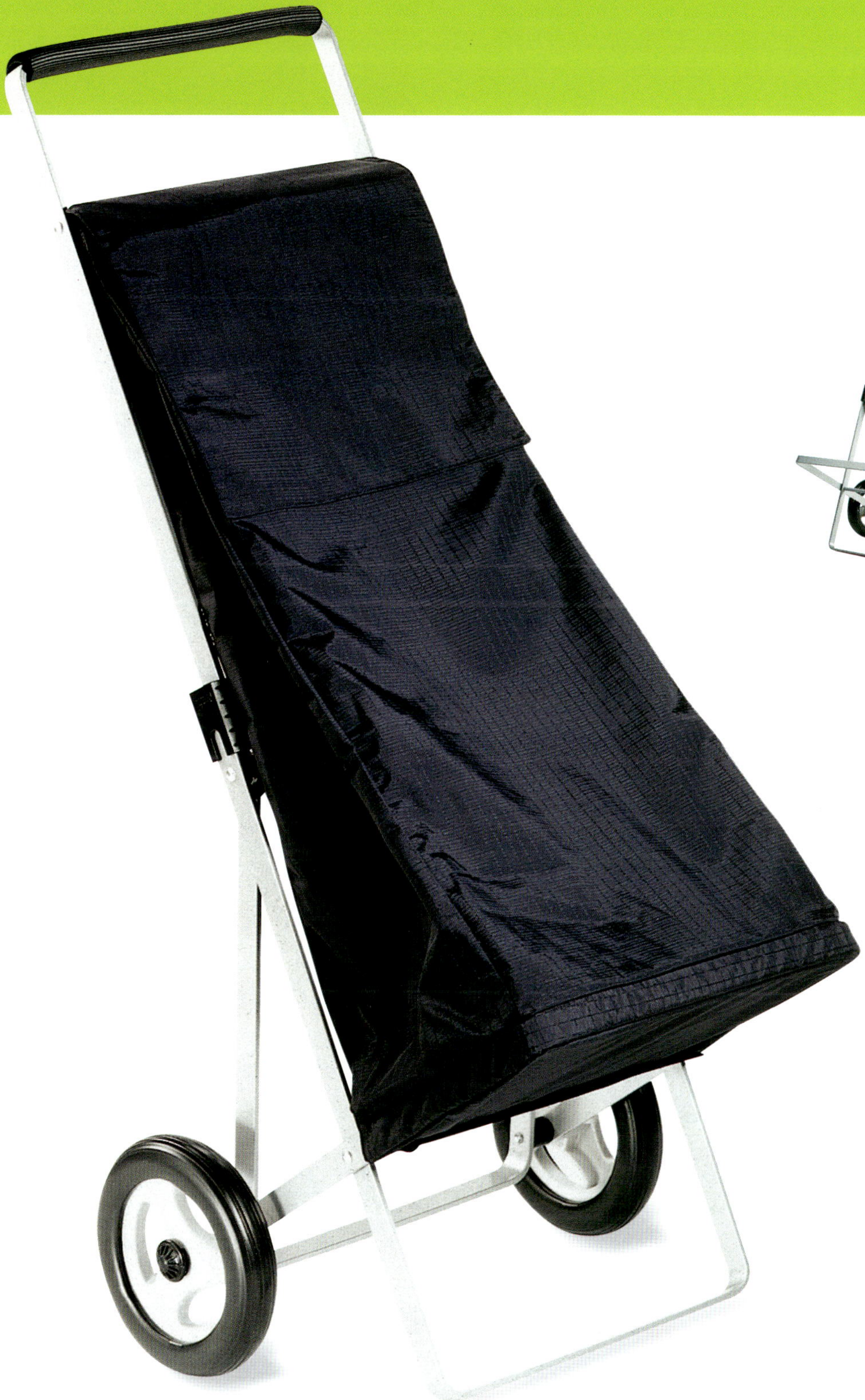

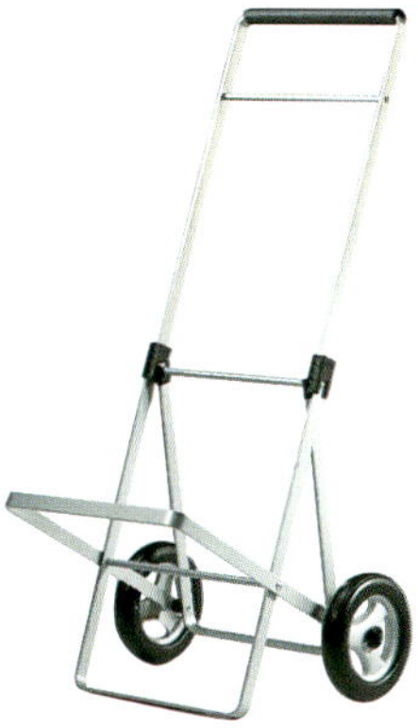

OUTDOOR + MOBILITÄT | SILBER
OUTDOOR + MOBILITY | SILVER

HERSTELLER . MANUFACTURER
Bree Collection GmbH & Co. KG
D-30916 Isernhagen

DESIGN . DESIGNER
Bree Designteam

Simply : Tasche . bag

„Simply" überrascht. Jeder kennt sie als Tüte aus Papier, vermutet aber nicht das Material, aus dem sie tatsächlich ist: eine Tüte aus pflanzlich gegerbtem Rindsleder.
Hier ist es erstmals gelungen, festes Leder auf einen halben Millimeter herunterzuspalten. Dadurch bekommt es ganz neue Eigenschaften: leicht wie Papier, zusammenfaltbar wie eine Papiertüte, aber weich, reißfest und beständig wie Leder. Das erhöht und verändert die bekannte Papiertüte in ihrer Wertigkeit. „Simply" könnte damit in ihrer Klarheit zu einem Kultobjekt werden.

.

"Simply" takes you by surprise. What seems at first glance to be a paper bag is in fact made of a quite different material: a bag made of naturally tanned cowhide.
For the first time, leather has been skived down to a thickness of just half a millimetre. As a result, it takes on completely new properties: it is as light as paper and can be folded together like a paper bag, but is as soft, tear-proof and long-lasting as leather. This enhances and changes the value of the commonplace paper bag. With its lucid qualities, "Simply" could become a cult object.

Jury . Judges panel

In sehr charmanter Weise wurde das Thema „heißgeliebte Papiertüte" aufgegriffen und in eine ganz andere Wertigkeit transferiert. Innovativ das ganz dünn verarbeitete, dennoch stabile Leder. Die Tüte wird wertig, sie wird sehr lange benutzt werden.
Für Männer und Frauen geeignet.

.

Very charmingly, the topic of the beloved paper bag has been taken up and transferred into a quite different value sphere. What is innovative about it is the way the leather has been worked: extremely thin yet robust. This lends value to the bag, it will be used for a very long time. Suitable for men and women.

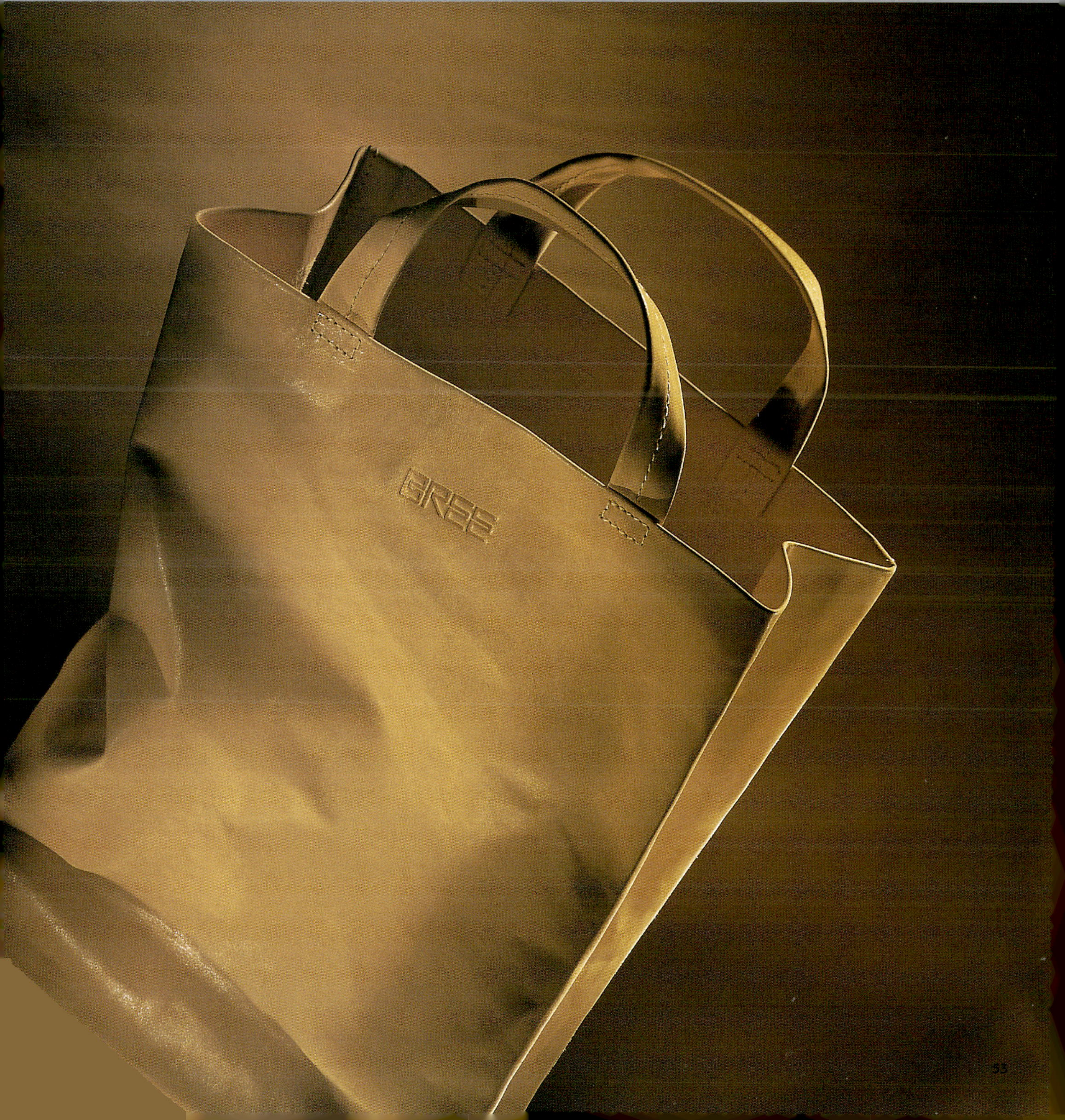
BREE

OUTDOOR + MOBILITÄT | SILBER

OUTDOOR + MOBILITY | SILVER

HERSTELLER . MANUFACTURER
Dima Freizeitmöbel GmbH
D-73278 Schlierbach

VERTRIEB . DISTRIBUTOR
Fischer Möbel GmbH
D-73278 Schlierbach

DESIGN . DESIGNER
Kurt Ranger Design
D-70469 Stuttgart

Domido : Tisch . table

„Domido" ist nicht nur ein formschöner, sondern auch ein sehr flexibler Tischbaukasten. Je nach Bedarf lassen sich damit fünf verschieden große Tische bauen: vom Kleintisch mit einer Länge von 90 Zentimetern bis zur großen, 270 Zentimeter langen Festtafel. Dazu werden einfach mehrere Grundplatten zusammengesteckt. „Males" und „Females" schaffen die notwendige Verbindung. Das sind handliche Edelstahlelemente, mit denen die einzelnen Tischplatten aneinander gekoppelt werden. So lassen sich Tische mit glatter Oberfläche schaffen. Zusätzlich kann an den Tischenden jeweils noch ein Element angedockt werden. Bei den Tischbeinen hat man die Wahl zwischen Teakholz oder Edelstahl.

"Domido" is not just an elegant modular table, but also a very flexible one. According to need, five different tables can be built from it, ranging from a small, 90 centimetre long table to a large banqueting table, 270 centimetres long. To create these different lengths, all that has to be done is to plug several basic table tops together. "Males" and "females" connect them together. These are handy stainless steel elements which allow individual table tops to be joined to each other. The resulting tables have a smooth surface. In addition, a further element can be docked on to each end of the table. For the table legs, there is a choice of teak or stainless steel.

Jury . Judges panel

Flexibilität pur. Eine Gartentafel, die sich leicht verlängern lässt. Sehr gute Qualität, sieht schön aus und das Material fühlt sich sehr gut an.

Pure flexibility. A garden table that can easily be extended. Very good quality, looks attractive, and the material is pleasant to the touch.

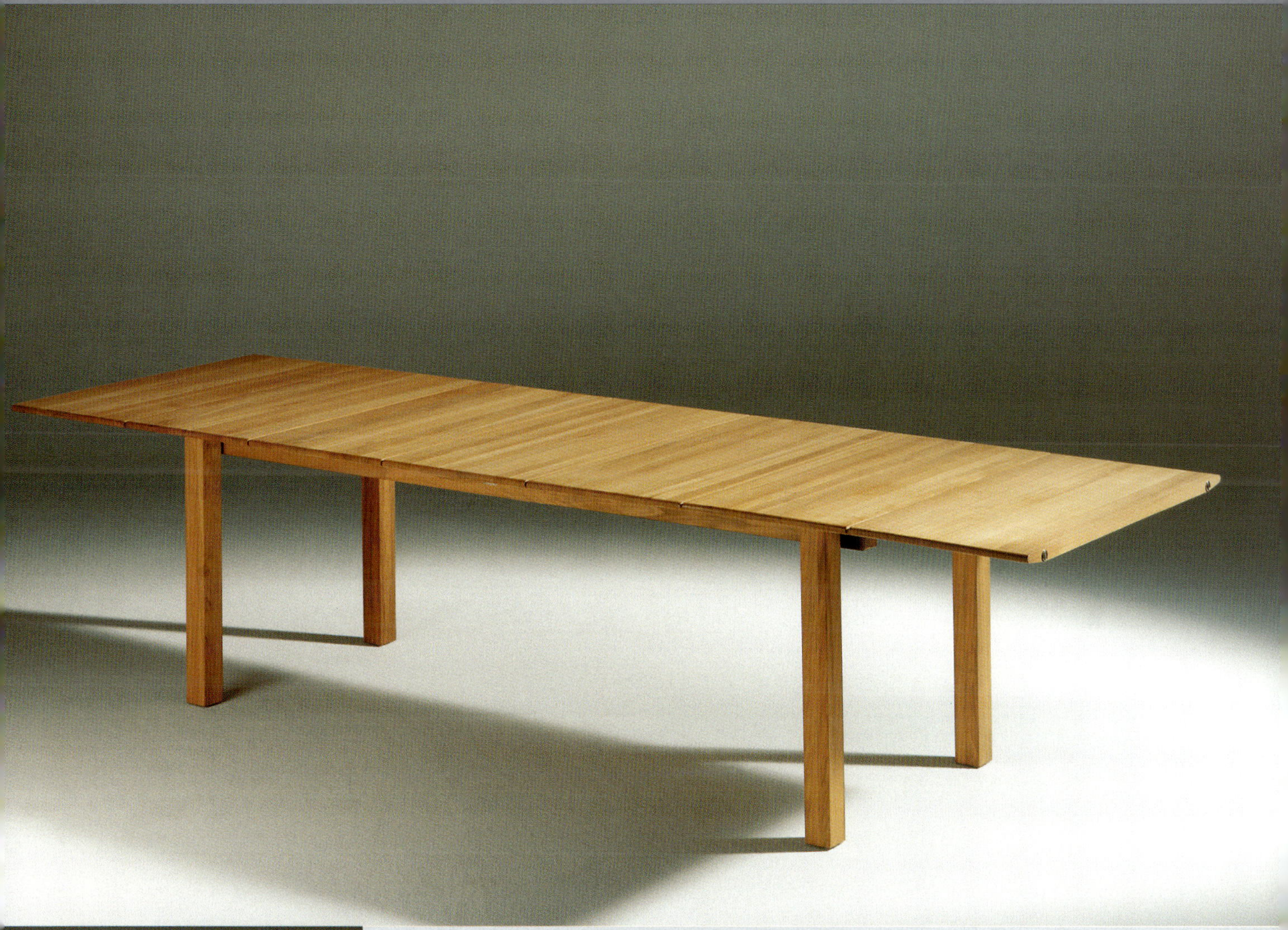

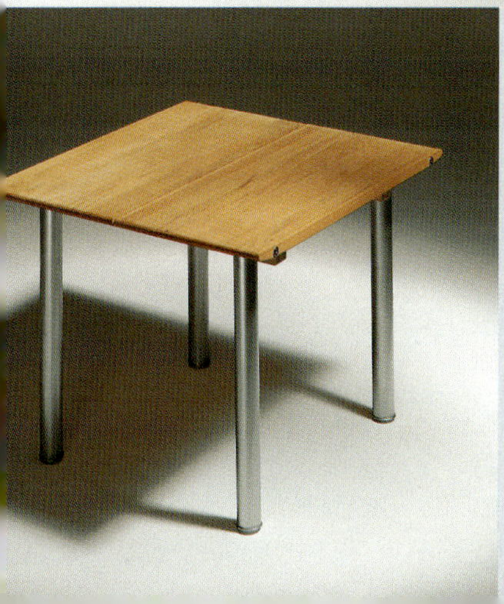

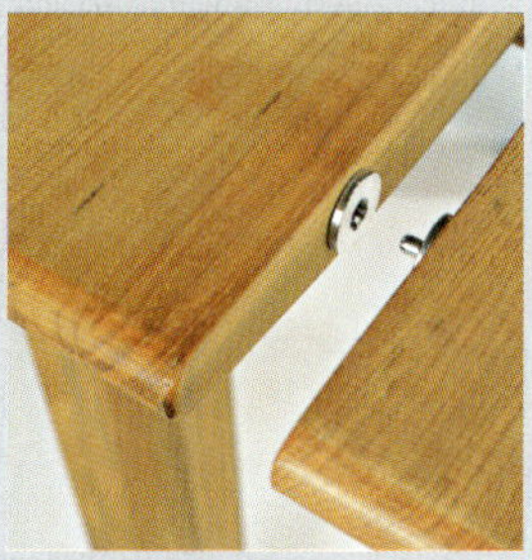

HERSTELLER . MANUFACTURER
Fischer GmbH
A-4910 Ried/Innkreis

VERTRIEB . DISTRIBUTOR
Fischer und Löffler Deutschland GmbH
D-94315 Straubing

DESIGN . DESIGNER
Factor Product GmbH
D-81667 München

Soma Tec F 9000 - F 7000 : Skischuh . ski boot

Der Skischuh für sportliche Skifahrer basiert auf der ergonomischen Erkenntnis, dass der Fuß beim Stehen üblicherweise leicht nach außen zeigt. Diese natürliche V-Stellung ist in einem normalen Skischuh nicht möglich. Dort werden die Füße in eine Parallelstellung gezwungen und die Knie damit leicht nach innen verdreht. Die patentierte Lösung des „Soma Tec" ermöglicht dagegen die natürliche V-Stellung, damit eine direktere Kraftübertragung und so ein sportlicheres und dynamischeres Skifahren bei geringerem Kraftaufwand.
Der Skischuh zeichnet sich durch klare Formen aus und ist damit ein bewusstes Gegenstück zur vorherrschenden organischen Formensprache im Skischuh-Design. Durch den neuen Entwicklungsansatz wurden die Technologie und das Design miteinander verschmolzen. Der Skischuh hat eine geschlossene, kompakte Form. Die Schnallen integrieren sich optisch in die Fläche und sind extrem flach.

This ski boot for sporty skiers is based on the ergonomic finding that a person's foot usually points slightly outwards when he is standing still. This natural V position is not possible in a normal ski boot. Normal ski boots force the feet into a parallel position, twisting the knees slightly inwards. By contrast, the solution patented in the Soma Tec allows this natural V position, and in doing so makes a more direct transmission of energy possible, which means more sporty and dynamic skiing using less energy.
The ski boot stands out by virtue of its clear forms, and is therefore a deliberate counterpart to the organic formal expression that dominates ski-boot design. The new development approach used here has merged technology and design. The ski boot has a closed, compact form. The buckles are integrated optically into the surface of the boot and are extremely flat.

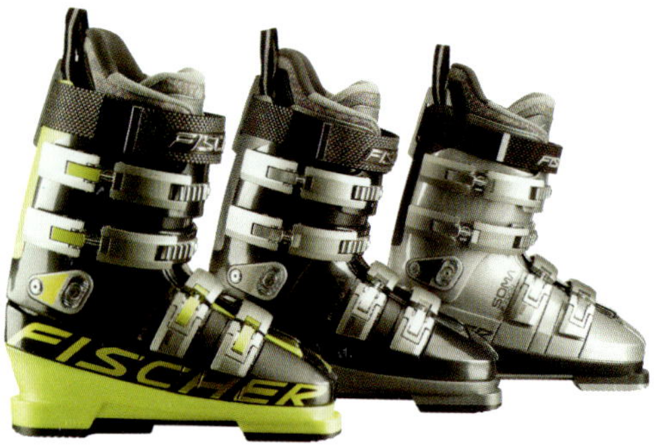

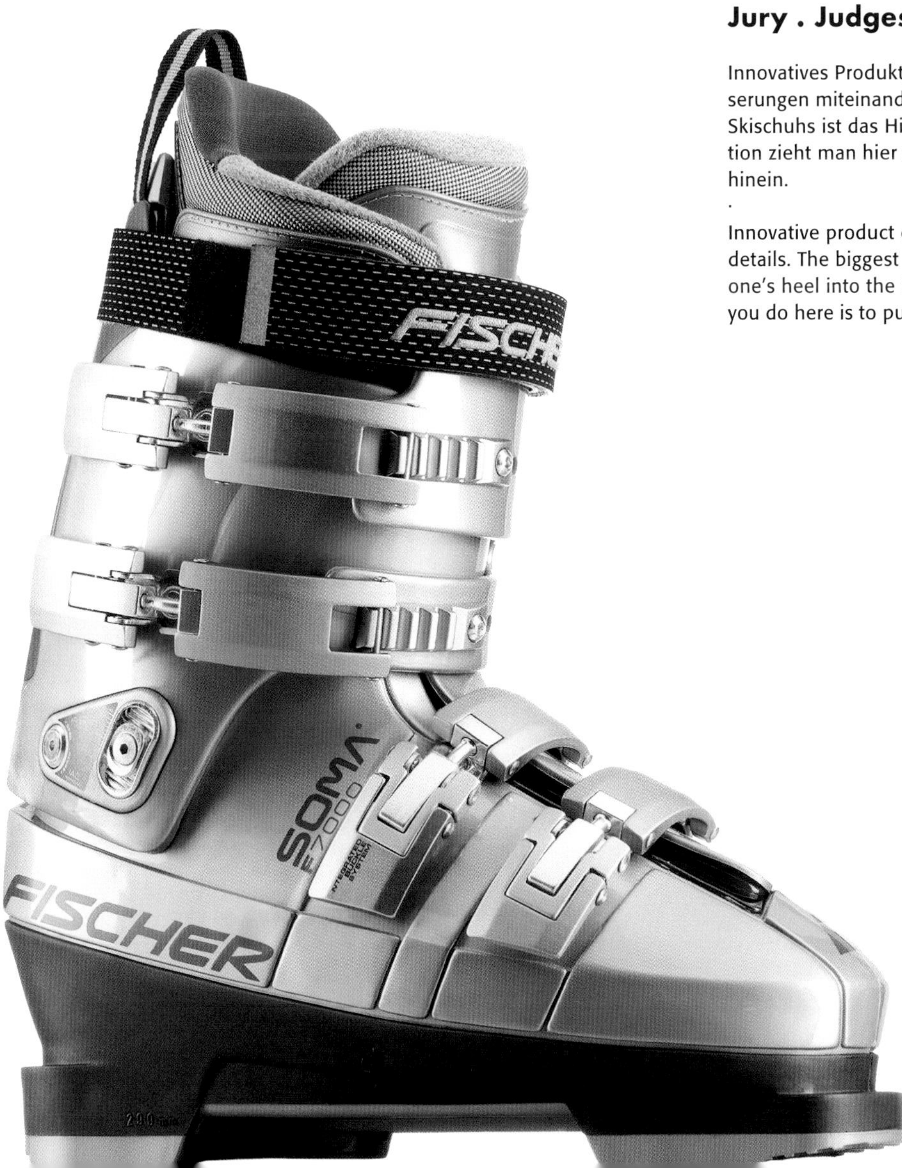

Jury . Judges panel

Innovatives Produkt, verbindet Sportlichkeit und Detailverbesserungen miteinander. Größtes Problem eines Vierschnallen-Skischuhs ist das Hineinschlupfen der Ferse. Dank einer Innovation zieht man hier einfach die Fersenkeile heraus und schlupft hinein.

.

Innovative product combining sporty qualities and improved details. The biggest problem with a four-buckle boot is getting one's heel into the boot. Thanks to an innovative feature, all you do here is to pull out the heel section and slide your foot in.

HERSTELLER . MANUFACTURER
Hollister's MotorCycles
Inh. Volker Sichler
D-78658 Zimmern-Horgen

VERTRIEB . DISTRIBUTOR
Hollister's MotorCycles
Inh. Volker Sichler
D-78658 Zimmern-Horgen

DESIGN . DESIGNER
Werksdesign
Volker Sichler

Hollister's Excite : Motorrad . motorcycle

Das Design von „Hollister's Excite" charakterisiert sich durch weiche, fließende und organische Formen. Das Motorrad ist vollständig CAD-konstruiert. Alle Teile sind dreidimensional gefertigt, beispielsweise fräst man die Felge am Stück aus einem 100 Kilogramm schweren Aluminiumblock.
Das Motorrad wird nur über Funk ohne Zündschloss gesteuert. Über diese Funkfernbedienung aktiviert man auch die Luftdruckstoßdämpfer vor der Fahrt, die auf Körpergewicht programmiert sind, und startet den programmierten MultiBike-Controller (MBC), der die gesamte Elektronik zentral steuert und überwacht.
Das handgefertigte Chassis ist aus Leichtaluminium. Eine spezielle Lackmischung führt bei jedem Lichteinfall zu überraschend neuen Farbspielen. Die Züge und die elektrischen Leitungen verlaufen unsichtbar.

.

The design of "Hollister's Excite" is characterized by soft, flowing, organic forms. The motorcycle is completely CAD-designed. All the parts are manufactured three-dimensionally. For example, the rims are milled in one piece from a 100-kilogram block of aluminium.
The motorcycle is radio-controlled only, without an ignition lock. Before driving, this radio remote control activates the pneumatic shock absorbers, which are programmed for body weight, and starts the programmed MultiBikeController (MBC), which centrally controls and monitors all the electronics.
The hand-made chassis is made of lightweight aluminium. A specially mixed paint means that there is a surprising play of colour every time light falls on the motorcycle. The cables and electrical wires are concealed.

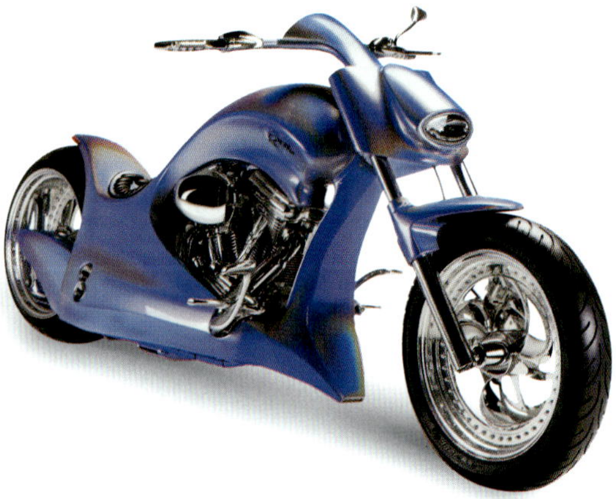

Jury . Judges panel

Die Jury zeichnet den Lenker aus; denn die Lenker-Details überzeugen, da hier zum ersten Mal Bremse, Handgriffe und Pneumatik einheitlich zusammengefasst werden. In seiner Perfektion heute auf dem Standard der besten Beispiele aus der Badmöbel- oder Badarmatur-Industrie.

.

The judges particularly single out the handlebar, because the details of its design are convincing, and brake, handgrips and pneumatics have been combined in a uniform way for the first time. This perfection places it on the same level as the best modern examples of bathroom furniture or bathroom fittings.

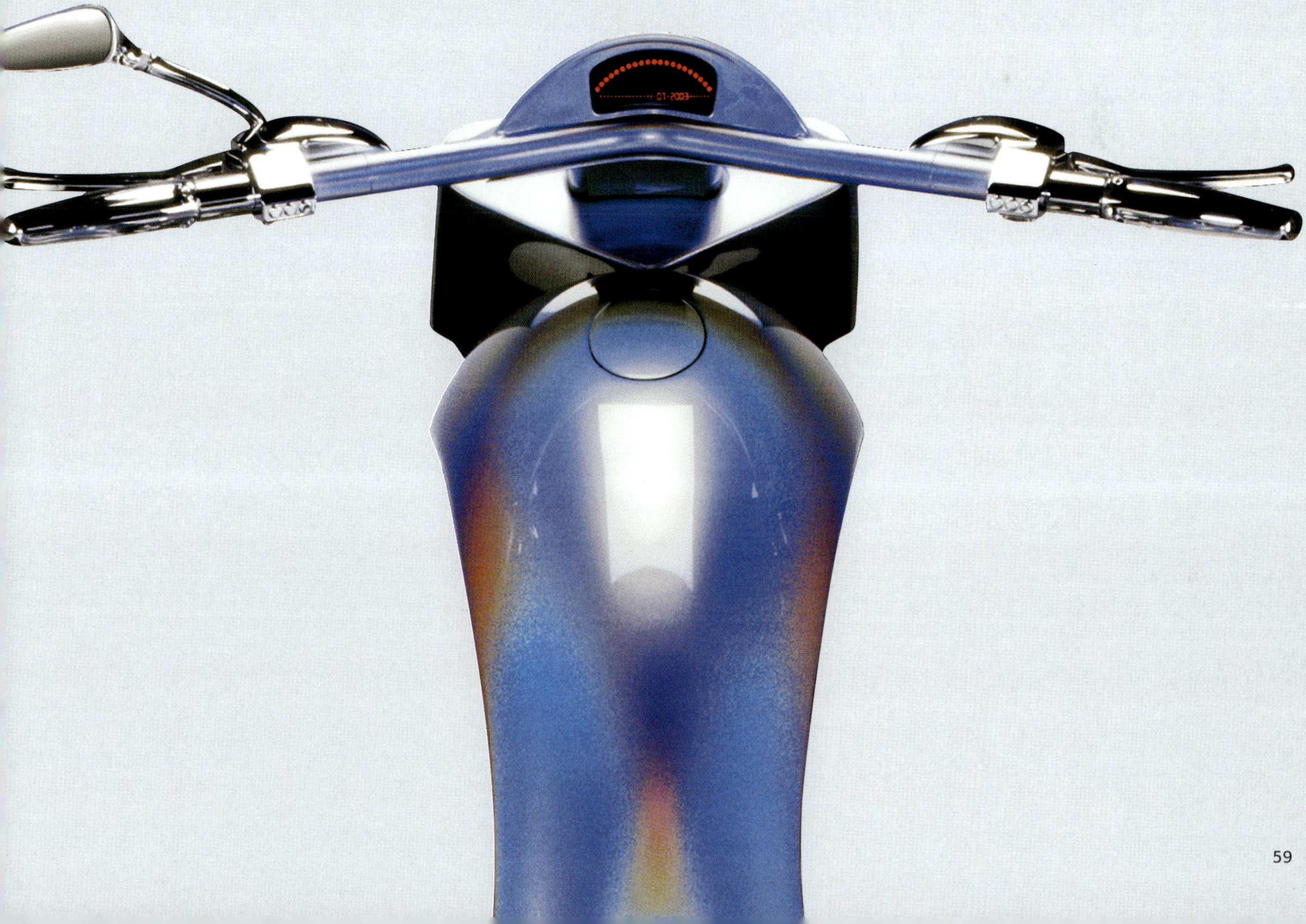

OUTDOOR + MOBILITÄT | SILBER
OUTDOOR + MOBILITY | SILVER

HERSTELLER . MANUFACTURER
Audi AG
D-85045 Ingolstadt

VERTRIEB . DISTRIBUTOR
Audi AG
D-85045 Ingolstadt

DESIGN . DESIGNER
Audi Design Team

Audi A 8

Diese Limousine aus dem Luxussegment zeichnet sich durch überzeugende Fahrdynamik, zukunftsweisende Technologie und Exklusivität in Design und Ausstattung aus. Die Schwerpunkte bei der Entwicklung des neuen Audi A 8 waren neben dem Design Leichtbau, Dynamik und Elektronik. Alle vier Bereiche eint die Idee der Leichtigkeit. Das bedeutet: Leichtigkeit in der Formgebung, in der Beschleunigung und im Gewicht, in der Fahrdynamik und schließlich in der Bedienung.
Das Design des Autos ist gekennzeichnet durch weit gespannte und deutlich überwölbte Flächen, eingefasst von präzise geführten Linien. Die Silhouette mit flacher Front und kurzen Überhängen sowie einer kräftig konturierten Heckpartie unterstreicht die Dynamik.
Bemerkenswert ist das geringe Gewicht der Aluminiumkarosserie in Audi-Space-Frame-Bauweise (ASF), der luxuriös gestaltete Innenraum und die herausragenden Fahreigenschaften.
Der Audi A8 ist nicht nur mit dem intuitiv erfassbaren und selbsterklärenden, integrierten Bedienkonzept Multi Media Interface (MMI), sondern auch mit der radargestützten Abstandsregelung Adaptive Cruise Control ausgestattet.

.

This luxury-class limousine stands out by virtue of its excellent handling, pioneering technology and exclusive design and features. When developing the new Audi A8, the main emphasis apart from design was on a lightweight construction, dynamism and electronics. Ease is common to all four aspects.
The result is lightness in terms of design and weight, coupled with effortless acceleration, handling and operation.
Its design is characterized by expansive, strikingly arched surfaces framed by precisely plotted lines. The silhouette of flat front and short body overhangs, together with the forceful contours of the tail, emphasize the car's dynamism.
Other remarkable features are the light weight of the aluminium body, built using the Audi Space Frame (ASF), its luxurious interior and its excellent handling characteristics.
The Audi A8 is equipped not only with Multi Media Interface (MMI), an intrinsically logical and self-evident operating concept, but also with radar-assisted Adaptive Cruise Control.

Jury . Judges panel

Audi zeigt in eindrucksvoller Weise, wie man ein innovatives Produkt machen kann ohne ein überlautes Produkt zu bekommen. Es geht sehr elegant mit Materialien um, es zeigt Dynamik, kaschiert die Größe und schafft so ein Designobjekt von hoher Faszination.

.

Here, Audi impressively demonstrates how to make an innovative product without getting an overloud product at the same time.
It presents its material qualities very elegantly, displays dynamism and disguises its size, and in this way creates an extremely fascinating design object.

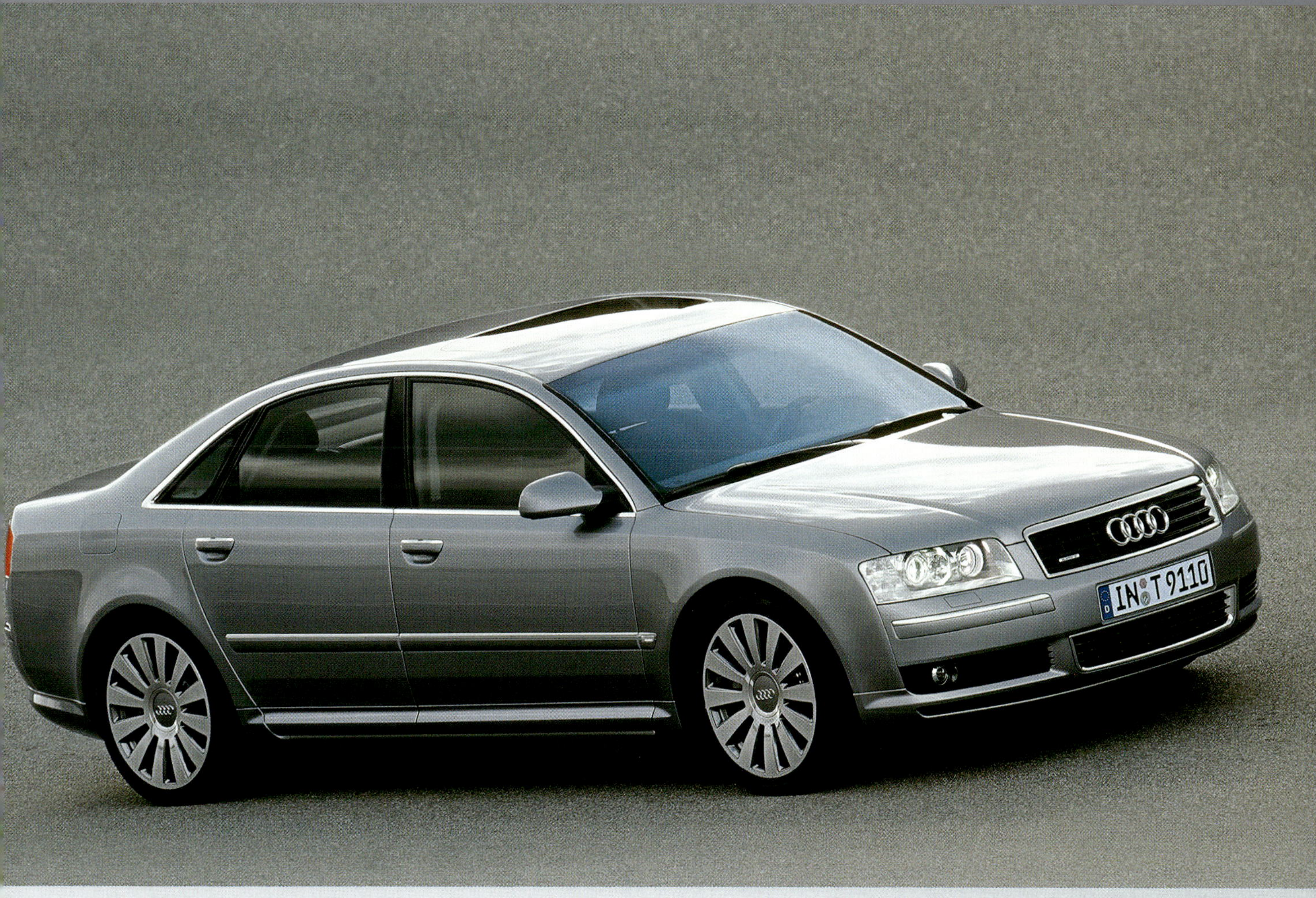

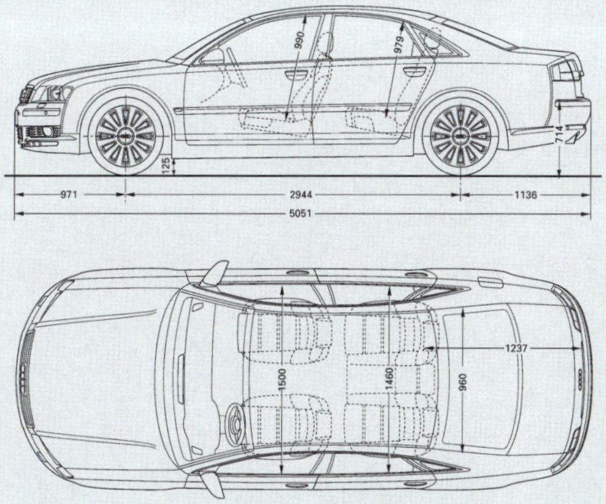
990
979
714
125
971
2944
1136
5051
1237
1500
1460
960

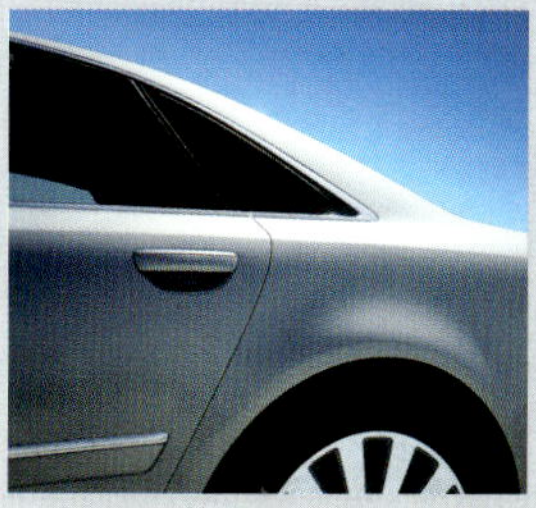

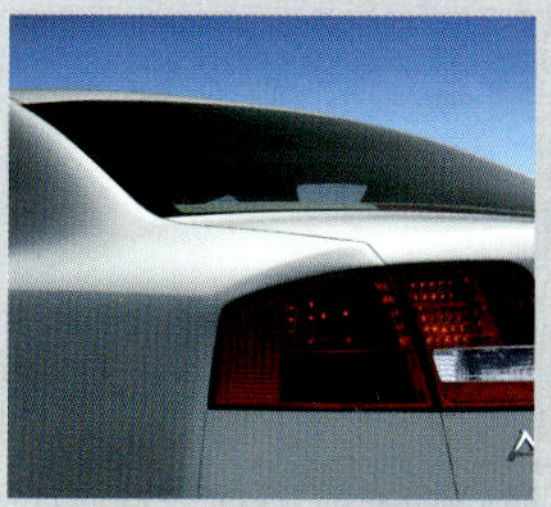

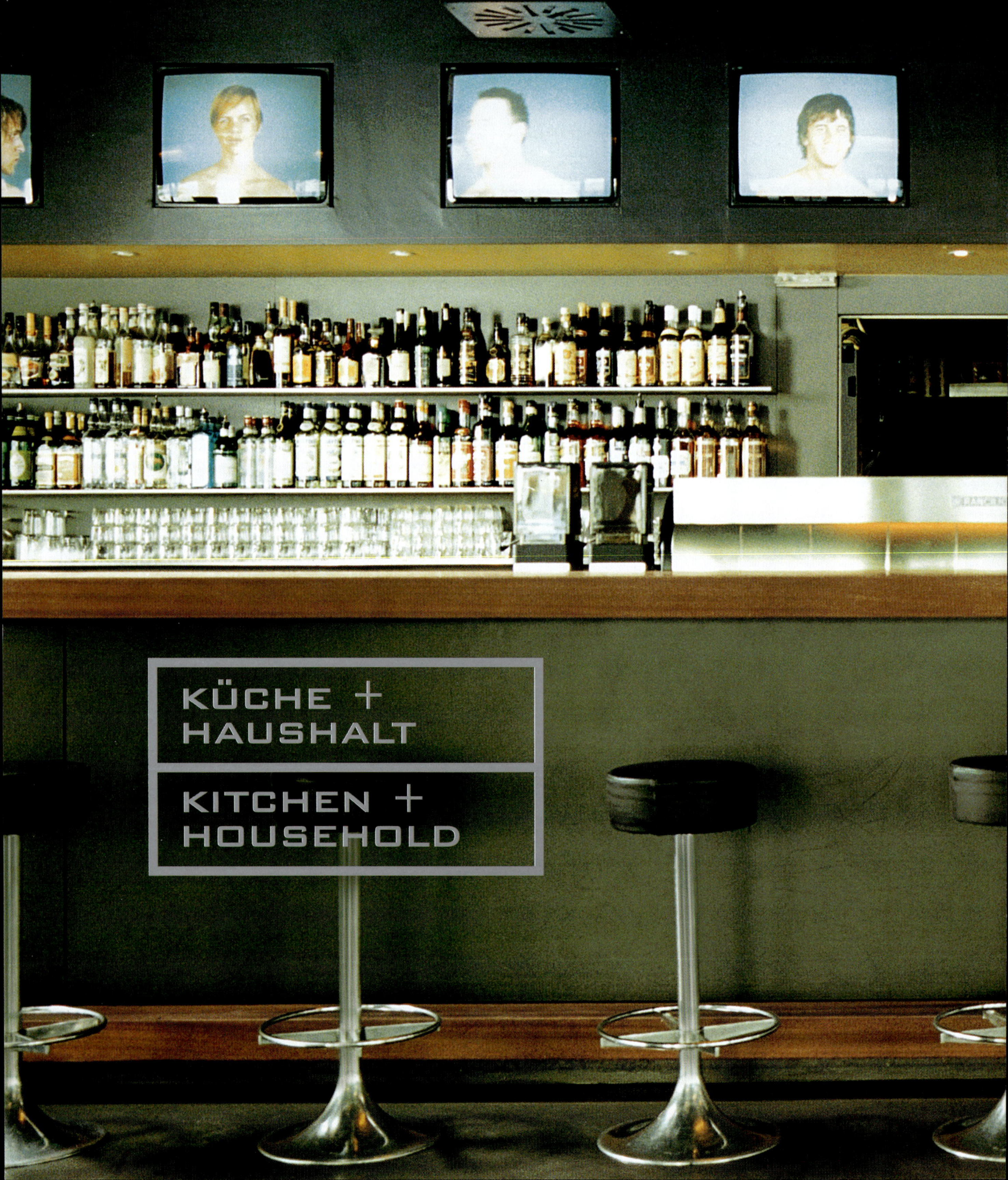

KÜCHE + HAUSHALT

KITCHEN + HOUSEHOLD

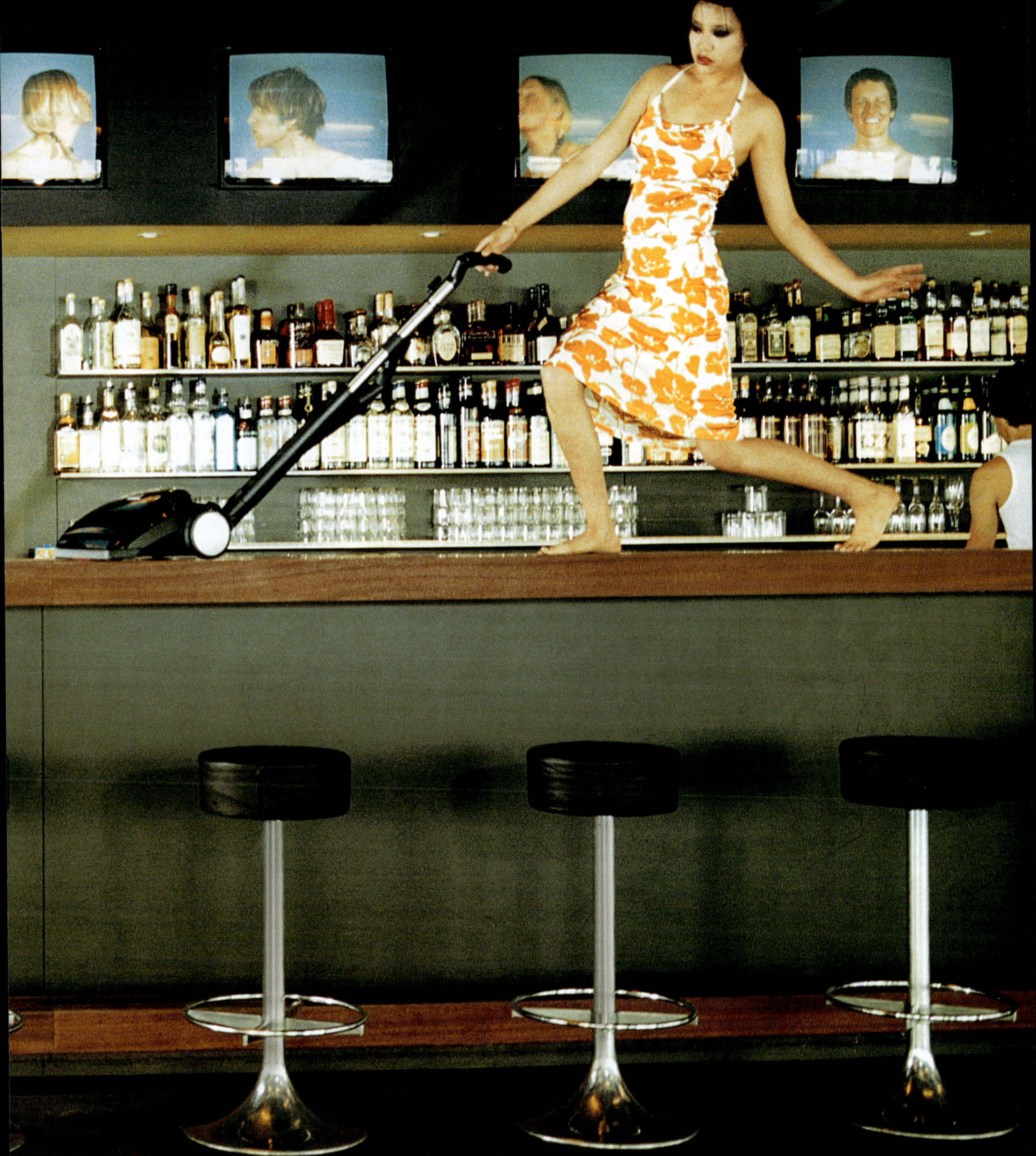

KÜCHE + HAUSHALT — GOLD

KITCHEN + HOUSEHOLD — GOLD

HERSTELLER . MANUFACTURER
Miele & Cie. GmbH & Co.
D-33332 Gütersloh

VERTRIEB . DISTRIBUTOR
Miele & Cie. GmbH & Co.
D-33332 Gütersloh

DESIGN . DESIGNER
Werksdesign
und
NPK Industrial Design
NL-2311 CD Leiden

Art. by Miele : Multifunktionsstaubsauger . multi-functional vacuum cleaner

Der Multifunktionsstaubsauger zeichnet sich durch eine hohe Designqualität und Funktionalität aus. Er besitzt nicht nur eine gute Saugleistung und Filtereigenschaft, sondern ist auch schnell betriebsbereit. „Art. by Miele" lässt sich flexibel einsetzen, das heißt er saugt verschiedene Bodenbeläge, Möbel und Polster. Dabei ist er auffallend wendig und sehr leicht und zielsicher zu bewegen.
Dank einem integrierten Saugschlauch mit Kombidüse müssen Zubehörteile nicht erst umständlich aufgesetzt oder zusammengesteckt werden. Die Kombidüse lässt sich sowohl als Fugendüse zum raschen Aufsaugen von Krümeln und Schnipseln als auch als Saugpinsel etwa zum Reinigen von Tastaturen nutzen. Schwer zugängliche Ecken, wie Raumecken und Heizkörpernischen, werden mit Hilfe des Teleskoprohres gesäubert.
Die hochwertigen Kunststoffe und die sehr gute Verarbeitung sorgen für Langlebigkeit. Der Multifunktionsstaubsauger ist in verschiedenen Farben und Dessins zu haben.

.

This multi-functional vacuum cleaner stands out by virtue of its high-quality design and functional features. Not only are its suction power and filtering properties good, but it is also ready to use very quickly. "Art. by Miele" can be used flexibly, cleaning different floor types, furniture and upholstery. At the same time, it is surprisingly manoeuvrable, and can be moved very easily and accurately.
Thanks to an integrated vacuum hose with a multi-purpose nozzle/brush, there is no need for time-consuming fitting or assembly of accessories. The multi-purpose nozzle/brush can be used both as a crevice tool for quickly vacuuming up crumbs and snippets and as a dusting brush, for cleaning keyboards for example. Corners that are difficult to reach, such as corners of rooms or gaps in radiators, can be cleaned with the help of a telescopic tube.
The high-quality plastic and excellent workmanship make this a long-lasting product. The multi-functional vacuum cleaner is available in various colours and designs.

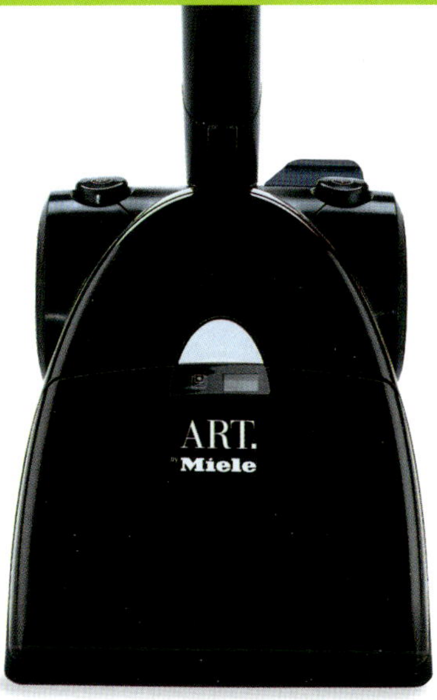

Jury . Judges panel

Eindrucksvoll, wie man eine innovative Handhabung verbunden mit Design in eine neue Dimension bringen kann. Das Produkt macht richtig Spaß, man kommt sehr leicht in alle Ecken; viele Hausmänner werden damit in Zukunft fröhlich zum Staubsauger greifen.

.

Impressive to see how innovative use combined with design can be applied to a new area. The product is a joy to use, you really can reach every corner with the greatest of ease. A lot of house husbands will be encouraged to gaily wield their vacuum cleaners in the future.

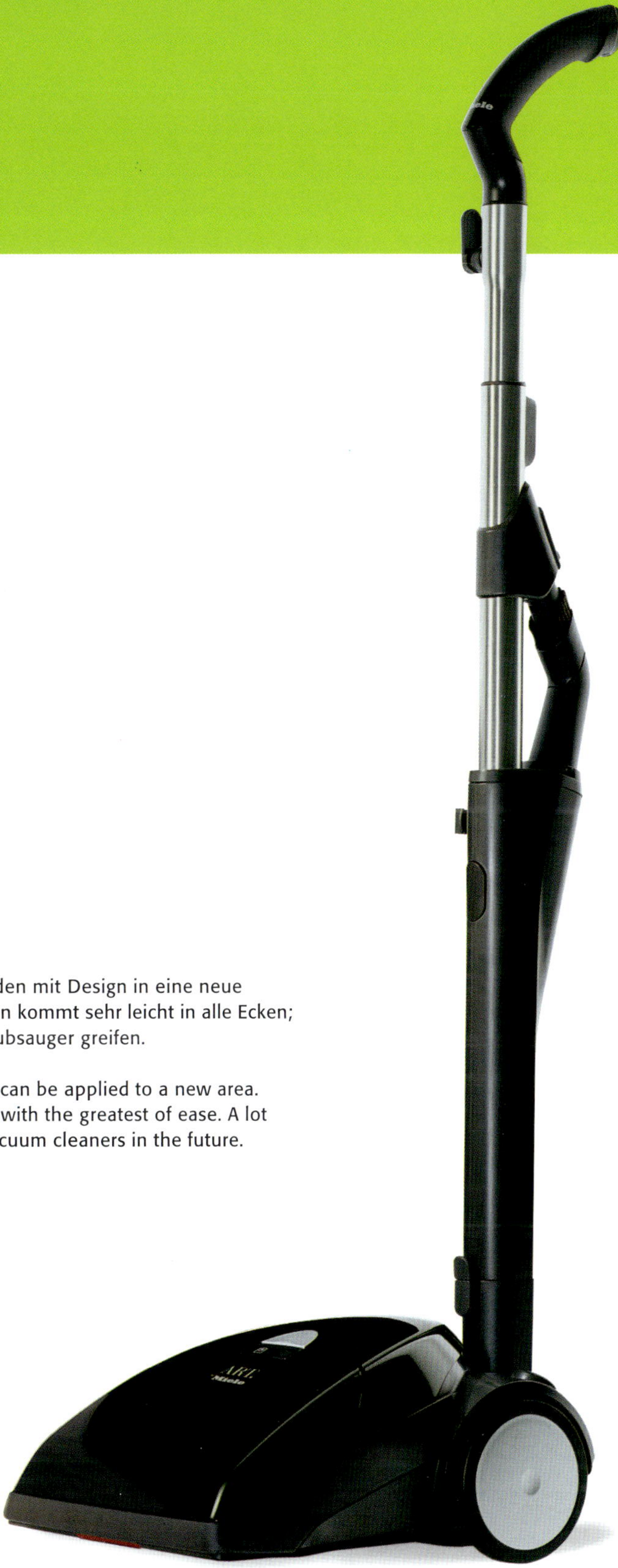

HERSTELLER . MANUFACTURER
AEG Hausgeräte GmbH
D-90429 Nürnberg

VERTRIEB . DISTRIBUTOR
AEG Hausgeräte GmbH
D-90429 Nürnberg

DESIGN . DESIGNER
AEG Industrial Design Center
Heinz-Peter Huxhold
D-90429 Nürnberg

Pyroluxe und Super Clean : Product Range . product range

Ein Herd, der sich selbst reinigt, wer wünscht sich das nicht? Mit „Pyroluxe", einem Selbstreinigungssystem für die Herdbaureihen Preference und Competence, ist das möglich. Zusammen mit dem Zubehör „Super Clean" erleichtert es wesentlich die Benutzung und Reinigung des Herdes.
Das Selbstreinigungssystem „Pyroluxe" verkohlt bei hohen Temperaturen die Verschmutzungen und lässt nur Asche zurück, die mit einem feuchten Tuch herausgewischt werden kann. Ein Katalysator verhindert nahezu vollständig Gerüche.
Der Vollauszug „Super Clean" erleichtert den Zugang zu den verschiedenen Ebenen im Backraum während des Back- oder Bratvorgangs. Jedes Blech kann in seiner vollen Größe herausgezogen werden. Die Beschichtung von „Super Clean" lässt sich mühelos und zeitsparend säubern, auch in der Spülmaschine. Bei der Gestaltung stand die benutzerfreundliche und ergonomisch sinnvolle Bedienung für Jung und Alt ebenso im Vordergrund wie leichte und schnelle Reinigung der Geräte.

.

Who would not want a cooker that cleans itself? This is what "Pyroluxe" offers, a self-cleaning system for the "Preference" and "Competence" cooker series. Together with its "Super Clean" accessories, it considerably eases using and cleaning the cooker.
The Pyroluxe self-cleaning system uses high temperatures to carbonize dirt, leaving only ash behind, which can be wiped up with a damp cloth. A catalyser almost completely prevents any odour.
The "Super Clean" fully extendable telescopic shelf system makes it easy to get at the various levels in the oven during baking or roasting. Each baking shelf can be pulled out to its full length. The "Super Clean" coating can be cleaned effortlessly and quickly, or even in the dishwasher. Its design was above all influenced by the desire for user-friendly, ergonomically sensible controls for young and old, and for appliances that can be cleaned easily and quickly.

Jury . Judges panel

Ein Produkt mit hoher Qualitätsanmutung, das Produktgrafik und Handhabung in vorbildlicher Weise miteinander verbindet und hier neue Wege geht.

.

A product with a high-quality feel, an exemplary and innovative combination of product graphics and operation.

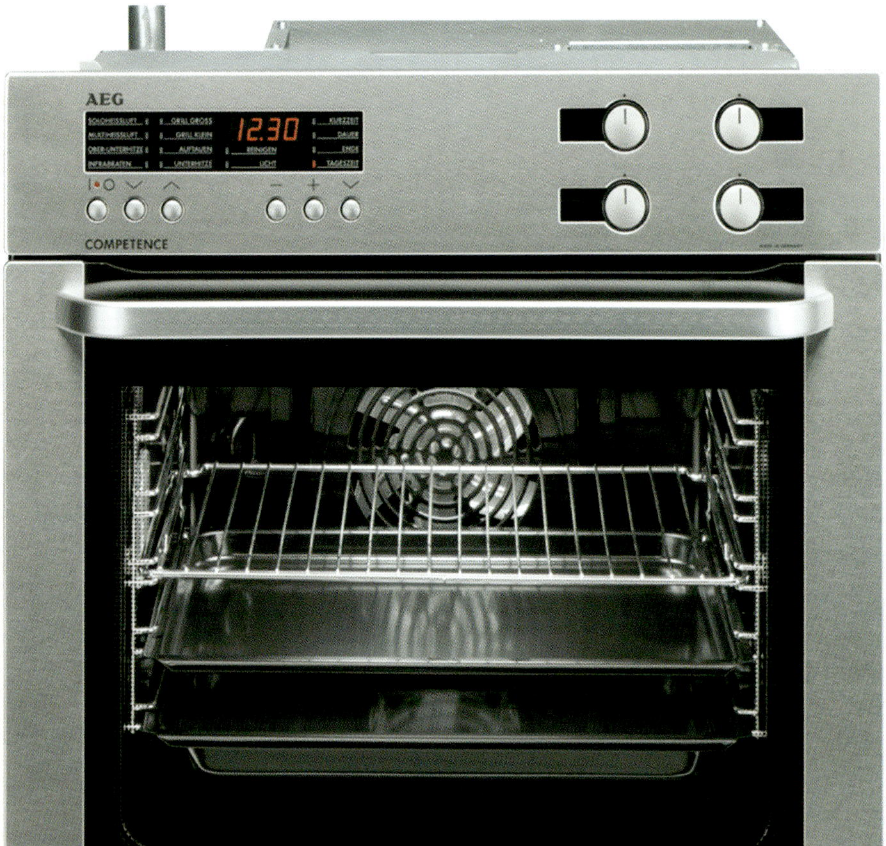
AEG
12.30
COMPETENCE

KÜCHE + HAUSHALT | SILBER
KITCHEN + HOUSEHOLD | SILVER

HERSTELLER . MANUFACTURER
Siemens Electrogeräte GmbH
D-81669 München

VERTRIEB . DISTRIBUTOR
Siemens Electrogeräte GmbH
D-81669 München

DESIGN . DESIGNER
Werksdesign
Gerd E. Wilsdorf

EK 78 H 55 : Kochfeld . hob

Auf diesem Kochfeld zerkocht nichts, läuft nichts mehr über – nicht einmal die Milch – und entflammt kein Fett in der Pfanne. Ein Kochsensor und ein Bratsensor machen dies möglich.
Der ausfahrbare Kochsensor nimmt dem Benutzer lästiges Nachregeln der Temperatur ab. Mit ihm kann man die Temperatur im Topf punktgenau kontrollieren.
Der Bratsensor verbirgt sich unter der Glaskeramik und regelt die Temperatur in der Pfanne optimal. Er verhindert die Entflammung von Fett und vermeidet, dass die Pfanne durch zu große Hitze beschädigt wird.
Das Kochfeld zeichnet sich durch ein zeitloses Design aus. Es ist klar, übersichtlich und funktional gestaltet.

Nothing can boil dry or boil over on this hob – not even milk – and no fat can catch fire in the frying pan. This is made possible by sensors to detect liquid and fat temperatures.
The retractable boiling temperature sensor does the tiresome job of adjusting temperature for the user. It allows temperature in the pot to be controlled accurately.
The hot fat sensor is concealed beneath the ceramic hob, providing perfect control of frying pan temperature. It prevents fat catching fire and prevents the frying pan from being damaged by excessive temperatures.
The hob stands out by virtue of its timeless design. Its layout is clear, logical and functional.

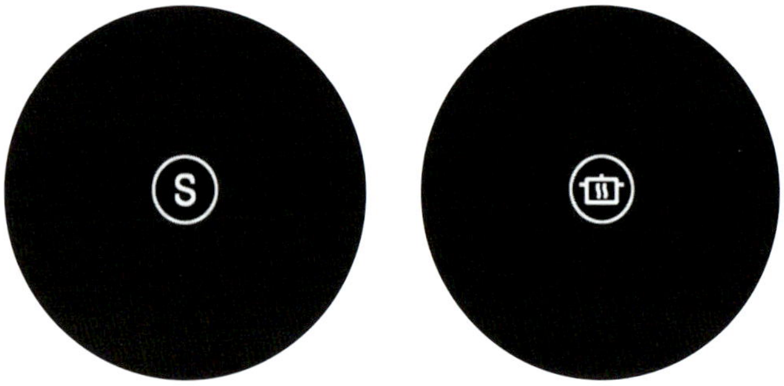

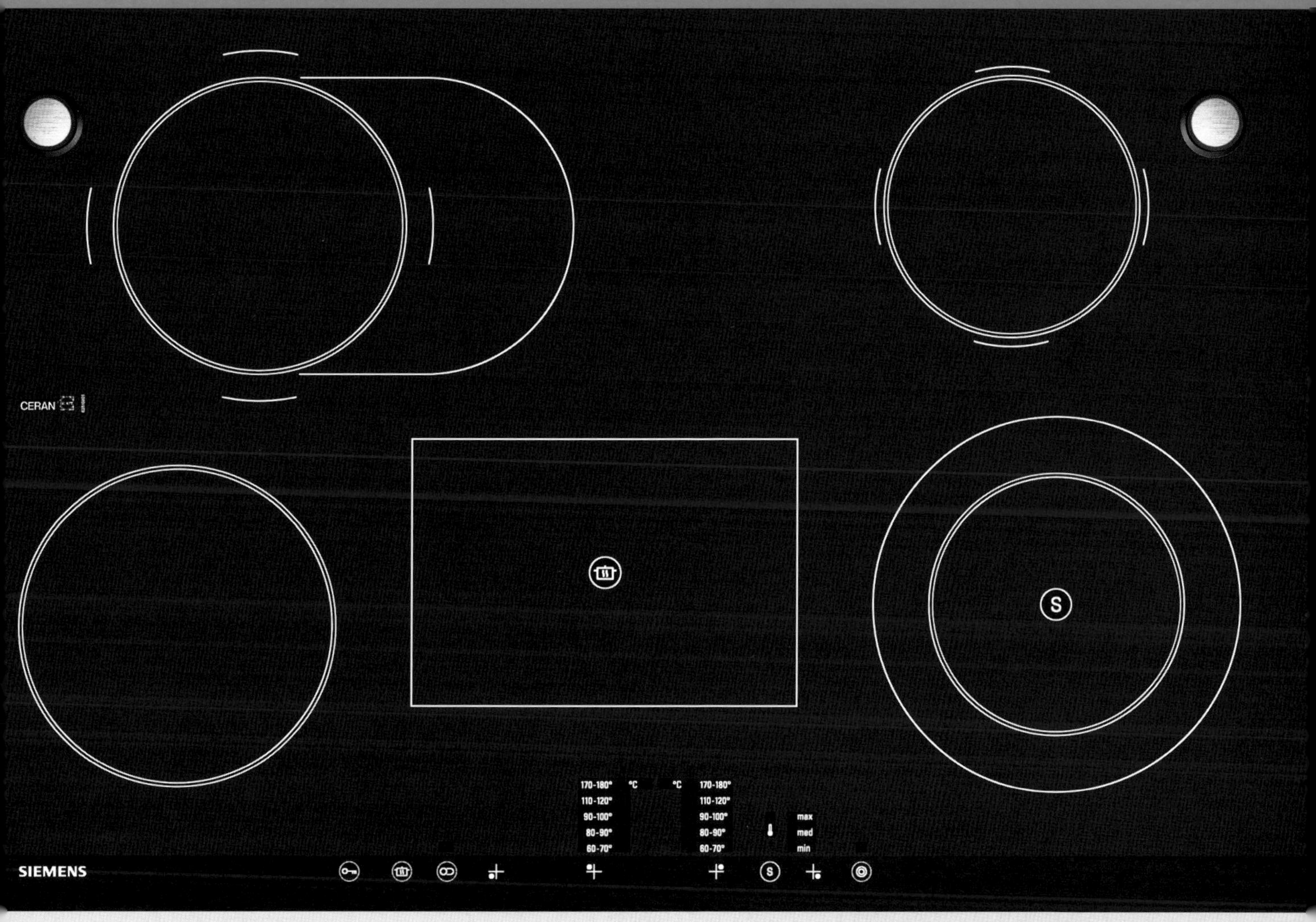

Jury . Judges panel

Ein sehr schlichtes, funktionales Produkt mit klarer Grafik. Sehr übersichtlich, man orientiert sich leicht. Sehr ansprechend zum Kochen. Empfehlenswert.

.

A very uncomplicated, functional product with clear graphics. Very logically arranged, you quickly find your way around. Very much makes you want to get cooking. To be recommended.

HERSTELLER . MANUFACTURER
Robert Bosch Hausgeräte GmbH
D-81669 München

VERTRIEB . DISTRIBUTOR
Robert Bosch Hausgeräte GmbH
D-81669 München

DESIGN . DESIGNER
Werksdesign
Hans-Reinhart Janssen
Robert Bosch Hausgeräte GmbH
D-89537 Giengen

EasyStore Bosch KTR 18420 : Tisch-Kühlautomat . refrigerator

Das neue Kühlgerät unterscheidet sich völlig von herkömmlichen Modellen: Alle Ablageschalen lassen sich durch Teleskopschienen bequem wie eine Schublade herausziehen. Das erleichtert gerade das Einlagern und den Zugriff von oben auf die Lebensmittel, bietet einen besseren Überblick über alle Vorräte, auch auf deren Verfallsdatum. Außerdem wurde eine um dreißig Prozent größere, durchgängige Nutzfläche geschaffen.
Flexibilität und Benutzerfreundlichkeit bestimmten die neue Gestaltung.
So sorgen flexibel einsetzbare Zubehörteile, wie Raumteiler, Flaschenhalter und Einhängeschale, in den verschiedenen Ebenen für Ordnung und Übersichtlichkeit. Weiteren Komfort bietet die in die Deckplatte integrierte Bedieneinheit. So lässt sich, ohne die Tür öffnen zu müssen, die Temperatur elektronisch präzise regeln. Die Leuchte unterhalb der Arbeitsplatte ermöglicht selbst bei voller Beladung des Geräts eine optimale Ausleuchtung.
.

This new refrigerator is completely different from conventional models: telescopic rails allow all the shelves to be pulled out with ease like drawers. This makes storing food away and getting it out of the refrigerator easier, and helps the user to keep better track of the provisions in the refrigerator and their use-by dates. In addition, usable space has been increased thirty percent and is unrestricted.
The new design is influenced by flexibility and user-friendliness. A flexible range of accessories such as partitions, bottle holders and suspended compartments provide for order and tidiness on every level. One further convenient feature is the control panel integrated in the cover panel. This allows temperature to be controlled electronically and precisely without opening the door. The lamp mounted on the under surface of the worktop provides for excellent illumination even when the appliance is full.

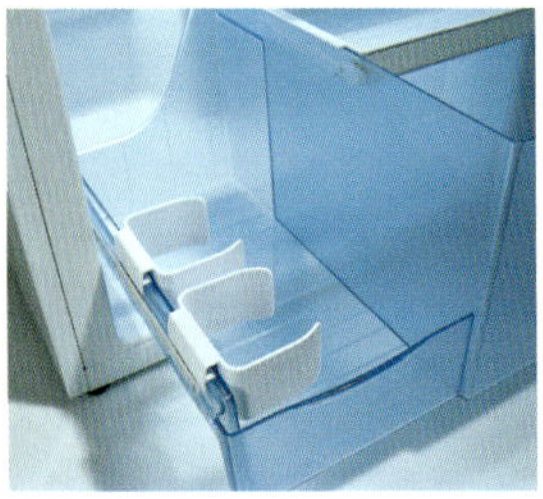

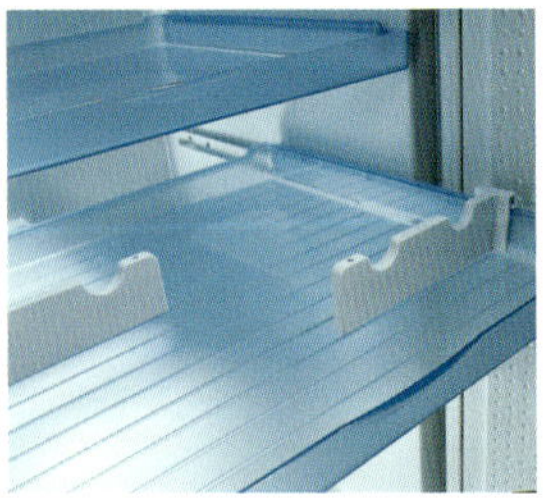

Jury . Judges panel

Sehr normales und schlichtes Äußeres, aber sehr guter Innenraum. Die Produkte kommen jetzt zu mir. Ich muss nicht mehr hinknien und erst andere Lebensmittel wegräumen, um etwas von hinten zu holen. Sehr schön auch die Elemente der Innenaufteilung, sodass die Produkte nicht verrutschen.

.

Very normal, plain exterior, but very good interior. Now the products come to me. I no longer have to kneel down to get things from the back of the fridge, or indeed get other food out of the way first. The internal partitioning elements are also very well done, preventing products from sliding around.

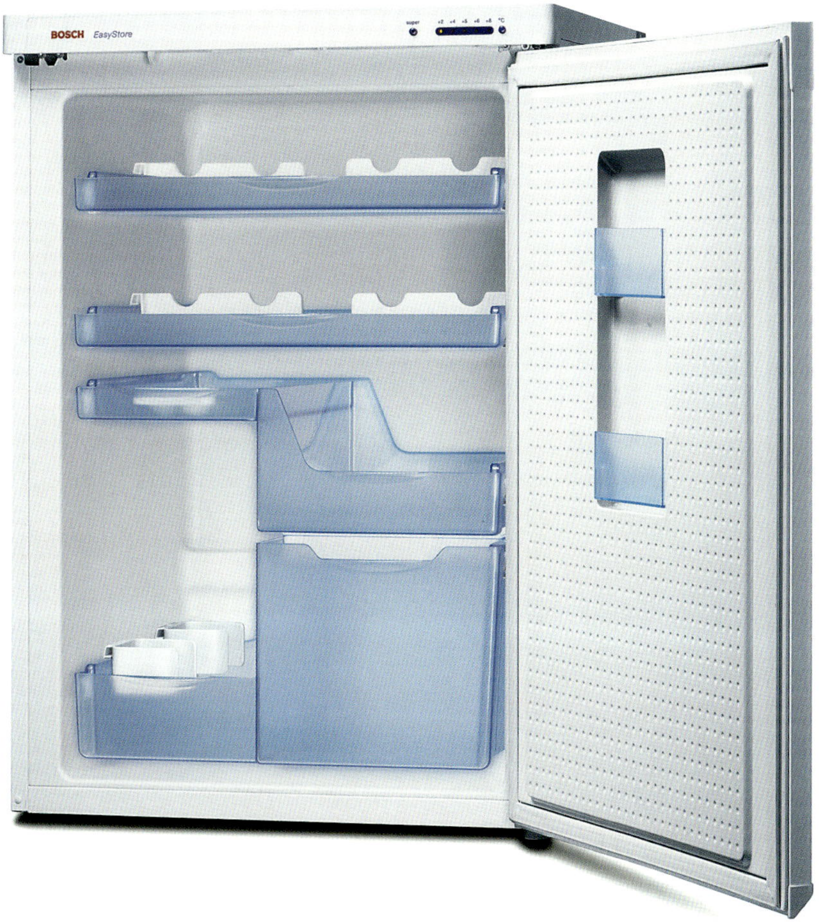

HERSTELLER . MANUFACTURER
Kahla/Thüringen Porzellan GmbH
D-07768 Kahla

VERTRIEB . DISTRIBUTOR
Kahla/Thüringen Porzellan GmbH
D-07768 Kahla

DESIGN . DESIGNER
Werksdesign
Cornelia Müller

Abra Cadabra : Porzellanset . chinaware

„Abra Cadabra" besteht nur aus drei Basisteilen: Schälchen, quadratische und rechteckige Platte. Dazu kommen zwei Zusatzteile: extralanges Tablett und eckiger Deckel. Alle Teile lassen sich unterschiedlich miteinander kombinieren und vielseitig einsetzen. So eignen sich die Schälchen zum Servieren von Süßigkeiten, Knabbereien oder Konfitüre bis hin zu Dips und Fingerfood. Selbst als Windlichter lassen sie sich einsetzen, besonders schön in der farbigen Version, da das Licht reflektiert wird. Die quadratische Platte dient als Tablett für Snacks oder Gläser und für die anderen Set-Teile. Der Deckel kann nicht nur als Abdeckung, sondern auch als Untersetzer oder zum Servieren von Snacks genutzt werden. Die Formensprache ist klar und übersichtlich auf dem Rasterprinzip aufgebaut. Alle Teile lassen sich dadurch nicht nur miteinander kombinieren, sondern passen auch problemlos in unterschiedliche Umgebungen. Das Material Porzellan hat durch seine Langlebigkeit einen hohen Gebrauchswert und ist spülmaschinenfest.

"Abra Cadabra" comprises just three basic parts: bowl, square and rectangular plate. Then there are two additional parts: an extra-long tray and a rectangular lid. All these parts can be combined with each other and used in different ways. The bowls can be used to serve confectionery, nibbles or jam, or even dips and finger food. They can even be used as lanterns, to particularly beautiful effect in the case of the coloured version, as light is reflected. The square plate can serve as a tray for snacks or glasses, and for the other components in the set. The lid can be used not only as a lid, but also as a coaster, or to serve snacks. Formal expression is clear, based on a grid pattern. Not only can all the parts be combined with each other, they also blend in harmoniously with a variety of environments. The long-lasting china used here has a high utility value, and is dishwasher-safe.

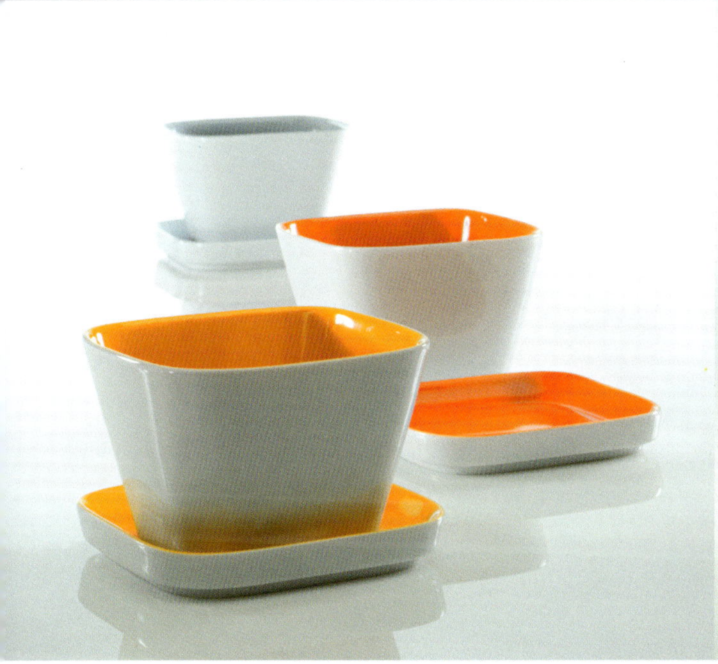

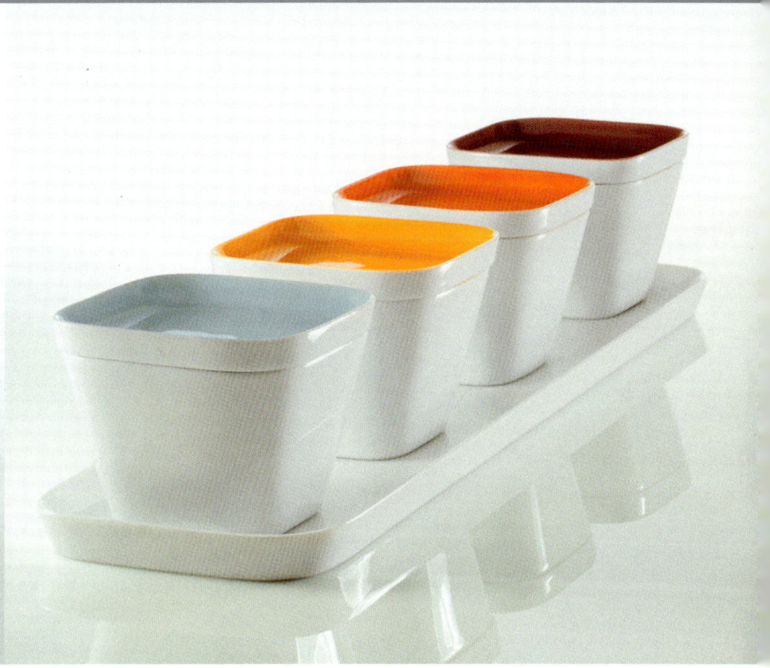

Jury . Judges panel

Sehr einfach und gleichzeitig elegant. Sehr nützlich. Man kann es für verschiedene Snacks oder was immer man in Schälchen tun kann, verwenden. Es ist eine Verknüpfung von Funktionalem mit einer sehr angenehmen Farbkombination. Eine sehr gute Verbindung.

.

Very simple, yet elegant at the same time. Very useful. You can use it for a variety of snacks or for whatever else can be put into bowls. It combines functionality with a very pleasant colour scheme. An excellent combination.

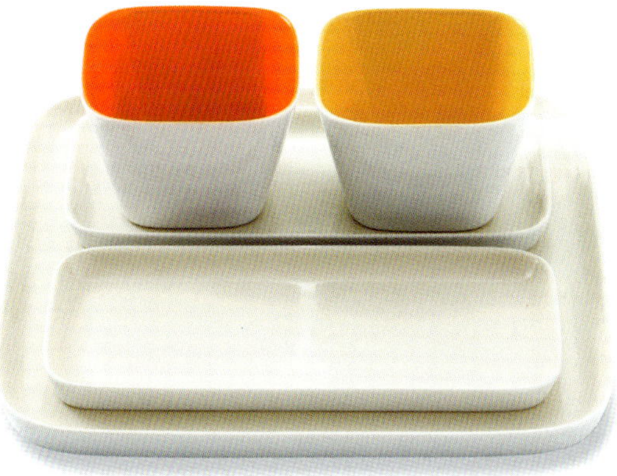

KÜCHE + HAUSHALT | SILBER

KITCHEN + HOUSEHOLD | SILVER

HERSTELLER . MANUFACTURER
Kahla/Thüringen Porzellan GmbH
D-07768 Kahla

VERTRIEB . DISTRIBUTOR
Kahla/Thüringen Porzellan GmbH
D-07768 Kahla

DESIGN . DESIGNER
Werksdesign
Barbara Schmidt

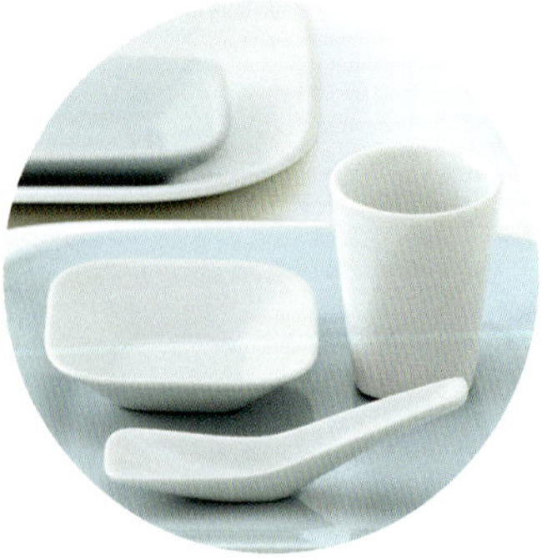

Cumulus : Geschirrprogramm . range of crockery

„Cumulus" ist ein verdichtetes Geschirrprogramm, das Freiräume für Experimentierfreude schafft. Es besteht aus wenigen Teilen mit beinahe unbegrenzten Möglichkeiten: quadratische und rechteckige Platten, Schalen und Schälchen, ein Amuse-Gueule-Löffel und drei Tassen. Jeder Nutzer kann die Funktionen selbst definieren. Von den Schalen und Platten kann gegessen werden, sie lassen sich als Serviergefäße einsetzen oder beispielsweise bei einem Buffet auf unterschiedliche Art auf- und nebeneinander gruppieren. Ihre Proportionen ermöglichen ganz verschiedene Konstellationen. Dank der ausgesparten Ecke liegt das Tablett gut in der Hand. Das Schälchen mit dem Ausgießer lässt sich als Milchkännchen oder für heiße Butter und Soße verwenden. „Cumulus" wirkt luftig und leicht. Es hat ungewöhnliche Formen mit gerundeten Ecken.
Das aus natürlichen Rohstoffen gebrannte Porzellan wird von Hand glasiert. Zusätzlich zur transparenten weißen Glasur wurde die Glasur „aqua" entwickelt, die an den Gefäßrändern dünner liegt und sich in ihrer Tiefe sammelt.

"Cumulus" is a compact crockery range which leaves room for experimentation. It comprises just a few parts with almost unlimited possibilities: square and rectangular plates, bowls and side-bowls, an "amuse-gueule" spoon and three cups. It is up to each user to define their functions himself. Food can be eaten from the bowls and plates, they can be used as serving vessels or be grouped on top of or next to each other – at a buffet, for instance. Their proportions allow quite different constellations. The tray fits well in the hand thanks to a recessed corner. The bowl with a spout can be used as a milk jug or for melted butter or gravy. "Cumulus" makes an airy, light impression. It has unusual shapes, with rounded corners.
The china is made of natural materials and glazed by hand. In addition to the transparent white glaze, an »aqua« glaze has been developed, which is applied more thinly at the edges and collects at the bottom of the vessels.

Jury . Judges panel

Formale Innovation bei einigen Geschirrteilen, beispielsweise bei den Schalen, dem Tablett und dem Schälchen mit Ausgießer. Einige andere sind noch etwas verbesserungsfähig.

·

Formal innovation in some of the pieces of china, such as the bowls, the tray and the bowl with a spout. Some of the other pieces could still be improved somewhat.

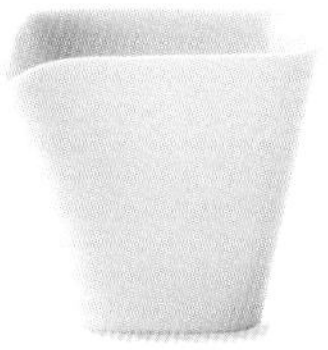

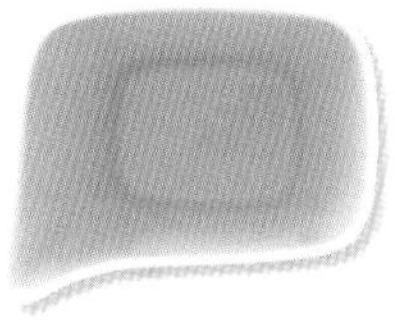

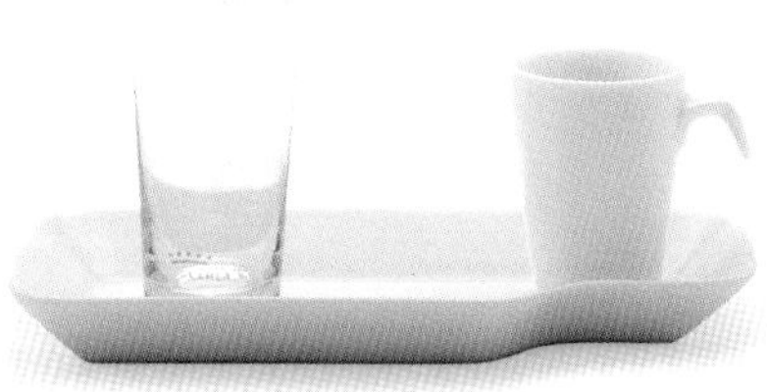

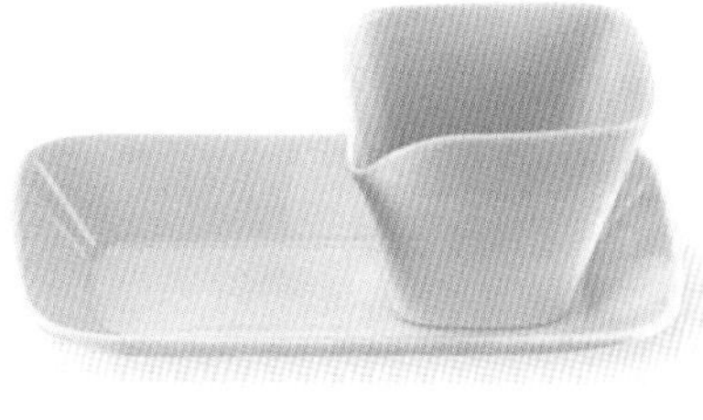

HERSTELLER . MANUFACTURER
WMF Württembergische Metallwarenfabrik AG
D-73312 Geislingen/Steige

VERTRIEB . DISTRIBUTOR
WMF Württembergische Metallwarenfabrik AG
D-73312 Geislingen/Steige

DESIGN . DESIGNER
Designpraxis Diener
D-89079 Ulm

Top Tools : Schöpf- und Schaumlöffelserie . series of ladles and skimmers

„Top Tools" sind professionelle Arbeitsgeräte zum Kochen, Schöpfen und Servieren. Alle Geräte sind sorgfältig gemäß ihrer jeweiligen Funktionen gestaltet worden. Der voluminöse Griff liegt gut in der Hand. Nicht nur durch die klare funktionale Formgebung, sondern auch durch das Material – matt gebürsteter Edelstahl – passen sich „Top Tools" hervorragend der Edelstahloptik eines modernen Küchenambientes an. Zum Sortiment gehört auch eine dazu passende Hängeleiste, auf der fünf Arbeitsgeräte Platz finden.

.

"Top Tools" are professional implements for cooking, ladling and serving. All the individual items have been carefully designed taking their respective functions into account. The bulky handle fits well in the hand. Not only the "Top Tools" clear, functional design, but also their material – matt brushed stainless steel – mean that they harmonize excellently with the stainless steel look of a modern kitchen environment. The range also includes a matching hanging rail, which has room for five of these implements.

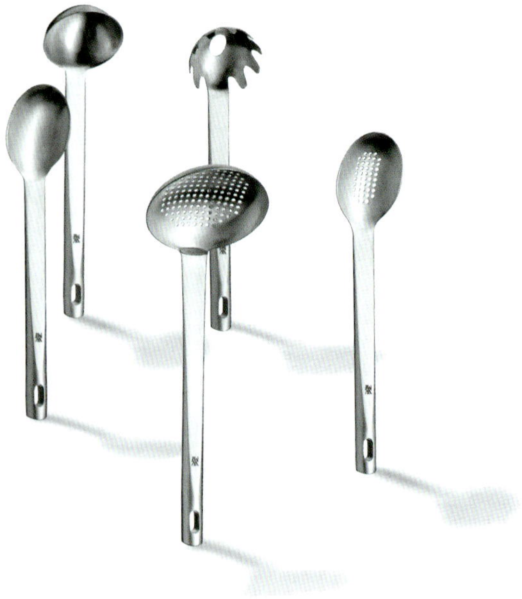

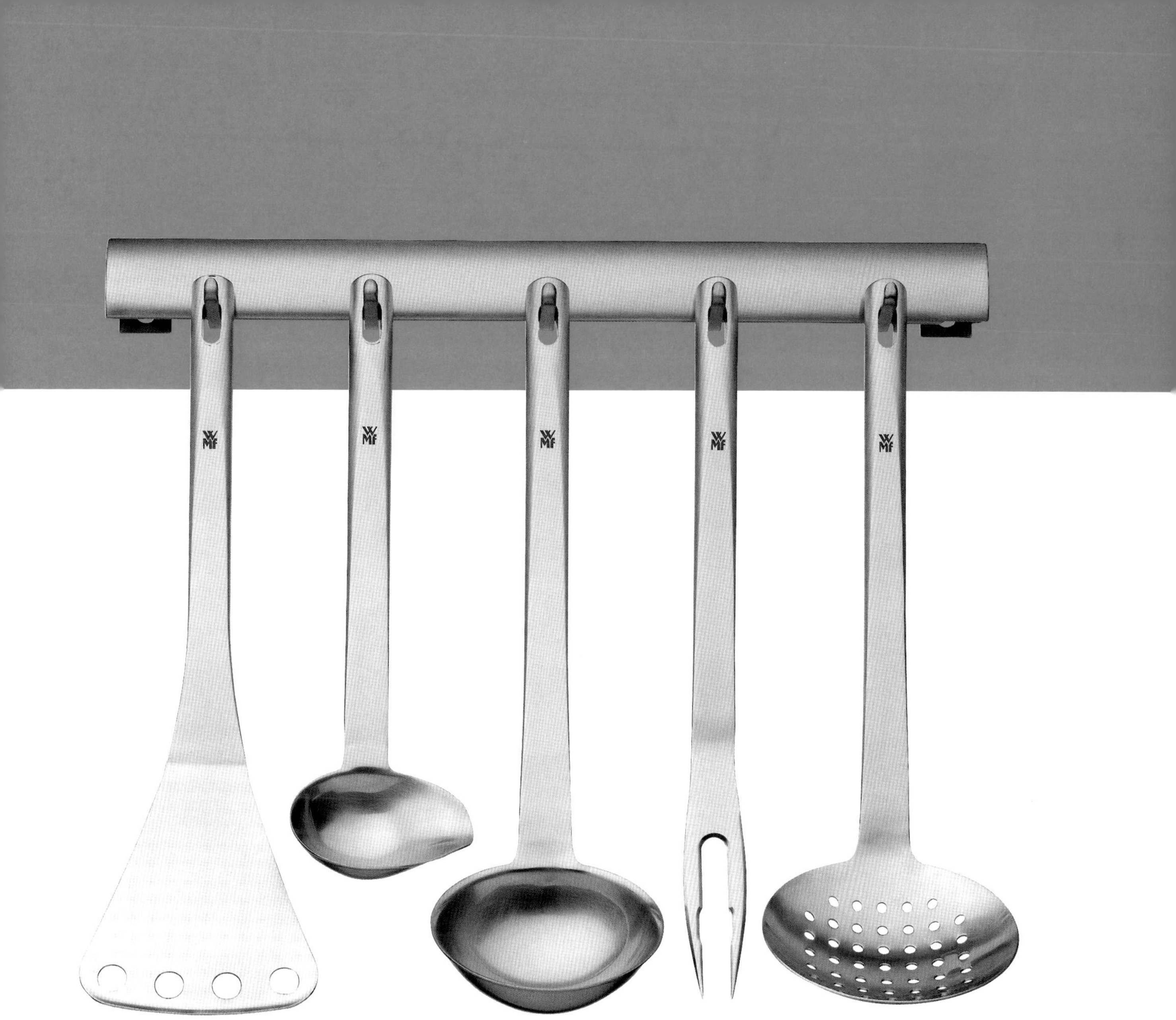

Jury . Judges panel

Das Zubehör macht Sinn, weil man verschiedene Formen hat für verschiedene Tätigkeiten in der Küche. Die Geräte sind leicht zu handhaben und verfügen über einen guten Griff, ohne dass sie schwer sind. Haptik und Optik vereint.

.

The accessories make sense, because there are different forms for the different activities in the kitchen. The items of equipment are easy to use and have good handles without being heavy. A combination of tactile and visual qualities.

HERSTELLER . MANUFACTURER
Robert Herder GmbH & Co. KG
Windmühlenmesser
D-42697 Solingen

VERTRIEB . DISTRIBUTOR
Robert Herder GmbH & Co. KG
Windmühlenmesser
D-42697 Solingen

DESIGN . DESIGNER
Werksdesign
Giselheid Herder-Scholz

Parmoulin : Parmesan- und Käsemesser . parmesan and cheese knife

„Parmoulin" gehört zu der Serie „Unikate aus Tradition". Stahl und Holz vereinen sich hier zu hoher Funktionalität. Das Parmesan- und Käsemesser ist nicht ganz so dünn geschliffen wie sonst die Windmühlenmesser – denn zum Brechen von Parmesan braucht man eine stabile Klinge –, aber es ist allemal dünn und scharf genug, um zähe Käserinde und Hartkäse mühelos zu durchteilen.
Die stark geschwungene Form der Klinge erlaubt es, beim Schneiden abzurollen. Die Klinge ist von Hand blaugepließtet, im traditionellen Solinger Schleifhandwerk die feinste Bearbeitung der Klingenoberfläche. Den Griff gibt es in österreichischem Pflaumenholz, spanischem Olivenholz oder Schwarzwälder Birnenholz. Letzteres wird mit heißem Wasserdampf gedämpft, wodurch das Holz eine stabile innere Struktur bekommt. Der ovale Griff liegt angenehm in der Hand. Die geschwungene Form des Messers, die der gewinkelten Stellung der Hand beim Schneiden nachempfunden ist, bewirkt den leichten Umgang.

The "Parmoulin" is part of the "Unikate aus Tradition" series. Here, steel and wood have been combined to make a highly functional product. The Parmesan and cheese knife has not been ground as thin as the other knives made by Windmühlenmesser – for a sturdy blade is needed to break Parmesan – but it is at all events thin and sharp enough to cut through tough cheese rind and hard cheese with no effort. The pronounced curve of the blade means that cutting can be done with a rocking motion. The blade has been honed by hand, using the finest blade surface processing that Solingen's long tradition of hand-grinding has to offer. The handle is available in Austrian plum, Spanish olive or Black Forest pear. The latter is treated with hot steam, which gives the wood a stable inner structure. The oval handle fits pleasantly in the hand. The curved form of the knife, which alludes to the angled position of the hand when cutting, makes it easy to use.

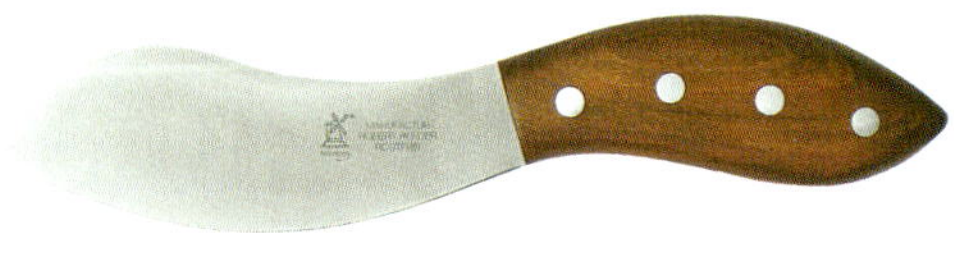

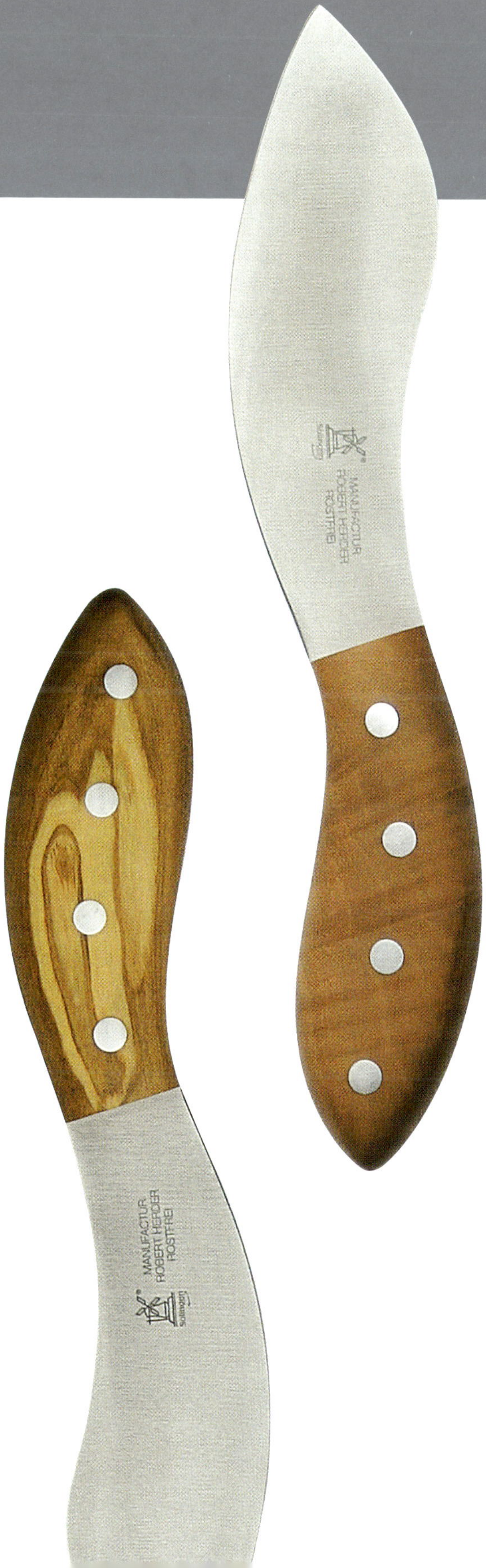

Jury . Judges panel

Ein Produkt für Liebhaber von sehr guten Küchengeräten, zu dem man eine emotionale Beziehung entwickelt und das man für immer behalten möchte. Deshalb macht es Sinn, dass man es zusammen mit Öl zur Pflege des Holzes bekommt. Sehr gutes Design. Die haptische Qualität von Holzgriffen ist unübertroffen.

A product for lovers of excellent kitchen tools, to which you can develop an emotional relationship and which you would like to keep forever. This is why it makes sense to supply the knife together with oil to treat the wooden handle. Excellent design. The feel of wooden handles cannot be beaten.

KÜCHE + HAUSHALT | SILBER
KITCHEN + HOUSEHOLD | SILVER

HERSTELLER . MANUFACTURER
W. F. Kaiser und Co. GmbH
D-65582 Diez/Lahn

VERTRIEB . DISTRIBUTOR
W. F. Kaiser und Co. GmbH
D-65582 Diez/Lahn

DESIGN . DESIGNER
Werksdesign

La Forme Quadro : Springform . springform

Eine Weltneuheit ist die quadratische Springform. Traditionelle Torten in neuer quadratischer Form und extra hohe Blechkuchen bringen Vielfalt und Abwechslung auf die Kaffeetafel und erlauben es, die Kuchen neu zu portionieren. Beim Backen muss aber nicht erst mühsam umgerechnet werden, denn „La Forme Quadro" ist für alle Springformrezepte der Standardgröße geeignet. Der Boden der quadratischen Springform ist schnittfest. „La Forme Quadro" lässt sich leicht zusammensetzen und einfach reinigen.

.

This square springform is a world first. Traditional gateaux in a new square shape or square sponge cakes bring variety and change to the tea table, and allow cakes to be portioned in a new way.
But no tiresome recalculation of baking recipes is needed, for "La Forme Quadro" is suitable for all standard-size springform recipes. The base of this square springform is scratch-resistant. "La Forme Quadro" is easy to put together and clean.

Jury . Judges panel

Ein Produkt mit einer hochinnovativen Idee. Es erschließt sich nicht auf den ersten Blick, aber wer es ausprobiert, erkennt, dass es sehr durchdacht ist. Manchmal dauert es Jahrzehnte, bis ein längst überfälliges Detail innovativ erfunden wird und so auch wieder neue Wege im Design geht.

.

A product based on a highly innovative idea. It is not immediately clear how it works, but once you have tried it out you realize how well designed it is. Sometimes it takes decades before someone has the idea for a long overdue detail, and a new design path is taken.

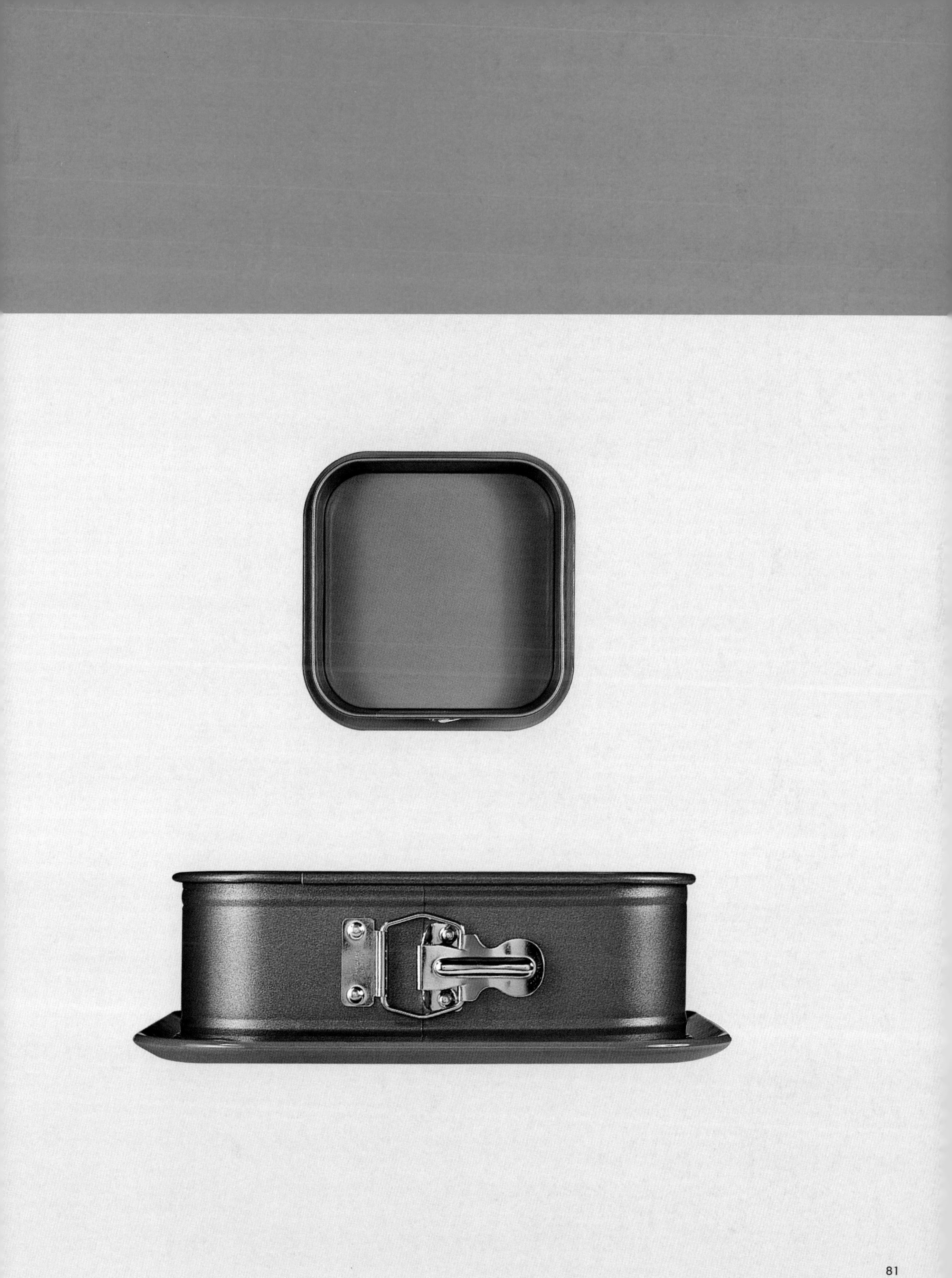

HERSTELLER . MANUFACTURER
KEV Konstruktion, Entwicklung und Vertrieb
Edgar + Andreas Renz GbR
D-75417 Mühlacker

VERTRIEB . DISTRIBUTOR
KEV Konstruktion, Entwicklung und Vertrieb
Edgar + Andreas Renz GbR
D-75417 Mühlacker

DESIGN . DESIGNER
Werksdesign
Andreas Renz

Aquaflip : Flaschenkasten-Untersatz . bottle crate base

Wer kennt das nicht. Man will beispielsweise eine Mineralwasserflasche aus dem Getränkekasten holen, sieht aber nicht gleich, welche noch voll ist. Meist erst nachdem man einige hochgehoben hat, greift man endlich eine volle. Darüber hat sich auch der Designer geärgert.
Deshalb kam er auf die Idee des „Aquaflip". Dieser Flaschenkasten-Untersatz aus mattem Kunststoff zeigt auf einen Blick, welche Flaschen noch voll sind. Eine Feder, die unter einer Kappe verborgen ist, hebt die leeren Flaschen auf ein höheres Niveau im Kasten und macht sie so leicht von den vollen unterscheidbar. Um den Untersatz kostengünstig herzustellen, ist er auf geringen Materialeinsatz ausgelegt und wird als Kunststoff-Spritzgussteil gefertigt.

.

We all know the problem. We want to get a bottle of mineral water, say, out of the crate, but cannot see at first glance which bottles are still full. It is usually not until we have lifted a few that we at last find a full bottle. This was something that annoyed the designer, too.
That is why he has come up with the idea of the "Aquaflip". This matt plastic bottle crate base shows at a glance which bottles are still full. A spring, concealed beneath a cap, pushes empty bottles higher in the crate, making it easy to distinguish them from full ones.
To allow this base to be manufactured inexpensively, it has been designed to require a minimum of material, and is produced as an injection-moulded plastic part.

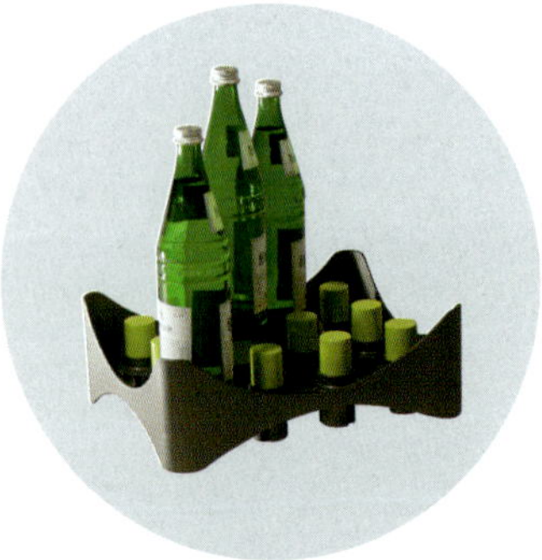

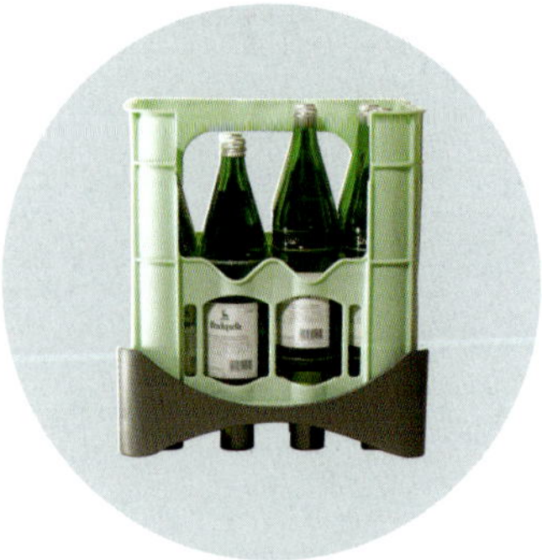

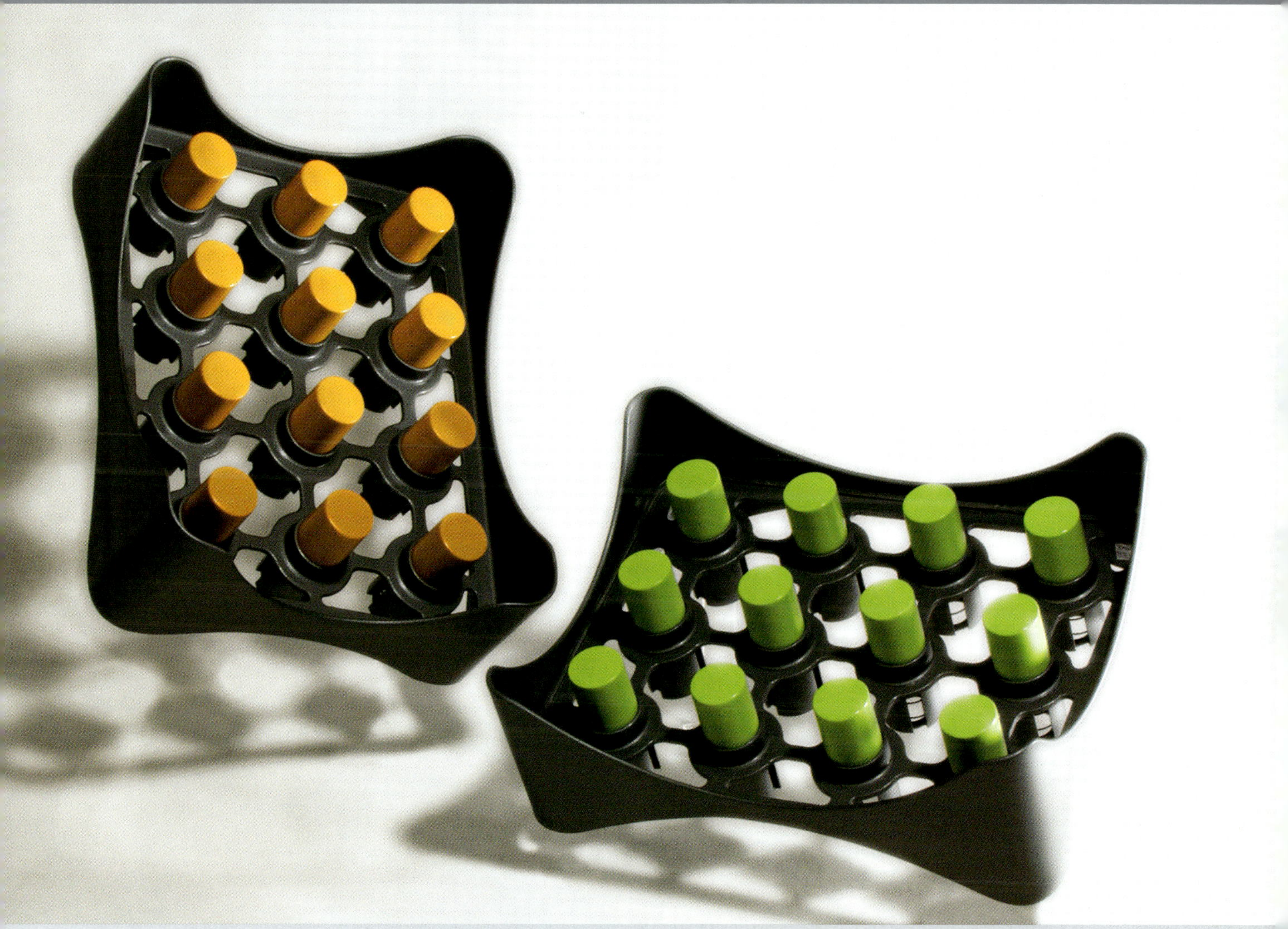

Jury . Judges panel

Flasche voll oder Flasche leer ist keine Frage mehr. Sehr wichtig auch, wenn viele Leute in einem Restaurant oder in einer Bar arbeiten, und man kein System für volle oder leere Flaschen hat. Mit diesem Kontrolluntersatz kann man sofort sehen, ob Flaschen voll oder leer sind.

.

No more guessing which bottle is full, which is empty. That is very important if people are working in a restaurant or bar, and there is no system for full or empty bottles. This base keeps a check on things, and allows you to see straight away whether bottles are full or empty.

HERSTELLER . MANUFACTURER
M. H. Wilkens & Söhne GmbH
D-28309 Bremen

VERTRIEB . DISTRIBUTOR
M. H. Wilkens & Söhne GmbH
D-28309 Bremen

DESIGN . DESIGNER
Kerstin Nolte
D-22587 Hamburg

Tema : Tischaccessoires . table accessories

Die Gestaltung der Vasen und Schalen, die ausgezeichnet wurden, ist auf das Wesentliche reduziert. Sie zeichnet sich durch klare Formen und kühle Eleganz aus, was durch den zeitlos schönen Edelstahl und die mattierten Oberflächen noch unterstrichen wird. Auf Dekore und angesetzte Elemente wurde bewusst verzichtet.
Das Design basiert auf dem Grundmotiv des Dreiecks verbunden mit harmonischen Rundungen. Das führt zu einem spannenden Kontrast zwischen elegantem Schwung und geometrischer Strenge.
Die Kollektion besteht aus verschieden großen Vasen, oval und in Kugelform, sowie Schalen. Gläser- und Flaschenteller, Tablett, Leuchter, Dosen und ein Krug vervollständigen das Programm.

·

The design of the vases and bowls commended here has been reduced to the bare essentials. It stands out by virtue of its clear forms and cool elegance, which is emphasized still further by the timelessly beautiful stainless steel and matt surfaces. Decoration and add-on elements have deliberately been avoided.
The basic motif of the design is a triangle, combined with harmonious rounded corners. The effect is that of a fascinating contrast between elegant dynamics and geometric severity.
The collection is made up of various large vases, both oval and spherical, and bowls. The product range is rounded off by coasters for glasses and bottles, a tray, candlestick, boxes and a jug.

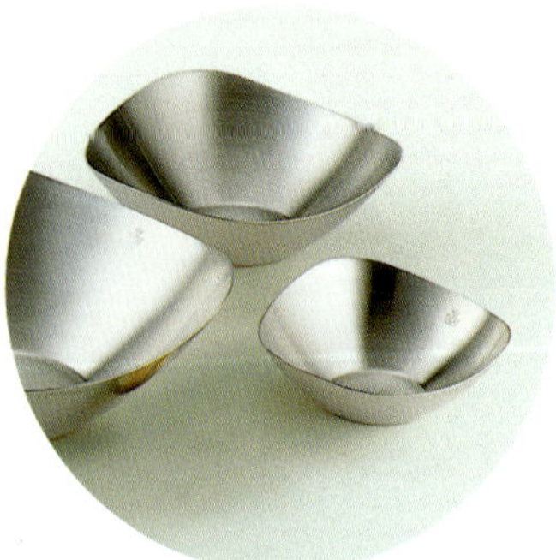

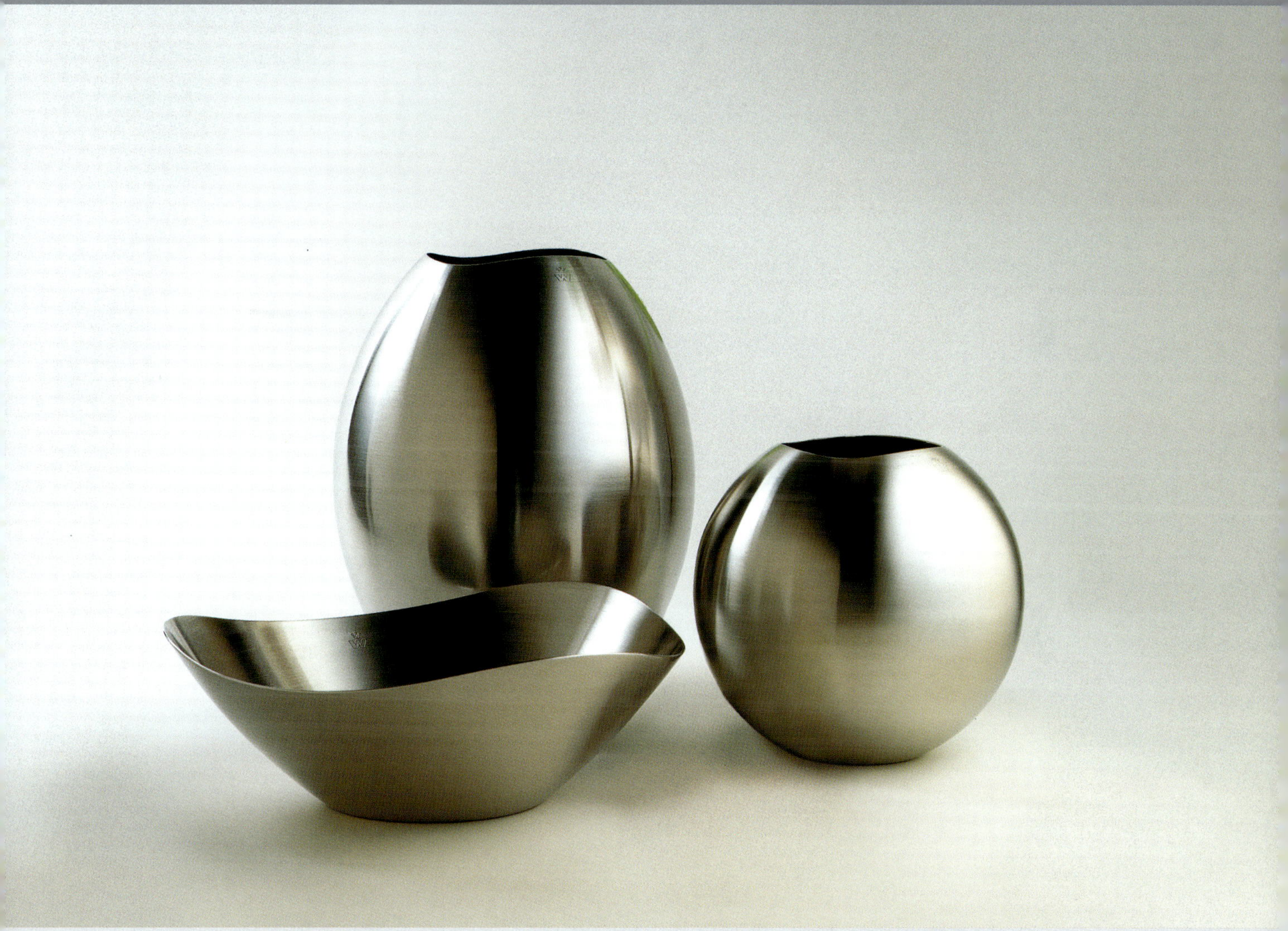

Jury . Judges panel

Ein schönes Beispiel für Balance zwischen Hightechmaterial und poetischer Form. Es sind einige Ansätze zu sehen, wo man aus traditionellen Formen herausgeht und trotzdem das Erscheinungsbild dank der Verbindung mit Edelstahl sehr, sehr neu und innovativ wirkt.

.

A nice example of balance between high-tech material and poetic form. There are the first signs here of an approach based on traditional forms where the resulting appearance, thanks to the combination with stainless steel, is very, very new and innovative in its effect.

KÜCHE + HAUSHALT | SILBER

KITCHEN + HOUSEHOLD | SILVER

HERSTELLER . MANUFACTURER
Alexander Hunger
Art & Wohndesign
D-88250 Weingarten

VERTRIEB . DISTRIBUTOR
Alexander Hunger
Art & Wohndesign
D-88250 Weingarten

DESIGN . DESIGNER
Werksdesign
Alexander Hunger

Modell 1001 XL : Schale . bowl

Die Schale besteht aus Edelstahl, der fein gebürstet wird. Sie wird mit Hilfe von Lasertechnik hergestellt. Die Grundform ist ein Rechteck, das in die Dreidimensionalität verformt wurde.
Die ungewöhnliche schlichte Form gibt der Schale ihren Charakter und lässt sie in Balance sein, das heißt sie schaukelt sich immer wieder in ihre Ursprungsstellung zurück. Durch die Einschnitte, die ihr einen eigenen Charakter geben, ist sie luftdurchlässig. Deshalb eignet sie sich besonders gut auch zur Aufbewahrung von Obst.
Die Schale ist in drei verschiedenen Größen, je nach Aufstellungsort, zu haben: eine Schale, die auf dem Tisch steht, dann eine größere Schale, die auf den Boden gestellt wird, und die größte, die im Garten ihren Platz findet.

.

The bowl is made of finely brushed stainless steel. It is manufactured with the help of laser technology. Its basic form is a rectangle which has been worked into a three-dimensional shape.
Its unusually plain shape gives the bowl its character and lends it its balance. In other words, it always rocks back to its original position. The cuts made in it are a distinguishing feature and allow air to pass through it. For this reason, it is also particularly suitable as a fruit bowl.
The bowl is available in three sizes, depending on where it is to be put: a bowl for the table, then a larger one for the floor, and the largest one for the garden.

Jury . Judges panel

Ein schönes, attraktives Objekt, mit dem man spielen kann. Faszinierend, dass hier aus einer einfachen Fläche, einem Rechteck, durch Einschneiden und Verformen eine ganz präzise Form entstanden ist. Extrem ausgewogen, tariert sich immer wieder von selbst aus.

A beautiful, attractive object that you can play with. It is fascinating how, starting from a simple rectangular surface, cutting and working can give rise to such a precise form. Extremely balanced, always finds its way back to an even keel.

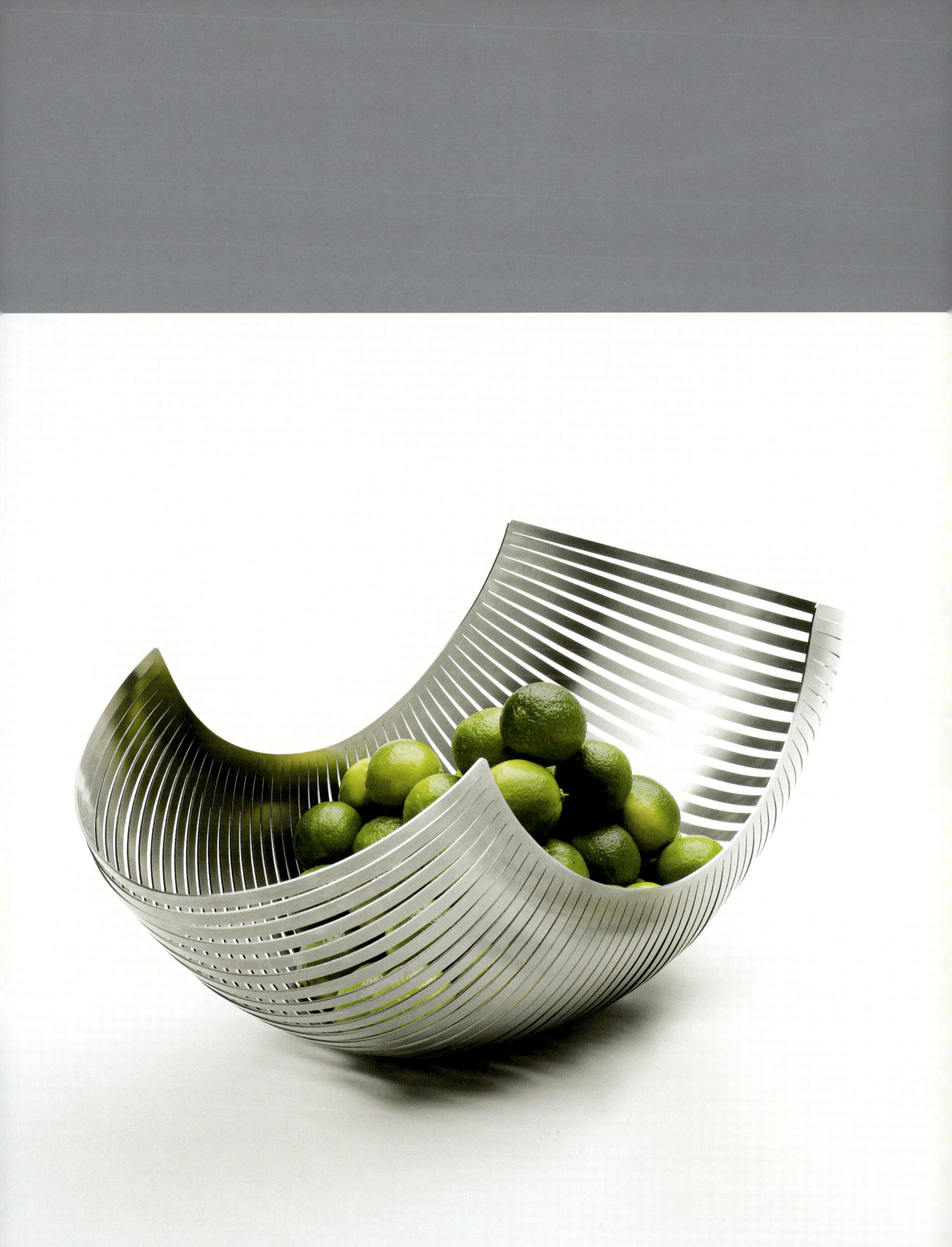

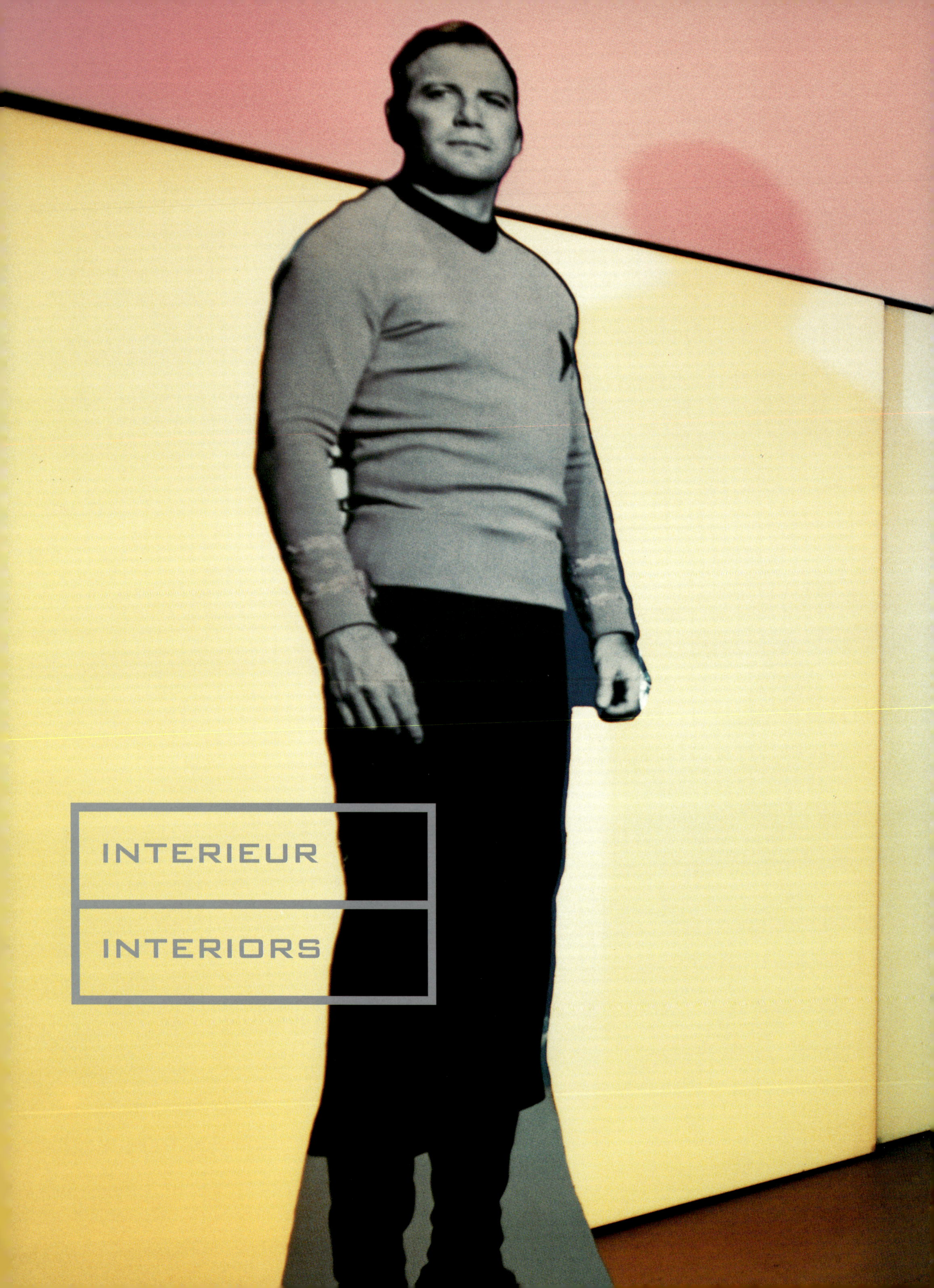

INTERIEUR

INTERIORS

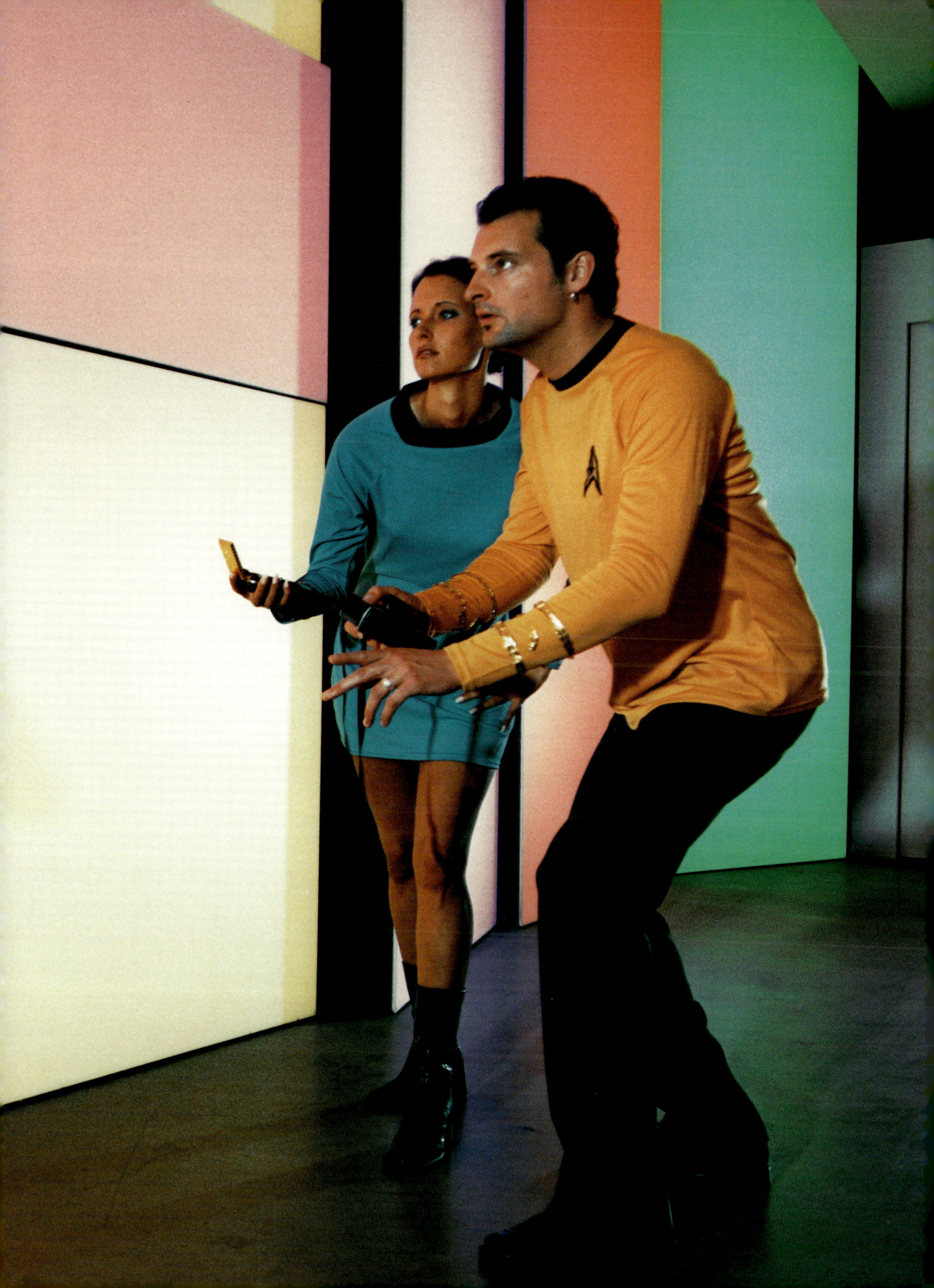

INTERIEUR GOLD
INTERIORS GOLD

HERSTELLER . MANUFACTURER
Zumtobel Staff GmbH
A-6850 Dornbirn

VERTRIEB . DISTRIBUTOR
Zumtobel Staff Deutschland Vertriebs GmbH
D-32657 Lemgo

DESIGN . DESIGNER
Werksdesign

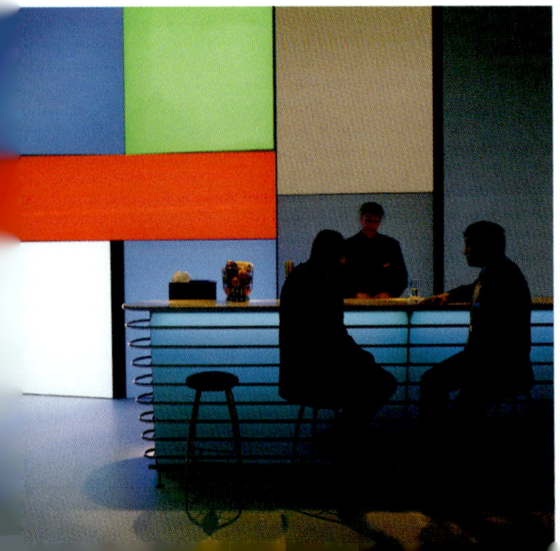

Active Light Wall

Der Mensch wird durch natürliches Licht beeinflusst, das je nach Tages- und Jahreszeit wechselt. „Active Light Wall" nimmt dieses Wechselspiel auf, indem es Intensität, Farbe und Richtung des Lichts ändert. An die Stelle statischer Beleuchtungskonzepte treten hier abwechslungsreiche, subtil wirkende Lichtdramaturgien. Mit intelligenten Luxmate Lichtmanagementsystemen werden gewünschte Raumstimmungen erzeugt, die nach Bedarf stimulieren, motivieren, aktivieren oder entspannen.
Dank des besonderen, matten Diffusormaterials der über die Vorderseite des Wall-Elements gespannten Folie verteilt sich das farbige Licht gleichmäßig über die gesamte Fläche ohne störende Spiegelungen wie bei Glas und Acryl. „Active Light Wall" erregt Aufmerksamkeit und ist ideal in Verkaufs- und Präsentationsräumen. Sie kann nach Bedarf aufgestellt und flexibel im Raum positioniert werden. Es gibt verschiedene Varianten: in der schmalen Version an die Wand angelehnt beziehungsweise fest montiert oder von der Decke abgependelt, in der tiefen Ausführung frei stehend als Einzelelement oder zur imposanten Lichtwand aufgetürmt.

Human beings are influenced by natural light, which changes depending on the time of day or year. "Active Light Wall" takes up this interplay idea, changing the intensity, colour and direction of the light. Here, static lighting concepts have been replaced by varied, subtle dramaturgies of light. Intelligent Luxmate light management systems allow room moods to be generated as desired, whether to stimulate, motivate, activate or relax.
Thanks to the special matt material of the diffuser sheet stretched over the front surface of the wall element, the coloured light is distributed evenly over the entire surface, without the annoying reflections found with glass or Plexiglas. "Active Light Wall" attracts attention and is ideal for sales or presentation rooms. It can be set up as required and positioned flexibly in a room. It is available in different models: in its slim version, it can be leaned against or firmly attached to a wall or suspended from a ceiling, while its wide version can stand on its own as an individual element or be stacked to form an impressive wall of light.

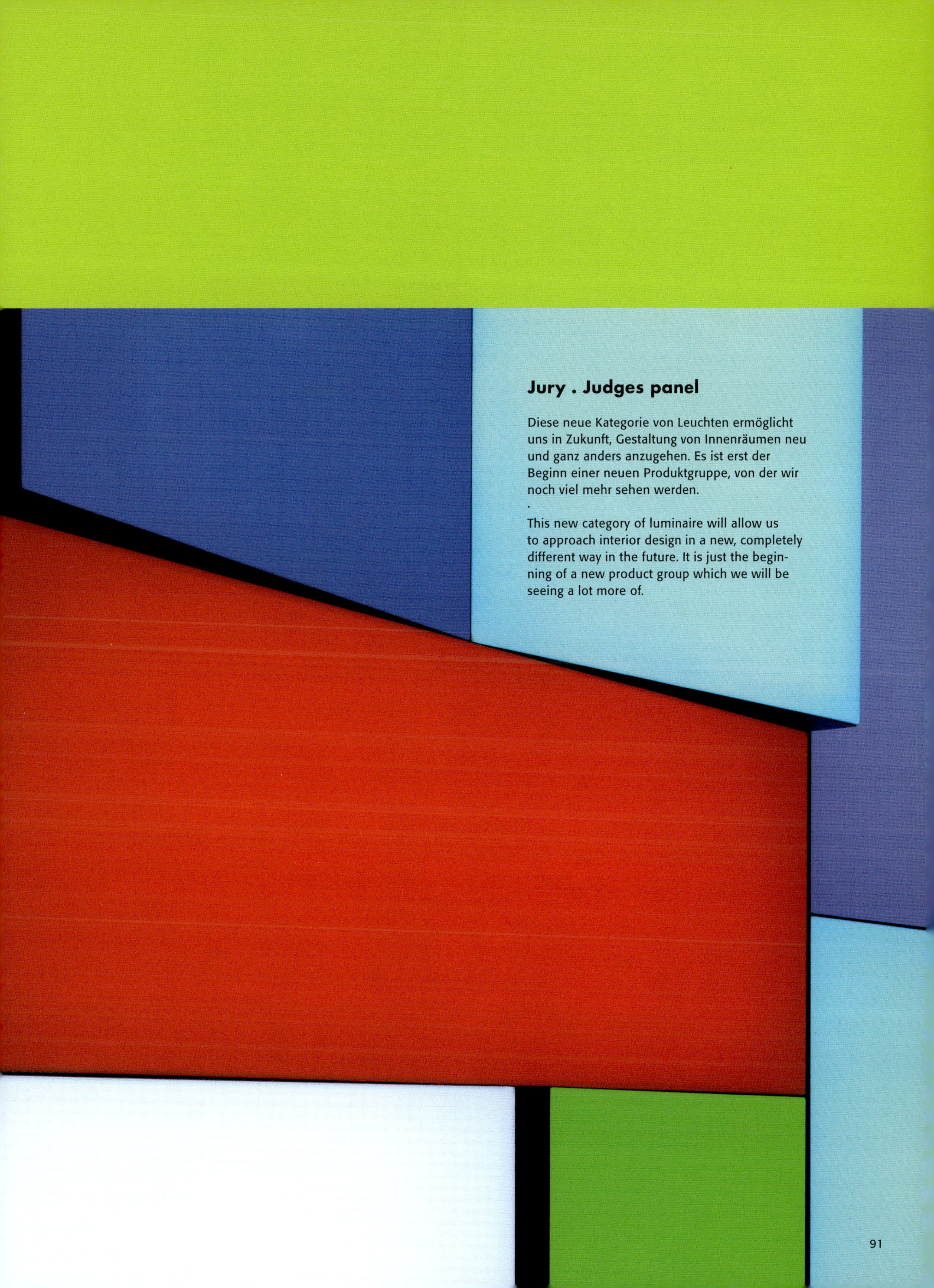

Jury . Judges panel

Diese neue Kategorie von Leuchten ermöglicht uns in Zukunft, Gestaltung von Innenräumen neu und ganz anders anzugehen. Es ist erst der Beginn einer neuen Produktgruppe, von der wir noch viel mehr sehen werden.

.

This new category of luminaire will allow us to approach interior design in a new, completely different way in the future. It is just the beginning of a new product group which we will be seeing a lot more of.

INTERIEUR | SILBER
INTERIORS | SILVER

HERSTELLER . MANUFACTURER
Zumtobel Staff GmbH
A-6850 Dornbirn

VERTRIEB . DISTRIBUTOR
Zumtobel Staff Deutschland Vertriebs GmbH
D-32657 Lemgo

DESIGN . DESIGNER
Zumtobel Staff Deutschland Vertriebs GmbH
D-32657 Lemgo

Phaos Kachel : LED Multicolorleuchte . multi-colour LED luminaire

„Phaos Kachel" verkörpert eine neue faszinierende Anwendung der modernen LED-Technologie. Hier bietet das Spiel mit den Farb-Lichtpunkten interessante Möglichkeiten. So kann die quadratische Kachel als Präsentationsfläche oder als farbdynamische Lichtwand eingesetzt werden, sie steht entweder für sich oder inszeniert ein Objekt.
Die rote, grüne und blaue Farbe der Leuchtdioden wird dabei elektronisch gemischt. Damit lässt sich sowohl eine statische als auch eine dynamische Beleuchtung mit über 16 Millionen Farben realisieren.
Die mattierte, homogene Glasfläche wird auf Trägern und Wänden montiert oder einfach in den Raum gelegt – alleine oder in vielerlei Formationen. Die „Phaos Kachel" zeichnet sich durch anspruchvolles Design mit zuverlässiger Verarbeitung aus. Die Gehäuse sind in Aluminium-Druckguss beziehungsweise -Strangguss gefertigt.

.

"Phaos Kachel" incorporates a new, fascinating application of modern LED technology. In this product, play with coloured points of light offers interesting possibilities. The square "tile" can be used as a presentation surface, for example, or as a dynamic wall of coloured light, it can either stand alone or be used to highlight an object.
The red, green and blue colour of the LEDs is mixed electronically. The result is a static or dynamic illumination effect with more than 16 million colours.
The matt, homogenous glass surface is fixed to supports and walls or simply laid down flat in the room – on its own or in a variety of formations. "Phaos Kachel" stands out by virtue of its sophisticated design and reliable workmanship. The housings are manufactured of die cast or continuous cast aluminium.

Jury . Judges panel

Ein Produkt in einer sehr schlichten Formgebung aber sehr stark in seiner ästhetischen Ausstrahlung. Schafft Atmosphäre.
Sehr schmale Leuchte, trotzdem gutes Licht.

.

A product whose design is very unassuming, but whose aesthetic impression is very strong. Creates atmosphere. Very slim luminaire, but good light nevertheless.

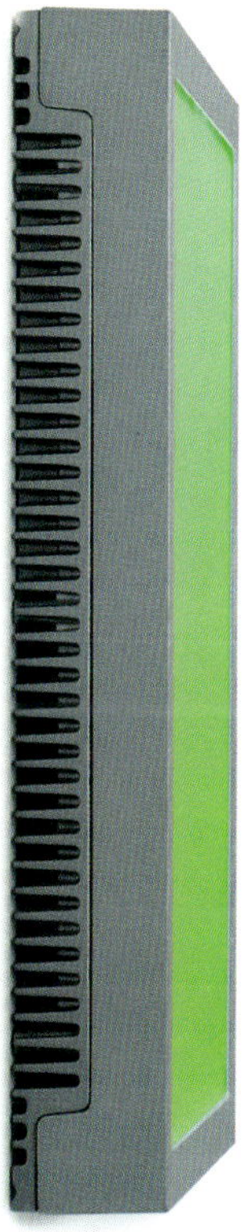

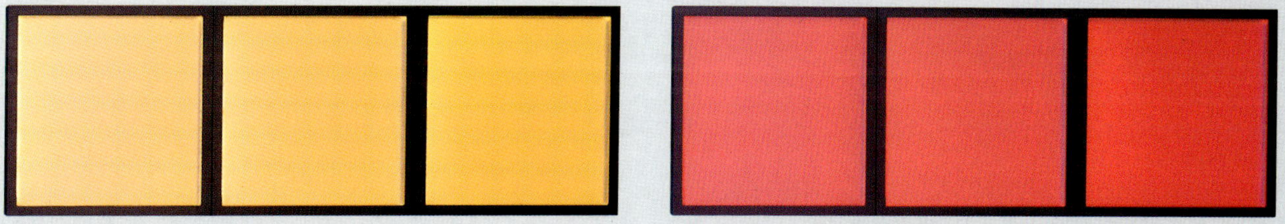

HERSTELLER . MANUFACTURER
Carpet Concept Objekt Teppichboden GmbH
D-33719 Bielefeld

VERTRIEB . DISTRIBUTOR
Carpet Concept Objekt Teppichboden GmbH
D-33719 Bielefeld

DESIGN . DESIGNER
TWO Design
Claudia de Bruyn, Achim Nagel
D-40878 Ratingen

Tec Pearl : Webteppichboden . woven carpet

Im Material von „Tec Pearl" vereinen sich die Qualitäten eines Webteppichs mit denen eines Hartbelages. Klarheit und formale Reduktion kommen vom Hartbelag, die sinnlichen Qualitäten einer handwerklich erzeugten Struktur stammen vom Webteppich. Ungewöhnlich ist das metallische Garn, das wie aufgereihte Perlen gleichmäßig auf der Fläche verteilt ist. Übliche Webteppiche haben ihre Farben und Muster im Polmaterial, ihre Stabilität bekommen sie aus Kette und Schuss. Bei „Tec Pearl" ist es umgekehrt: Das Flachgewebe erhält seine Farbe im stabilen Bereich von Kette und Schuss, darüber liegt als gedrehter Polfaden das metallische Garn. Dieses reflektiert das Licht und sorgt so für einen spannenden Wechsel aus Licht und Schatten.
„Tec Pearl" gibt es in zwei Strukturen: eine klare Oberfläche mit regelmäßigem Noppenbild und eine diffuse Oberfläche mit quadratischen, versetzten Noppen. Für den Teppichgrund stehen sechs verschiedene Farben zur Wahl: grau, blau, rot, schwarz, braun und natur.

.

The materials used in "Tec Pearl" combine the qualities of a woven carpet with those of a hard floor covering. It has the clarity and formal reduction of a hard floor covering and a woven carpet's sensuous qualities that go with a hand-crafted structure. One unusual feature is the metallic yarn, distributed evenly over the carpet's surface like a string of pearls. Conventional woven carpets take their colour and pattern from the pile material, while their stability comes from the warp and weft. With "Tec Pearl" it is the other way around: the flat weave gets its colour from the stable warp and weft, while the metallic yarn is superimposed on this as a twisted pile thread. This reflects light, and in doing so causes a fascinating play of light and shade.
"Tec Pearl" is available in two structures: a clear surface with a regular pattern of tufts and a diffuse surface with square tufts at staggered intervals. The carpet base is available in six different colours: grey, blue, red, black, brown and natural.

Jury . Judges panel

Interessante Balance zwischen einer Reichhaltigkeit und Tiefe mit einer sehr dezenten Oberfläche. Sehr schöne optische Eigenschaften, gute Qualität und hohe Strapazierfähigkeit für den Office-Bereich.

.

Interesting balance between richness and depth, with a very subtle surface. Very beautiful visual properties, good quality and very hard wearing for use in offices.

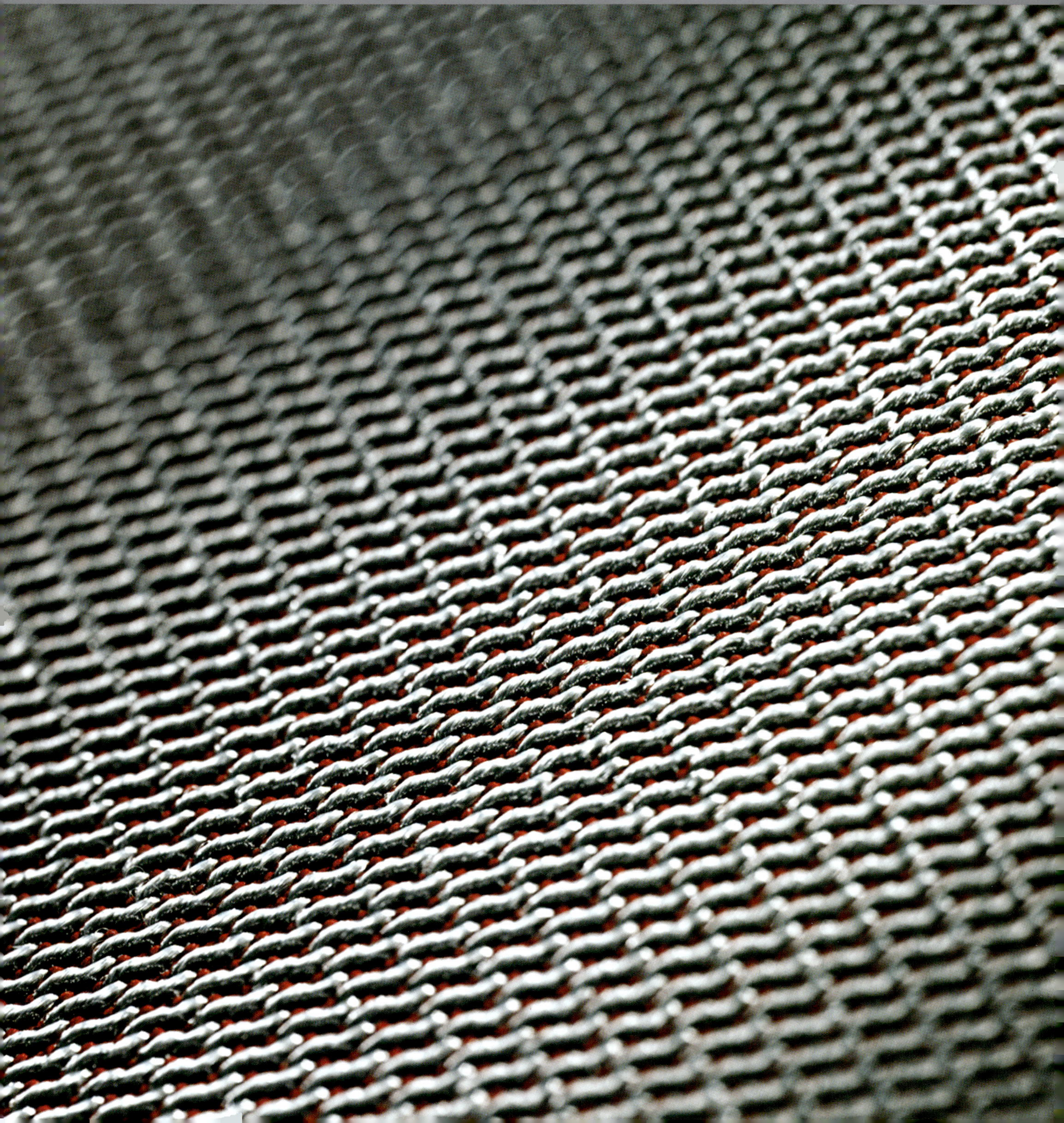

HERSTELLER . MANUFACTURER
Nils Holger Moormann
Möbel-Produktions- und Handels GmbH
D-83229 Aschau im Chiemgau

VERTRIEB . DISTRIBUTOR
Nils Holger Moormann
Möbel-Produktions- und Handels GmbH
D-83229 Aschau im Chiemgau

DESIGN . DESIGNER
Markus Boge & Patrick Frey
D-30451 Hannover

Kant : Tisch . desk

„Kant" sorgt für Ordnung auf dem Schreibtisch. Die Tischplatte ist in ihrem hinteren Teil eingeknickt und bietet dort die Möglichkeit Bücher, Ordner und anderes Arbeitsmaterial übersichtlich zu verstauen und zu archivieren. So ist die Arbeitsfläche frei für das Wesentliche. Benötigt man eines der Bücher oder einen Ordner, kann man schnell darauf zugreifen, da man den Überblick über das ganze Arbeitsmaterial hat.
Der Tisch setzt sich aus zwei Kreuzen und einer eingeknickten Arbeitsplatte zusammen, die mühelos in kurzer Zeit montiert sind. Die Tischplatte besteht aus weiß laminiertem Birkensperrholz, die Tischbeine aus Ahorn massiv.

"Kant" is the byword for a neat and tidy desk. A V-shaped fold runs along the back of the desk top, allowing books, folders and other work materials to be stowed and filed away there, where they can easily be found again.
This leaves the work surface free for work. If one of the books or a folder is needed, it can easily be retrieved, as everything that is needed for work can be seen at a glance.
The desk is made up of two X-braces and a desk top with a V-shaped fold running along one long side, which can effortlessly be put together in no time at all. The desk top is made of birch plywood with a white laminate coating, and the legs are made of solid maple.

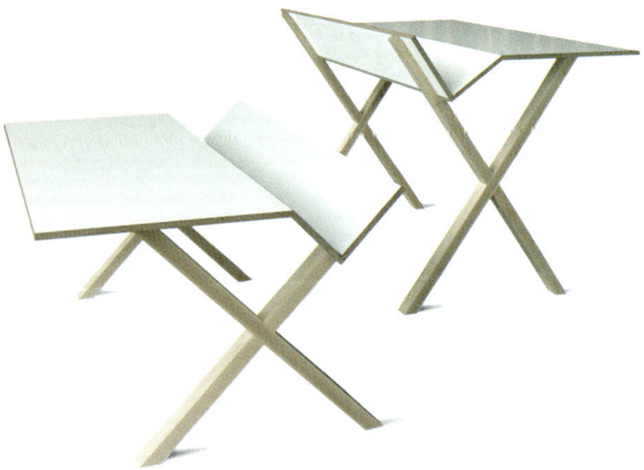

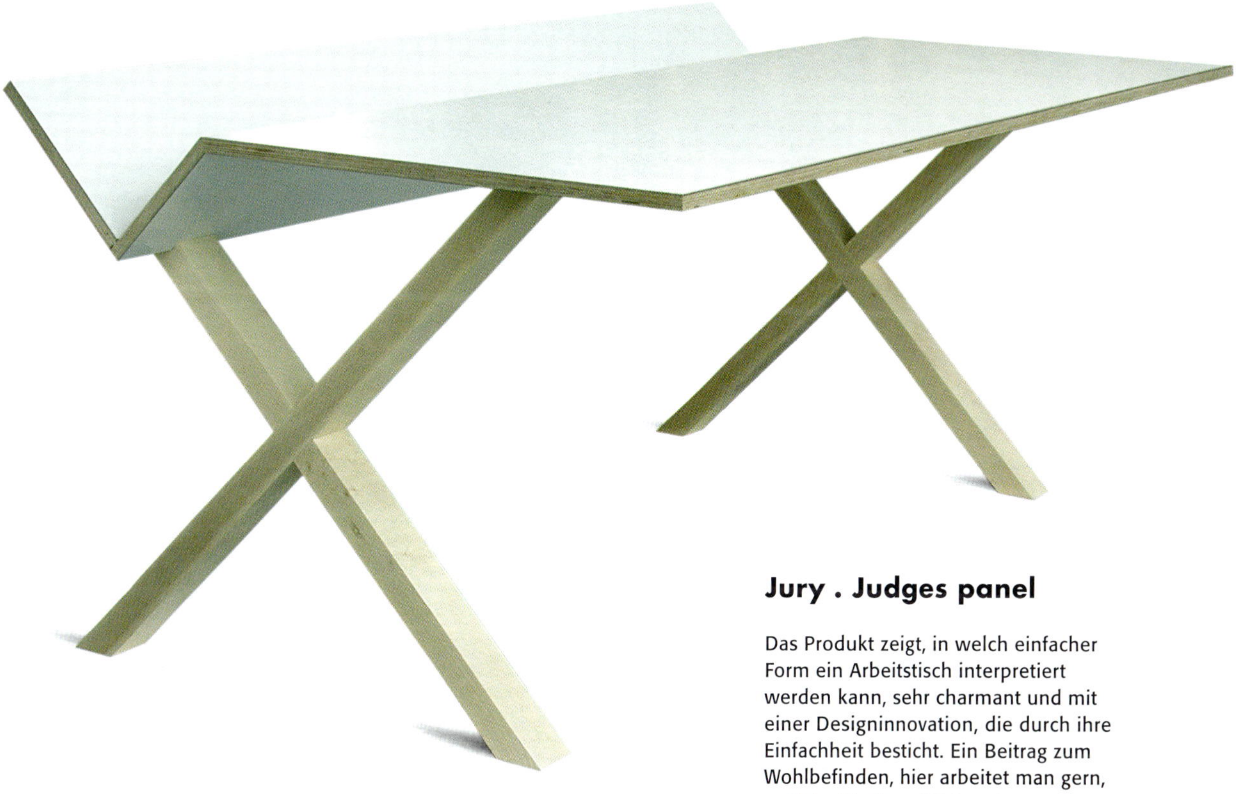

Jury . Judges panel

Das Produkt zeigt, in welch einfacher Form ein Arbeitstisch interpretiert werden kann, sehr charmant und mit einer Designinnovation, die durch ihre Einfachheit besticht. Ein Beitrag zum Wohlbefinden, hier arbeitet man gern, weil der Tisch groß ist und man alle seine Sachen da hat.

.

The product shows how simply a desk can be interpreted. Very charming, with a design innovation whose simplicity is captivating. A contribution to well-being. You enjoy working here because the desk is big and you have everything you need in front of you.

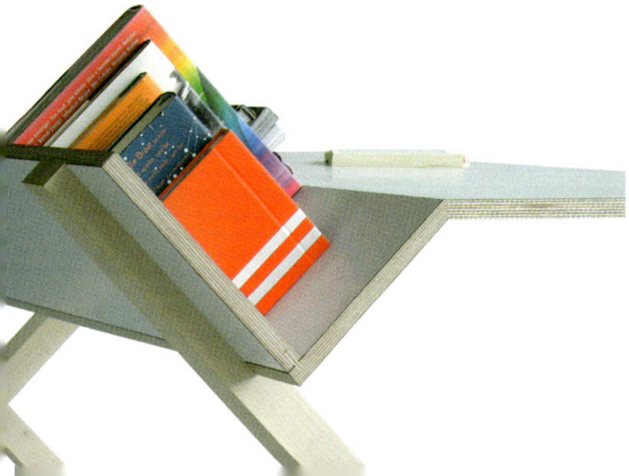

HERSTELLER . MANUFACTURER
Nils Holger Moormann
Möbel-Produktions- und Handels GmbH
D-83229 Aschau im Chiemgau

VERTRIEB . DISTRIBUTOR
Nils Holger Moormann
Möbel-Produktions- und Handels GmbH
D-83229 Aschau im Chiemgau

DESIGN . DESIGNER
Planzwei
Kai Ertel & Matthias Furch
D-70180 Stuttgart

Billy Clever : Regalsystem . shelving system

Eine Fläche wird auf verblüffend einfache Art sekundenschnell in ein Regal verwandelt, und das werkzeugfrei und selbsterklärend. Die Seitenflächen der einzelnen Regalebenen werden einfach hochgeklappt, sodass aus der flachen Platte ein Korpus entsteht. Zylinderstifte und Regalrückwände werden in vorgesehene Nuten nur eingesteckt. Auf diese Weise lassen sich Regalfächer in unterschiedlichen Höhen aufeinandersetzen.
Das Material ist MDF weiß beschichtet. Die Rückwand aus Acrylglas gibt es wahlweise in folgenden Farben: glasslook satiniert, grün, orange, rot und blau.

With astonishing simplicity, a surface is transformed into a shelf in a matter of seconds, and the whole process is self-explanatory and requires no tools. The side surfaces of the individual shelves are simply folded up, creating a shelf unit with sides out of a flat board. Cylindrical pins are inserted into concealed holes and the back panels of the shelving system are simply inserted into the grooves provided. This connecting system allows shelf compartments of different heights to be stacked on top of each other.
The shelving is made of white laminated MDF. The acrylic glass back is available in a choice of burnished glass look, green, orange, red or blue.

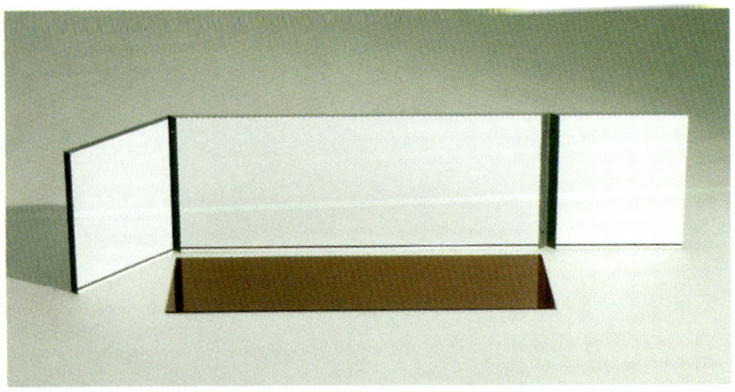

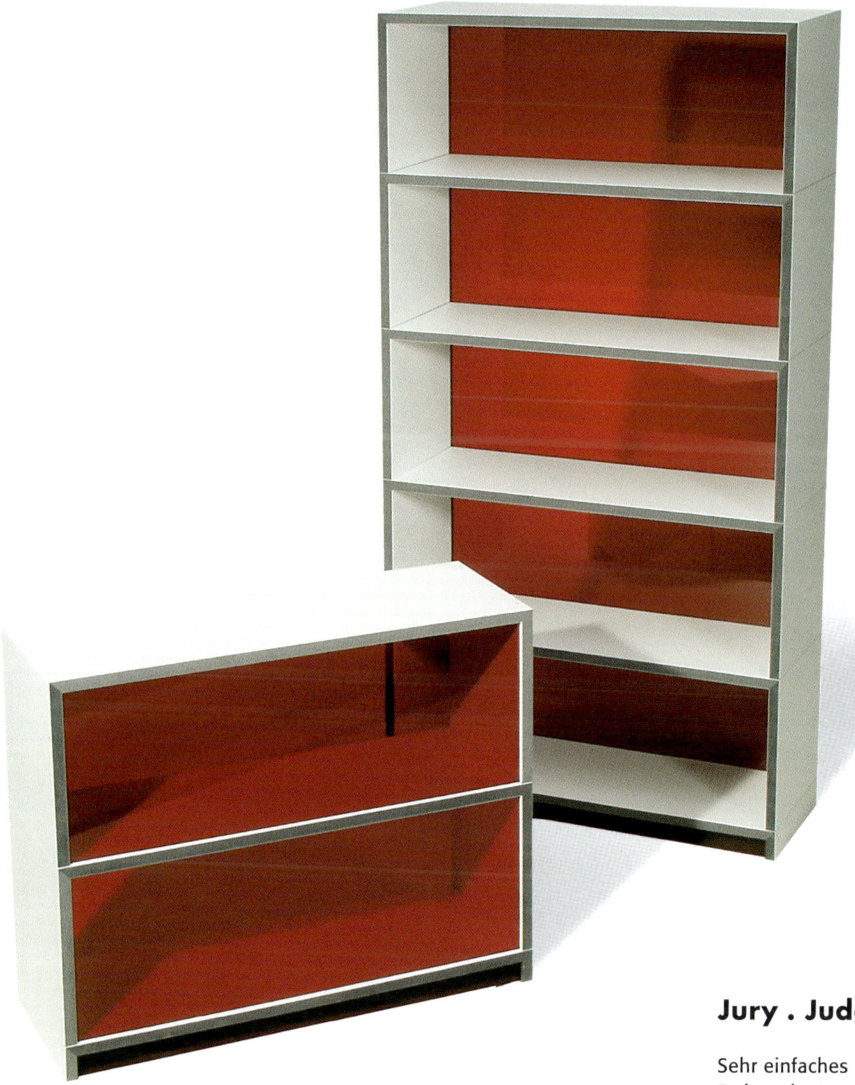

Jury . Judges panel

Sehr einfaches Produkt, leicht aufzubauen. Sehr schön die Farben der transparenten Rückwände. 30 Jahre lang war es dem Design verboten, Buntes zu gestalten. Dies ist ein Ansatz, wie auch Farben sehr positiv angewendet werden können.

·

Very simple product, easily assembled. The colours of the transparent back are especially attractive. Designers have been forbidden to use bright colours for thirty years. This approach shows how colours can be applied very positively.

INTERIEUR | SILBER
INTERIORS | SILVER

HERSTELLER . MANUFACTURER
Interlübke
Gebr. Lübke GmbH & Co. KG
D-33378 Rheda-Wiedenbrück

DESIGN . DESIGNER
Prof. Wulf Schneider & Partner
D-70188 Stuttgart

eo : Containersystem . container system

Das erste Systemmöbel, das auf Knopfdruck seine Farbe ändert. Je nach Stimmung oder Einrichtung kann der Nutzer jeden erdenklichen Ton auswählen. Möglich wird das durch die additive Farbmischung. Leuchtdioden (LED) im Inneren werfen rotes, grünes und blaues Licht auf die Rückwand des Containers. Von dort wird das Licht von den Scheiben aus mattiertem Glas aufgefangen. Den jeweiligen Anteil der drei Grundfarben mischt man einzeln bequem per Fernbedienung und kann so stufenlos jeden gewünschten Farbton erzeugen. Dabei lassen sich fünf Farbmischungen speichern.
Zu dem System gehören Container mit Türen und, halb so hoch, mit Schubkästen, die sich perfekt miteinander kombinieren lassen. Beide sind in zwei verschiedenen Tiefen, wandhängend oder auf Füßen, erhältlich. Dank verschiedener Einsätze lässt sich „eo" noch vielseitiger nutzen: Glasboden, Barboden, CD-Boden, Flaschengitter, Besteckeinsatz.

.

The first furniture system to change its colour at the touch of a button. Depending on mood or décor, the user can select any colour he likes. This is possible by virtue of an additive colour-mixing system. Light-emitting diodes (LEDs) inside the container project red, green and blue light onto the back of the container. From there, the light is caught by panes of matt glass. The proportions of the three primary colours are simply mixed by remote control, which means that any colour desired can be produced. Five colour blends can be stored.
The system includes containers with doors and other containers, half as high, with drawers. They can be combined perfectly with each other. Both are available in two different depths, and are available as wall-hanging or floor-standing versions. Various supplementary elements make "eo" even more versatile: glass shelves, cocktail bar shelves, CD shelves, bottle rack, cutlery drawer.

Jury . Judges panel

Ein sehr durchdachtes Produkt mit vielen innovativen Elementen, das neue Maßstäbe im Wohnbereich setzt.

.

A lot of thought has gone into this product with its many innovative elements. It sets new standards for the living room.

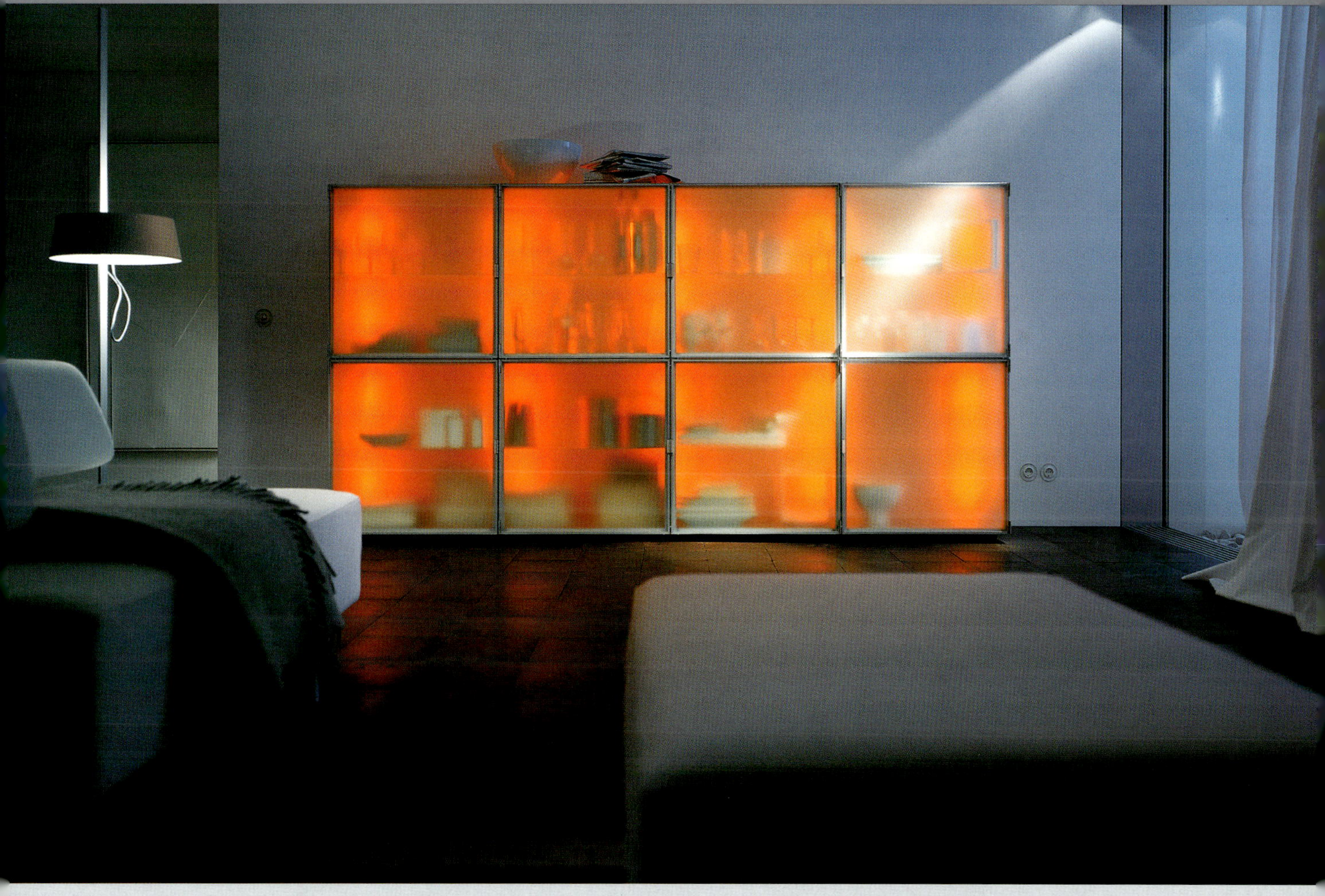

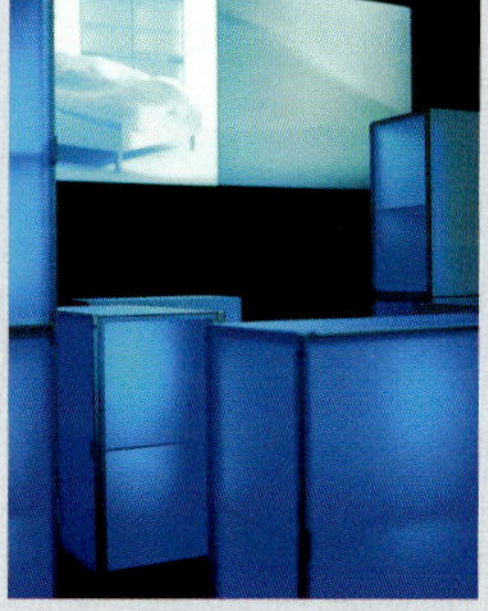

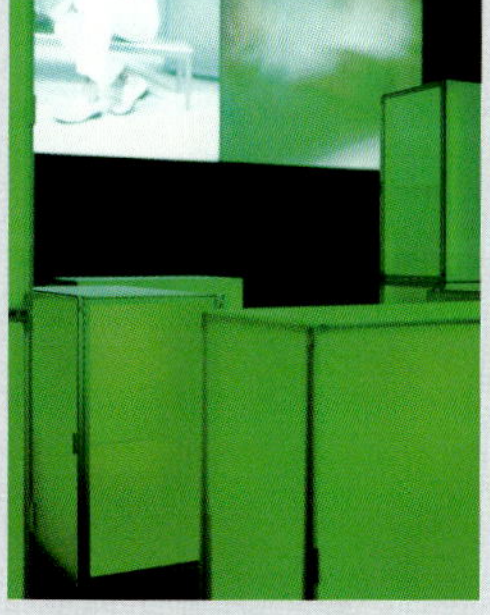

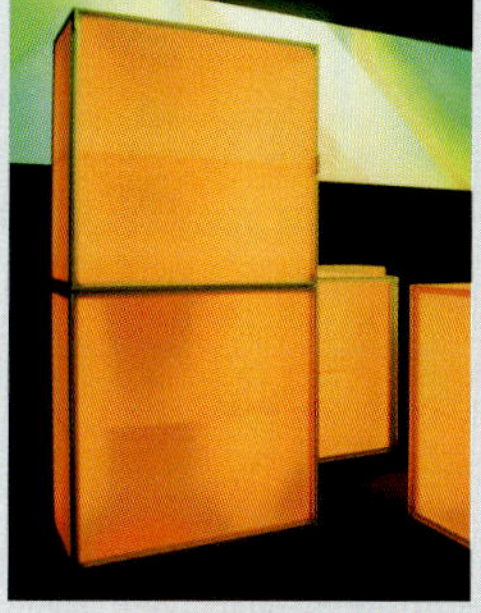

HERSTELLER . MANUFACTURER
Brühl & Sippold GmbH
D-95138 Bad Steben

DESIGN . DESIGNER
Siegfried Bensinger
D-21244 Buchholz i.d.N.

Orit : Liege . couch

Die Grundidee von „Orit" war, ein entspannendes Medienmöbel zum Musik hören, Lesen und Fernsehen zu entwerfen. Mittels einer Gaszugfeder lässt sich die Liege stufenlos von der Relaxposition bis zur planen Liegefläche absenken. Dazu sind sowohl das Rückenteil mit eigens ausgestatteter Kopfneigung als auch der Sitz- und Liegebereich eigenständig beweglich. Dadurch ergeben sich ganz individuelle Entspannungs- und Liegestellungen, die ergonomisch effektvoll den Kopf-, Nacken- sowie den gesamten Wirbelsäulenbereich unterstützen. Da die Liege formal auf ein Minimum reduziert und geometrisch klar gestaltet ist, integriert sie sich gut in verschiedene Interieurs. Das Design wird durch die ausgesuchte Materialauswahl noch unterstrichen: moderne, glänzende High-tech-Synthetik-Gewebe, hochwertige, pflegeleichte Microfasern und besonders langlebiges, geschmeidig-weiches Leder. Alle Bezüge sind komplett abziehbar und leicht zu reinigen.

The basic idea behind "Orit" was to design a relaxing piece of media furniture for listening to music, reading and watching TV. A pneumatic tension spring allows the couch to be lowered to any height, from a relaxing position to a fully reclining position. In addition, both the back of the couch, equipped with its own adjustable head rest, and the seating and reclining part can be moved independently of each other. This allows individual relaxation and reclining positions, which provide ergonomic support for head, neck and the whole length of the spine.
Formally speaking, the couch has been reduced to a minimum and has a geometrically clear design, and so it fits well into different settings. The design is emphasized still further by the choice of materials: modern, shiny, high-tech synthetic fabric, high-quality, easy-care microfibre and especially long-lasting, supple and soft leather. All coverings can be removed completely and are easy to clean.

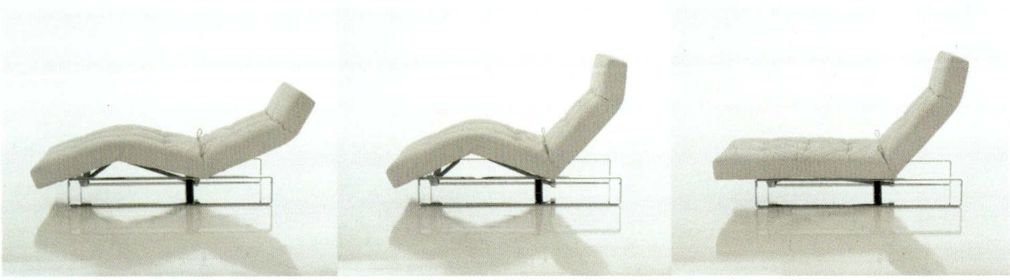

Jury . Judges panel

Besonders überzeugt, wie die Verstellmechanik gelöst ist. Gute Balance zwischen Funktionalität und Komfort. Ein schönes Produkt für den individuellen Bereich zum Entspannen, Fernsehen und Musik hören.

.

The way the adjusting mechanism has been done is particularly convincing. Good balance between function and comfort. An attractive product for the personal sphere – for relaxing, watching TV, and listening to music.

HERSTELLER . MANUFACTURER
Weber + Hätinger
D-72622 Nürtingen

VERTRIEB . DISTRIBUTOR
Weber + Hätinger
D-72622 Nürtingen

DESIGN . DESIGNER
Bredel.de sign
Thomas Bredel
D-70178 Stuttgart

K 12/18 : Regalsystem . shelving system

Das vielseitig einsetzbare, modulare Möbelsystem für den Privat- und Objektbereich zeichnet sich durch eine klare und ausgewogene Formensprache aus. Ein innovativer patentierter Plattenverbinder ermöglicht einfachste Handhabung und trotzdem extrem hohe Stabilität.
„K 12" hat 10 Millimeter starke Kompaktplatten in verschiedenen Farben mit schwarzem Kern. Bei „K 18" sind es 16 Millimeter dicke Multiplexplatten mit verschiedenen Furnier- und Schichtstoffoberflächen. Verchromte Vierkantprofile aus Aluminium verbinden die Platten. Die Verbindung erfolgt über Inbusschrauben in den Vorderkanten. Diese können in schwarz, nahezu unsichtbar, in Edelstahl oder einfach nur als Schrauben mit verchromten Stopfen als markantes Gestaltungsmittel gewählt werden. Zu dem System gehören außerdem in die Regalfächer einsteckbare Container mit Klappen, Türen oder Schubkästen. Fronten in verschiedenen Materialien setzen weitere individuelle Akzente.

.

This versatile, modular furniture system for the home and the office stands out by virtue of its clear, balanced formal expression. An innovative, patented connecting system for the boards makes it extremely easy to put together, while at the same time making it extremely robust.
"K 12" has 10 millimetre thick compact boards in various colours, with a black core. In the case of "K 18" the boards are made of 18 millimetre thick multiplex, available in a variety of veneered or laminated plastic surfaces. The boards are connected together by chromium-plated, rectangular aluminium sections. They are joined together by turning an Allan key on the front edge. These keys are available in black, making them almost invisible, in stainless steel, or simply as screws with chromium-plated plugs, as a striking design feature. The system also includes containers with fold-down flaps, doors or drawers which can be slotted into the shelf compartments. Front panels made of various materials offer the possibility of a further individual touch.

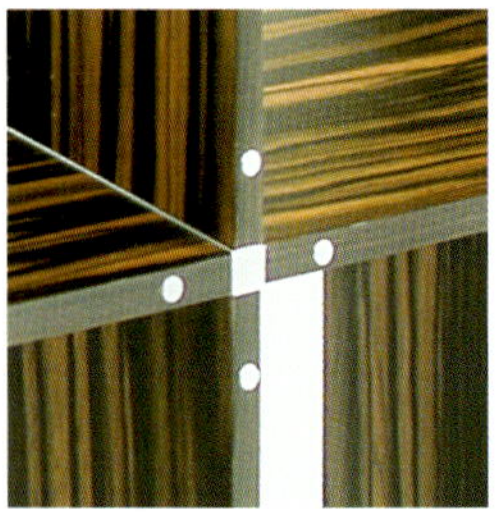

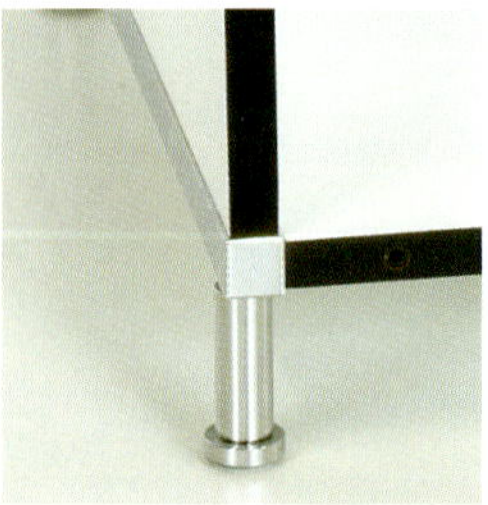

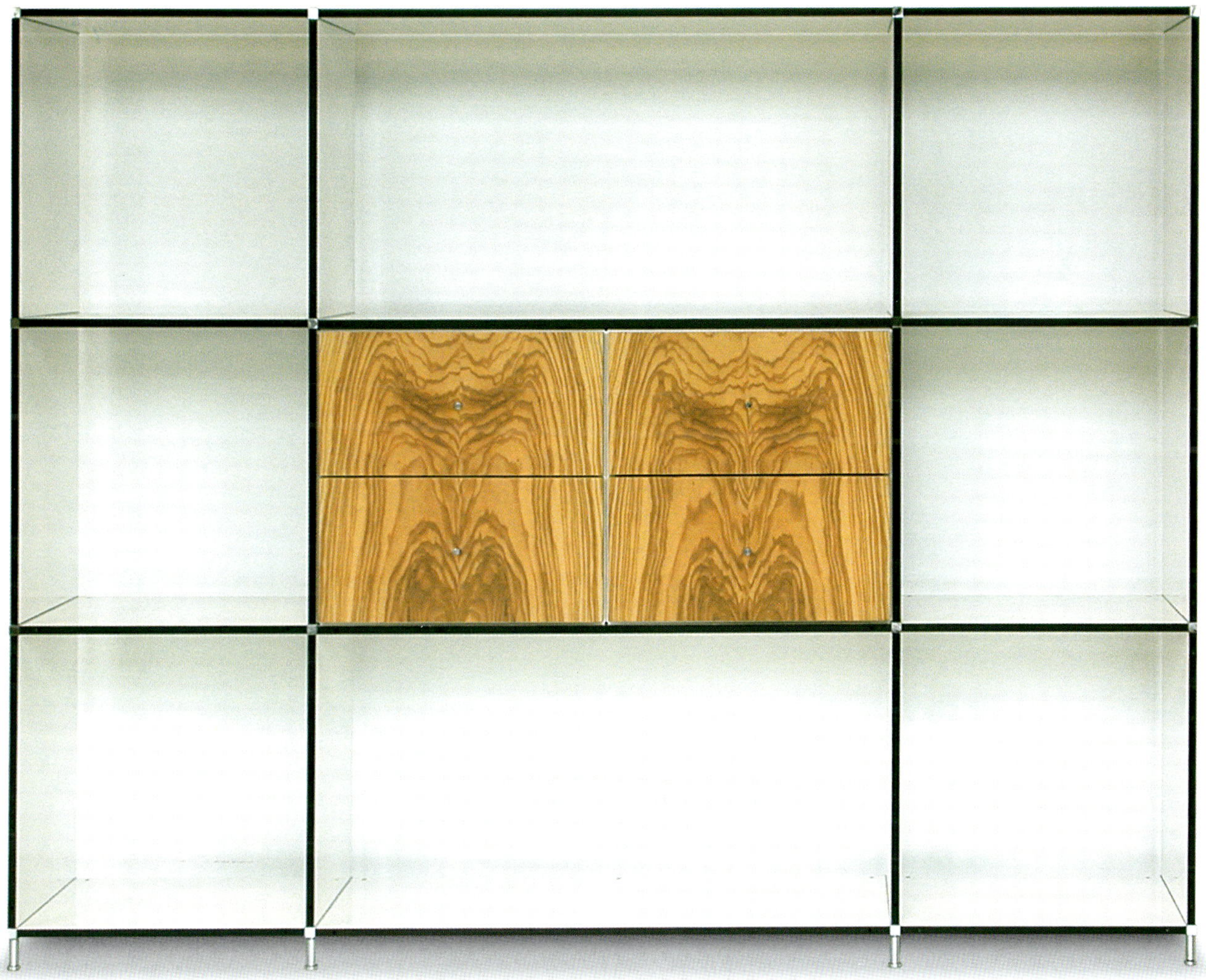

Jury . Judges panel

Das Regal besticht durch seine Leichtigkeit. Intelligente Knotenpunkte verbinden Regalböden und Seitenwangen. Kästen lassen sich zusätzlich einsetzen. Ein innovatives Produkt, das durch verschiedene Materialien ergänzt werden kann.

.

The shelf is captivatingly light. An intelligent system of nodes connects the shelves and sides. Containers can also be slotted in. An innovative product that can be supplemented by different materials.

HERSTELLER . MANUFACTURER
Brühl & Sippold GmbH
D-95138 Bad Steben

DESIGN . DESIGNER
Siegfried Bensinger
D-21244 Buchholz i.d.N.

Lobby : Polsterprogramm . range of upholstered furniture

Die freiflächig konzipierten kubischen Sitzobjekte bieten viel Komfort und vielfältige Nutzungsmöglichkeiten. Die raffinierte Versenkfunktion der Lehnen ermöglicht im Handumdrehen eine Umwandlung vom Sit-in zur Liegestätte. Die Lehnen lassen sich mit Hilfe einer Gasfeder stufenlos aufstellen und in jeder gewählten Position automatisch arretieren.
Drei verschiedene Varianten stehen zur Wahl. Das Grundmodell ist mit jeweils einer versenkbaren Rücken- und Armlehne ausgestattet. Damit lässt sich gut das zweisitzige Sofa mit zwei kleinen absenkbaren Rückenlehnen zu einer offenen Eckformation kombinieren. Das dritte, breitere Modell hat zwar wie das Grundmodell Rücken- und Armlehne, lässt sich aber mit ausgefahrener Rückenlehne von beiden Seiten besitzen.
Die Couchen stehen auf gebogenen Chromkufen. Oberflächenstrukturierte Gewebe, changierende Chenillebezüge oder samtweiche Microfasern, unterstreichen die freiflächigen Formate. Alle Bezüge sind komplett abziehbar und leicht zu reinigen.

These cubic seat objects designed for combination to form any surface offer a lot of comfort and a wide range of uses. The sophisticated mechanism for lowering the arm rests means that it can be changed in a thrice from somewhere to sit to somewhere to lie down. The arm rests can be lowered by a pneumatic spring to any position and locked in any position automatically.
Three different variants are available. In the basic model, the back and one arm rest can be lowered. This can be combined well with the two-seater sofa, whose two back rests can be lowered, to form an open corner seat arrangement. Like the basic model, the third, wider model also has back and arm rests, but the back also functions as a seat when it is lowered.
The couches stand on curved chrome runners. Their formats are emphasized by structured fabrics, shimmering chenille covers or silky soft microfibres. All coverings can be removed completely and are easy to clean.

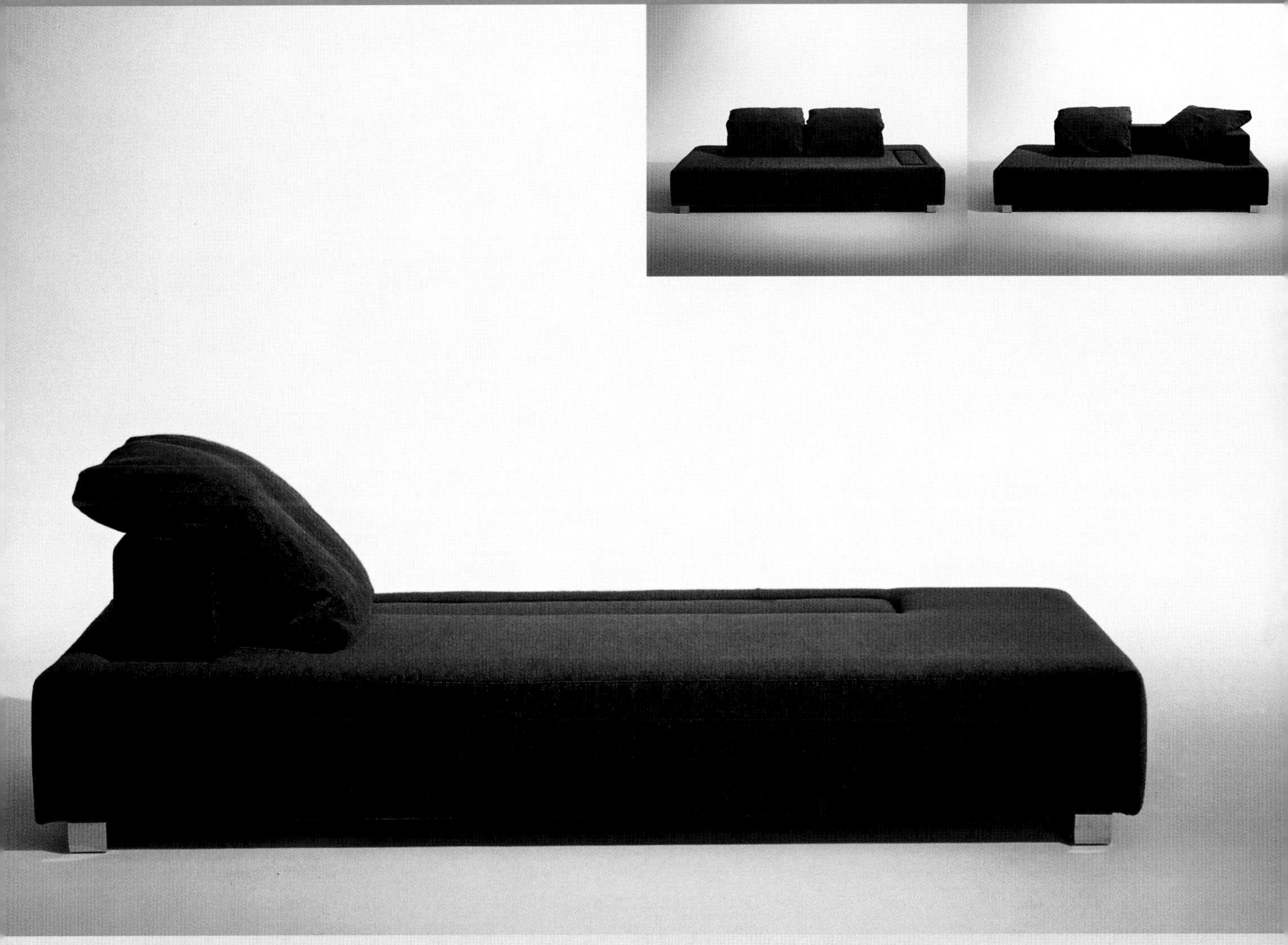

Jury . Judges panel

Aus der Fläche lässt sich durch Hochfahren der Rückenelemente eine Sofalandschaft gestalten. Ein innovatives Produkt mit interessanten neuen Designperspektiven.

.

By raising the back elements, a sofa landscape can be created out of a flat surface. An innovative product with interesting new perspectives for design.

INTERIEUR	SILBER
INTERIORS	SILVER

HERSTELLER . MANUFACTURER
Walter Knoll AG & Co. KG
D-71083 Herrenberg

VERTRIEB . DISTRIBUTOR
Walter Knoll AG & Co. KG
D-71083 Herrenberg

DESIGN . DESIGNER
EOOS Design GmbH
A-1180 Wien

Jason Lite 1700 : Polsterstuhl . upholstered chair

„Jason Lite" wurde im Zuge der Erweiterung des erfolgreichen Jason-Stuhlprogramms entworfen. Der Polsterstuhl zeichnet sich durch filigrane Leichtigkeit und eine formal aufs Minimalste reduzierte Gestaltung aus. Außerdem bietet er einen überraschenden Sitzkomfort. Taschenfederkerne im Sitz und Torsionsfederkerne im Rücken sorgen für bequemes Sitzen. „Jason Lite" ist ein Polsterstuhl mit der Qualität eines Sessels.
Der stapelbare Stuhl ist in drei verschiedenen Farben zu haben. Wahlweise gibt es auch ein Modell mit lederbezogener Armlehne.

.

"Jason Lite" was designed when the successful Jason range of chairs was extended. The upholstered chair stands out by virtue of its filigree lightness and the extreme formal minimalism of its design. It is also surprisingly comfortable to sit on. Pocket springs in the seat and torsion springs in the back make sitting comfortable. "Jason Lite" is an upholstered chair with armchair qualities.
The stackable chair is available in a choice of three different colours. There is also a model with a leather covered armrest.

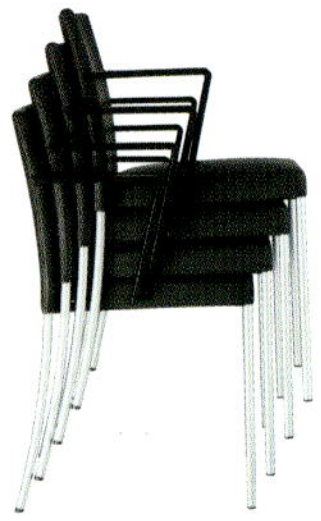

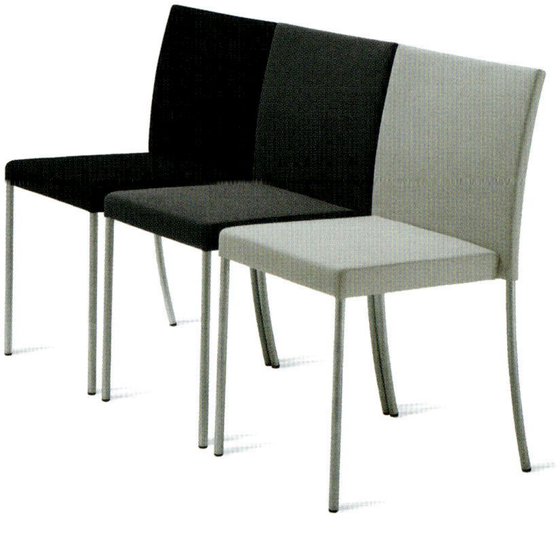

Jury . Judges panel

Ein Stuhl, der sowohl als Esstisch- als auch als Arbeitstischstuhl einzusetzen ist. Er besticht durch seine Leichtigkeit und überzeugt trotzdem durch einen Komfort, den man nicht vermuten würde, wenn man die leichten Formen sieht.

.

A chair that can be used both as a dining chair and for sitting at a desk. It is strikingly light and yet is undeniably comfortable – something which you would not expect when you see the lightness of its form.

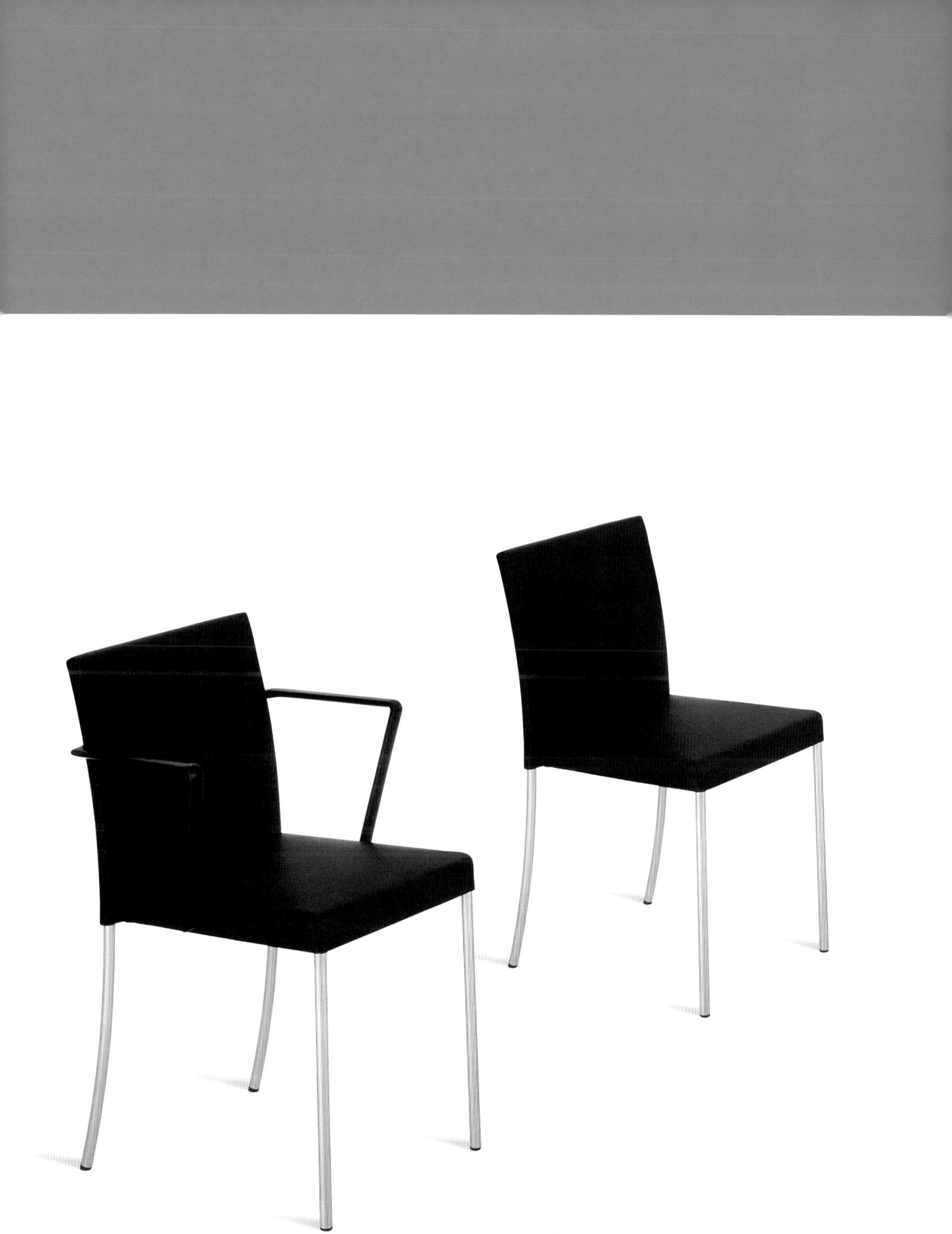

INTERIEUR | SILBER
INTERIORS | SILVER

HERSTELLER . MANUFACTURER
Walter Knoll AG & Co. KG
D-71083 Herrenberg

VERTRIEB . DISTRIBUTOR
Walter Knoll AG & Co. KG
D-71083 Herrenberg

DESIGN . DESIGNER
Pearsonlloyd
London E1 2AX

Kite 560 : Sofaprogramm . range of sofas

„Kite" wirkt erfrischend leicht. Das Sofaprogramm zeichnet sich durch klare Linien, wenig Volumen und konsequentes Design aus. Im Trend des Minimalen liegend, ist hier die Kunst der Reduktion auf die Spitze getrieben. Trotzdem ist das Sofa erstaunlich bequem und dauerhaft solide. Die dünnen, zierlichen Beine lassen die Sitzfläche beinahe schweben.
Das Programm besteht aus Sofa, Sessel und Polsterbank in ungewöhnlichen, fröhlichen Farben.

"Kite" makes a refreshingly light impression. This range of sofas stands out by virtue of its clear lines, lack of bulk and rigorous design. True to the minimalist trend, the art of reduction has been taken to extremes here. Nevertheless, the sofa is astonishingly comfortable and sturdily robust. On thin, delicate legs, the seating surface almost seems to float.
The range comprises a sofa, easy chair and bench, all in unusual, gay colours.

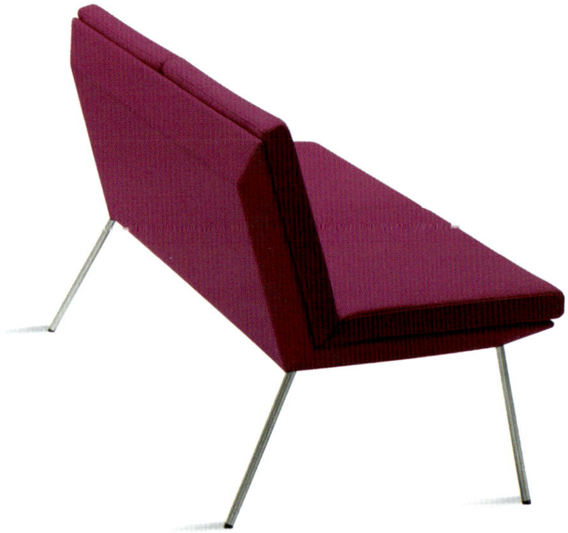

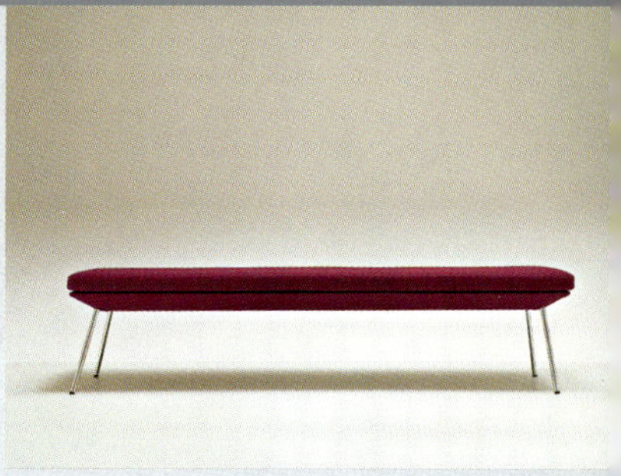

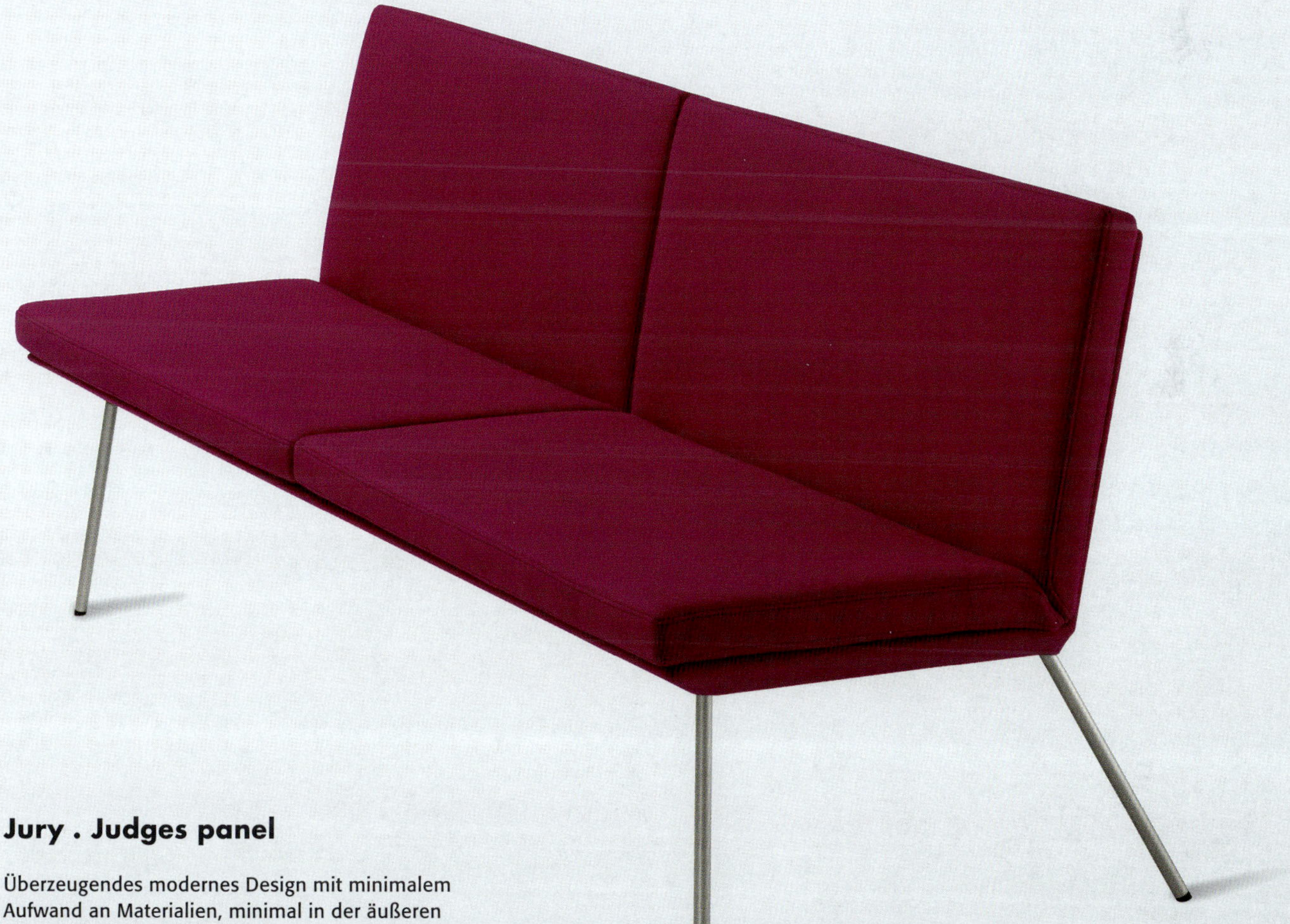

Jury . Judges panel

Überzeugendes modernes Design mit minimalem Aufwand an Materialien, minimal in der äußeren Erscheinung, aber zum Sitzen doch sehr komfortabel.

·

Convincing modern design with a minimal use of materials and a minimalist appearance, yet very comfortable to sit on.

INTERIEUR | SILBER
INTERIORS | SILVER

HERSTELLER . MANUFACTURER
Walter Knoll AG & Co. KG
D-71083 Herrenberg

VERTRIEB . DISTRIBUTOR
Walter Knoll AG & Co. KG
D-71083 Herrenberg

DESIGN . DESIGNER
EOOS Design GmbH
A-1180 Wien

Nelson 602/603 : Sofaprogramm . range of sofas

Das Sofaprogramm bietet eine Vielfalt individueller Lösungen dank einzelner Module, die miteinander kombiniert werden: verschiedene Mittel-, Eck- und Abschlusselemente mit zwei unterschiedlich breiten Armlehnen.
Wahlweise erhält jedes einzelne Modul eine ausklappbare Nackenrolle. Diese Funktion hat doppelten Nutzen: Neben dem Stützen des Kopfes ergibt das Entfalten eine größere Sitztiefe. Beides unterstützt das entspannte Sitzen.
Mit den Proportionen der Moderne und klassischer Qualität ist „Nelson" zeitlos. Das kubische Sofaprogamm mit den kraftvollen Proportionen wirkt großzügig und repräsentativ. Es ist deshalb für das private Wohnen ebenso einsetzbar wie für den öffentlichen Raum.

.

Thanks to individual modules – different middle, corner and end sections with arm rests in two different widths – that can be combined with each other, this range of sofas has many different individual solutions to offer.
As an optional extra, each individual module has a retractable bolster. This function is doubly useful, for apart from supporting the head, folding it away provides more seating depth. Both aspects help relaxation when sitting.
With its modernist proportions and classical qualities, "Nelson" is timeless. The cubist range of sofas with its dynamic proportions is generous and prestigious in its effect. It can therefore be used both in the home and in public space.

Jury . Judges panel

Mit einem einfachen Dreh lässt sich die Sitztiefe verstellen. Ein intelligentes und dazu sehr repräsentatives Produkt.

.

A simple turn is all that is needed to adjust seating depth. An intelligent and also extremely prestigious product.

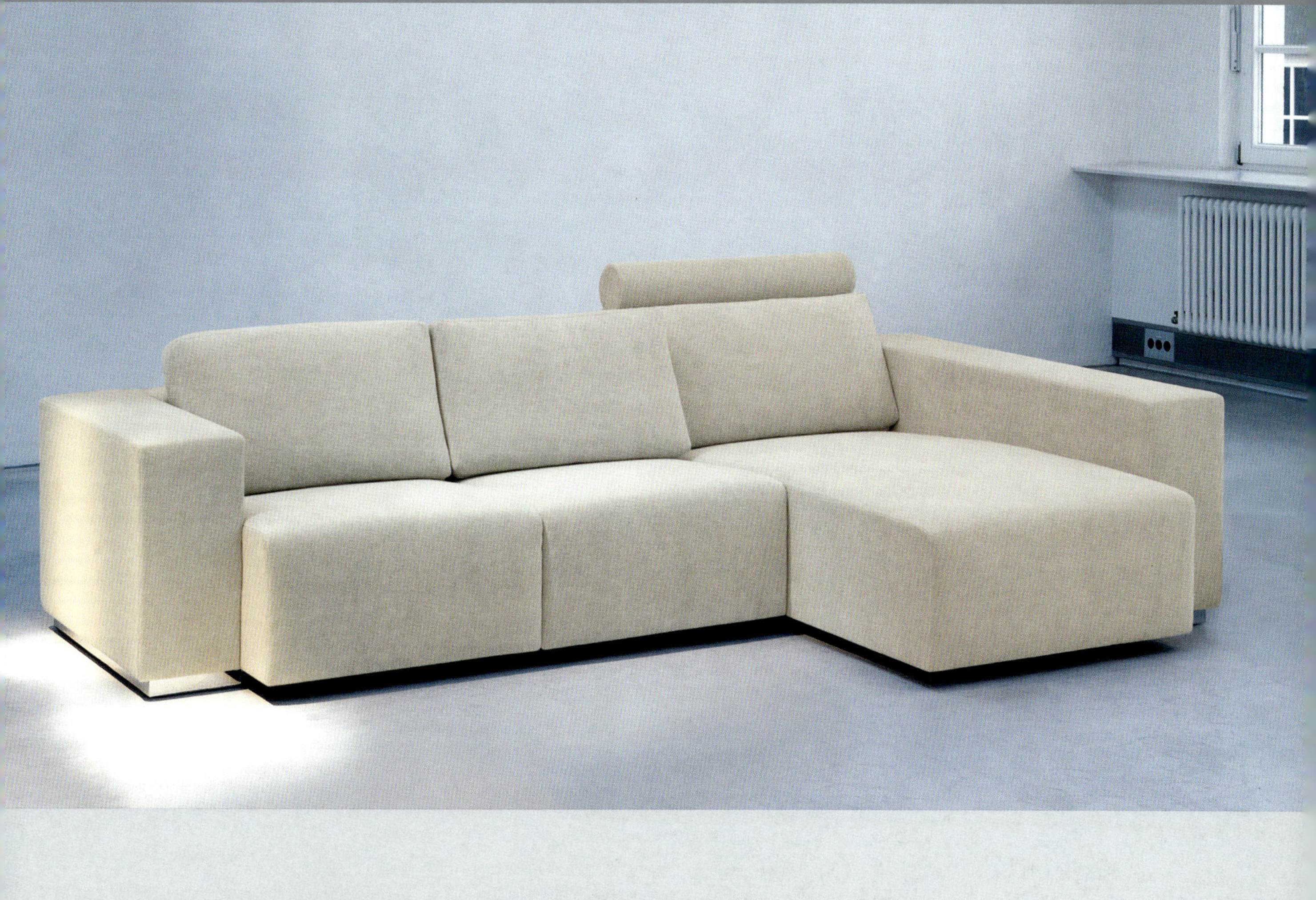

HERSTELLER . MANUFACTURER
Wagner-Design
D-60314 Frankfurt/Main

VERTRIEB . DISTRIBUTOR
Wagner-Design
D-60314 Frankfurt/Main

DESIGN . DESIGNER
Werksdesign
Wolf Udo Wagner

02 : Liege . couch

Eine Liege zum Relaxen und Entspannen von Körper und Geist. Der Körper liegt gerade, das Blut kann zirkulieren, die Gelenke sind entspannt, der Mensch findet seine „innere Mitte" und seine Balance. Die Aussparung im Kopfteil lässt Luft an den Kopf und die Gehirnzellen, während der Nutzer sich in einer ausgestreckten Position ausruht. Die Gestaltung ist auf das Wesentliche reduziert, die Liege wirkt sehr leicht und funktional.
Die 22 Millimeter starke Liegefläche aus GFK (Glasfiberlaminat) ist hochbelastbar. Es ist eine Sandwichkonstruktion aus 13 Schichten. Zum Abschluss wurde die Liege poliert und gebürstet.
Die GFK-Technologie kommt ursprünglich aus dem Flugzeug- und Schiffsbau. Ihre Charakteristika sind Leichtigkeit und hohe Stabilität.
Das Gestell besteht aus Edelstahl, das Kopfpolster ist aus Polyurethanschaum geformt. Die Verbindungen von Gestell und Liegefläche sind als sichtbare Details gestaltet.

.

A couch for body and soul to relax and hang loose. Your body lies straight, blood can circulate, joints are relaxed, and you can find your "inner core" and your balance. The recess in the head section allows air to get at the head and the brain cells, while the user relaxes in a stretched position. Design has been reduced to the bare essentials, and the impression the couch makes is very light and functional.
The 22 millimetre thick couch surface, made of GFK (glass fibre reinforced plastic), can take high loads. It is a sandwich construction made up of 13 layers. At the end of the process, the couch is polished and brushed.
GFK technology originally comes from aircraft and shipbuilding. It is characterized by lightness and extreme sturdiness.
The frame is made of stainless steel, while foamed polyurethane has been used for the cushion. The connections between frame and couch surface have been left as visible details.

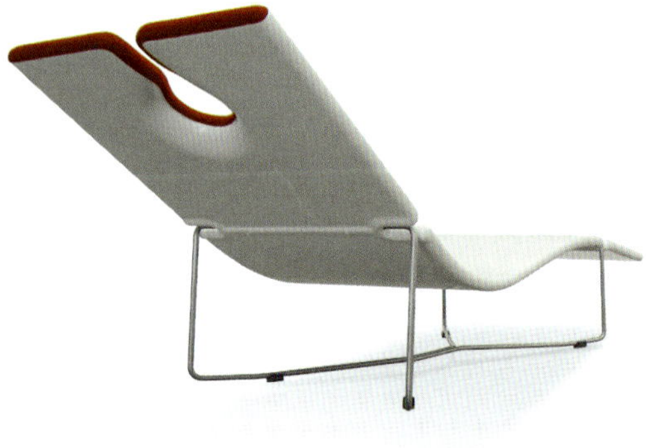

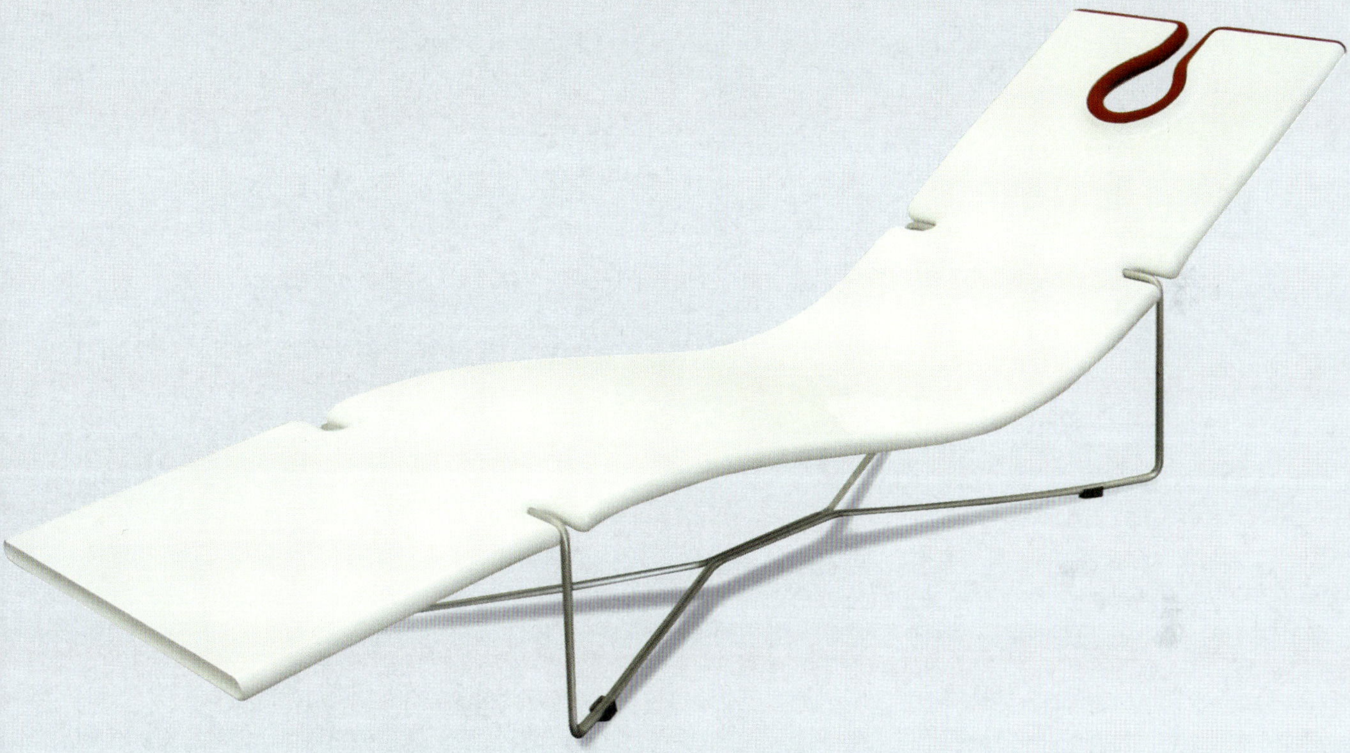

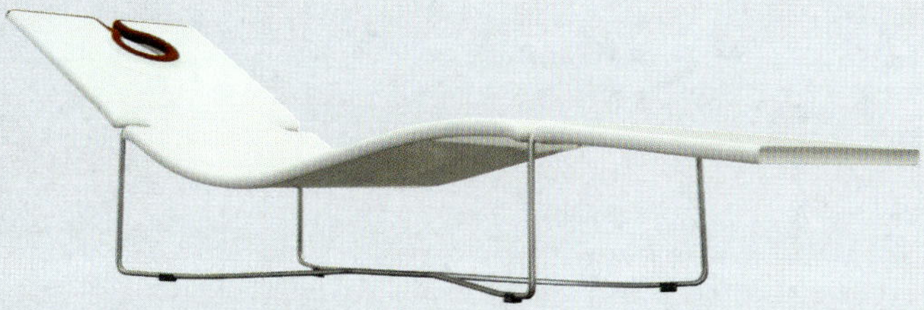

Jury . Judges panel

Eine sehr innovative Liege, die neue Materialien mit einer sehr geradlinigen Form verbindet. Ein innovatives Produkt.

.

A very innovative couch, combining new material qualities with a very linear form. An innovative product.

HERSTELLER . MANUFACTURER
Steuler Fliesen GmbH
D-75417 Mühlacker

VERTRIEB . DISTRIBUTOR
Steuler Fliesen GmbH
D-75417 Mühlacker

DESIGN . DESIGNER
Werksdesign

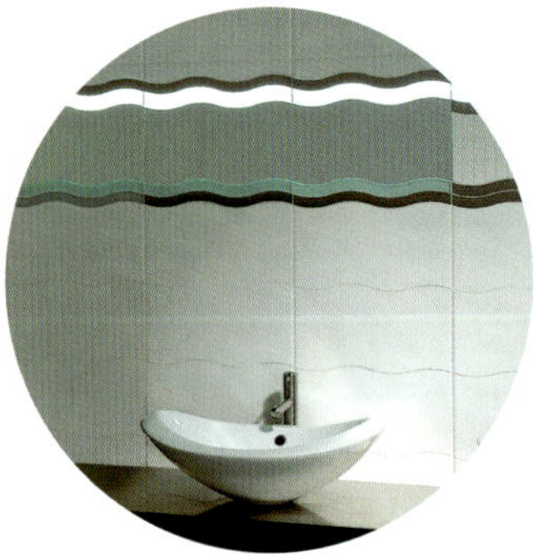

Flow : Fliesenserie . tile range

Müssen Fliesen denn immer rechteckig sein? Diese Frage brachte Designer auf „Flow". Geflieste Flächen sind sehr stark vom Fugenbild geprägt. „Flow" löst hier nun rechte Winkel und gerade Linien in eine harmonische Wellenbewegung auf, was den Gedanken an fließendes Wasser im Badezimmer widerspiegelt.
Die Grundfliese bleibt im Winkel, sie wird so graviert, dass sie der Fliesenleger an der Gravurlinie brechen kann und so eine oder zwei wellenförmige Kanten entstehen. Wird der Anschluss an herkömmliches Fliesenraster gewünscht, ist dies somit weiterhin möglich.
Zu der Fliesenserie gehören Bordüren in Wellenform aus warmen und kalten Materialien, wie mattiertes Glas, Edelstahl und Wengé-Holz, sowie geschwungene Mosaikbänder. Völlig neu sind die wellenförmigen, LED-beleuchteten Glasbordüren in blau, gelb und weiß. Sie geben dem Badezimmer als Ort der Erholung und Entspannung stimmungsvolles Licht und lassen alles fließen: Wasser, Fuge, Licht, Energie ...

Do tiles always have to be rectangular? It was this question that led designers to "Flow". Tiled walls are very much dominated by the pattern of joints. It is here that "Flow" does away with right angles and straight lines in favour of a harmonious wave-like movement, reflecting the idea of flowing water in the bathroom.
The basic tile is still square to the wall, but is engraved in such a way that the tiler can break it along this engraved line, creating one or two wave-like edges. This means that it is still possible to link up to a conventional tiling pattern, if so desired.
The tile series also includes wavy borders made of warm and cold materials, such as matt glass, stainless steel and wenge wood, as well as wavy bands of mosaic. One completely new feature is the wavy, LED illuminated glass borders in blue, yellow and white. They give the bathroom, as a place of recuperation and relaxation, atmospheric light and get everything flowing: water, joint, light, energy ...

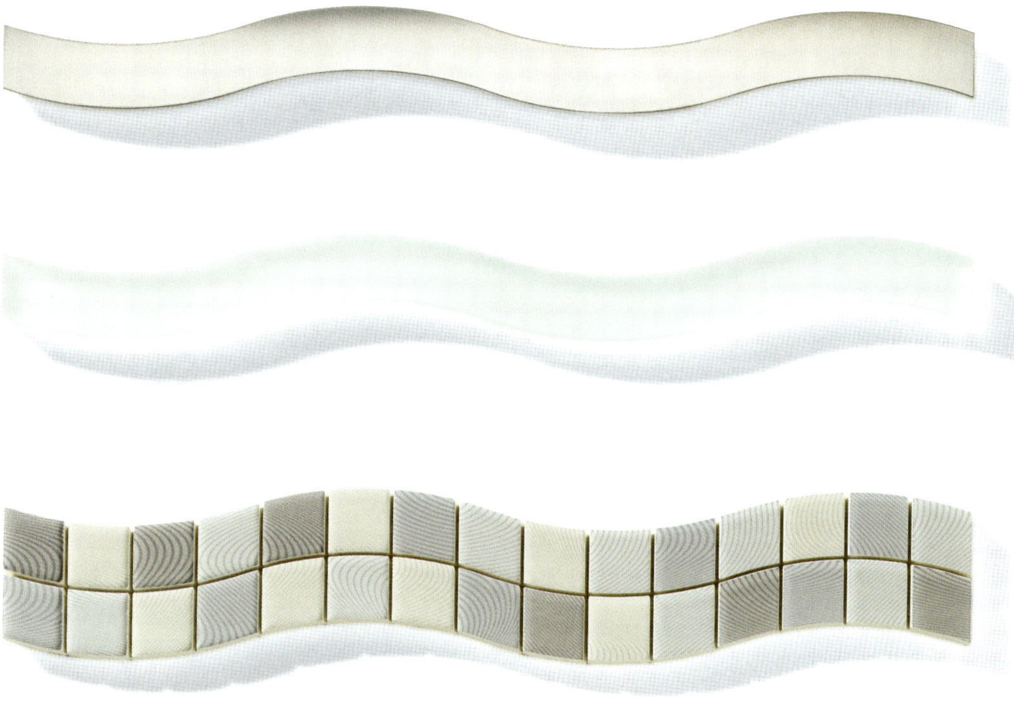

Jury . Judges panel

Sehr schön und interessant diese neue Art, Fliesen zu nutzen und dabei auch Licht, Glas, Holz und Metall einzusetzen. Eine gute Idee. Die Jury regt jedoch an, sehr vorsichtig mit den Farben und Details zu sein, weil nicht alles geht. Manche der Beispiele sind dekorativ aber vom Designstandpunkt nicht wirklich gut entwickelt.

.

This new way of using tiles, making use of light, glass, wood and metal at the same time, is very attractive and interesting. A good idea. However, the judges would suggest being very careful with colours and details, because some things do not work. While some of the examples are decorative, from a design perspective they have not really been well developed.

HERSTELLER . MANUFACTURER
Duscholux GmbH
D-69198 Schriesheim

VERTRIEB . DISTRIBUTOR
D+S Sanitärprodukte GmbH
D-69198 Schriesheim

DESIGN . DESIGNER
Werksdesign
und
Code.2.design
Michael Schmidt
D-70176 Stuttgart

Life Evolution : Duschtrennwand . shower partition

„Life Evolution" steht in der Tradition der türlosen Duschwände, die die Firma bereits vor über zehn Jahren eingeführt hat. Die neue Duschtrennwand hat eine Schneckenform mit einem großzügigen Dusch- und Eingangsbereich. In die Duschsäule in Bicolor-Optik – silber-matt und platinum-silber – ist eine Fitness- und Wellness-Ausstattung integriert: eine flache Kopfbrause, vier Seitenbrausen und eine Handbrause mit drei Strahlarten.
„Life Evolution" zeichnet sich durch hohe Designqualität und zeitgemäß klare Gestaltung aus. Kreis und Bogen prägen das Erscheinungsbild, kleine Radien an den Flächenübergängen betonen die geometrischen Formen zusätzlich. Die extrem gebogenen Sicherheitsglasscheiben geben dem Ganzen einen markanten optischen Schwung und bilden den Blickfang im Badezimmer.
Die Duschwanne ist aus Sanitäracryl gefertigt.

.

"Life Evolution" belongs to the tradition of doorless shower partitions, introduced by the company more than ten years ago. The new shower partition is spiral, with a generous shower and entrance area. The two-tone shower column, in matt silver and platinum silver, houses fitness and wellness attachments: a flat shower head, four lateral shower heads and a portable shower head with three types of jet.
"Life Evolution" stands out by virtue of its high-quality design and modern, clear lines. Circle and arch are its dominant features, and small circles at the transitions between surfaces give additional emphasis to these geometric forms. The extremely curved panes of safety glass give the whole ensemble a striking visual dynamism, and are an eye-catcher in the bathroom.
The shower tray is made of sanitary acrylic.

Jury . Judges panel

Eine Raumskulptur verbindet Ästhetik mit Funktionalität. Ein Produkt, das eine Plattform braucht, aber auch dem Benutzer eine Plattform bietet. Das Reinigen des Körpers wird vom Muss zum zelebrierten Vergnügen. Ein Brausenerlebnis der besonderen Art.

.

A room sculpture combines aesthetics and functionality. A product that needs a platform, as well as offering the user a platform. Cleansing the body is transformed from a chore into a celebration and pleasure. A shower experience of a quite special kind.

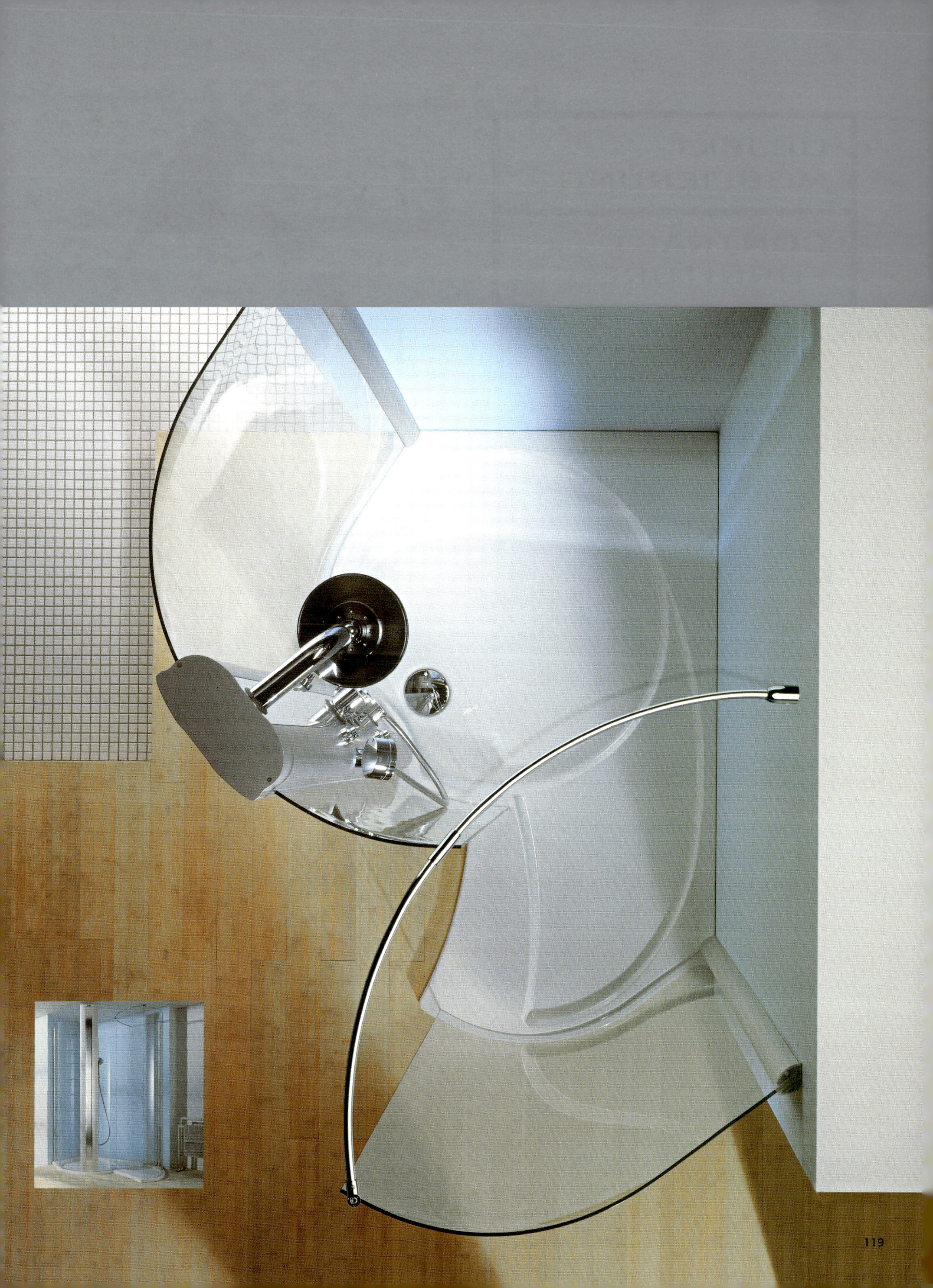

OBJEKT-MÖBLIERUNG

CONTRACT FURNITURE

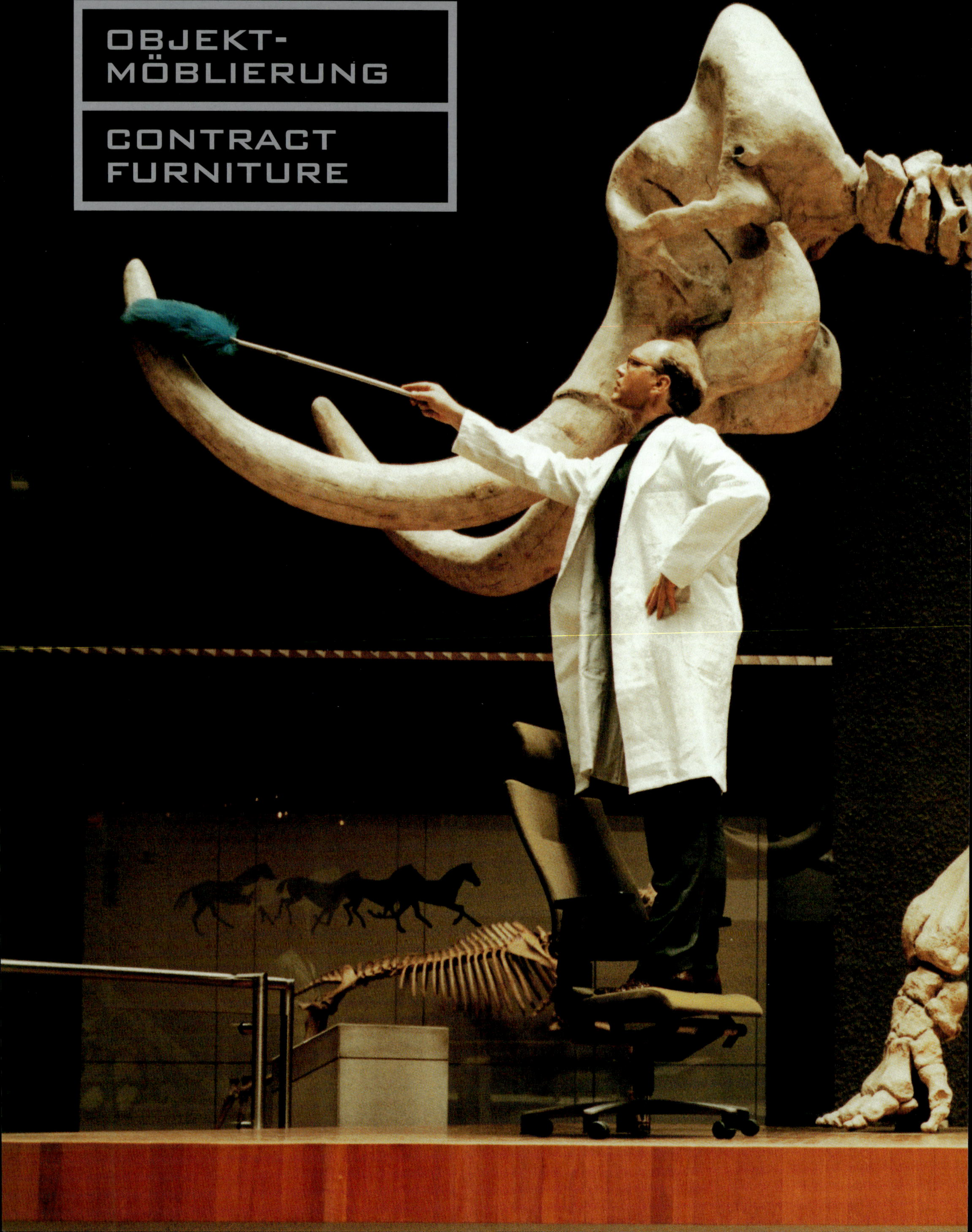

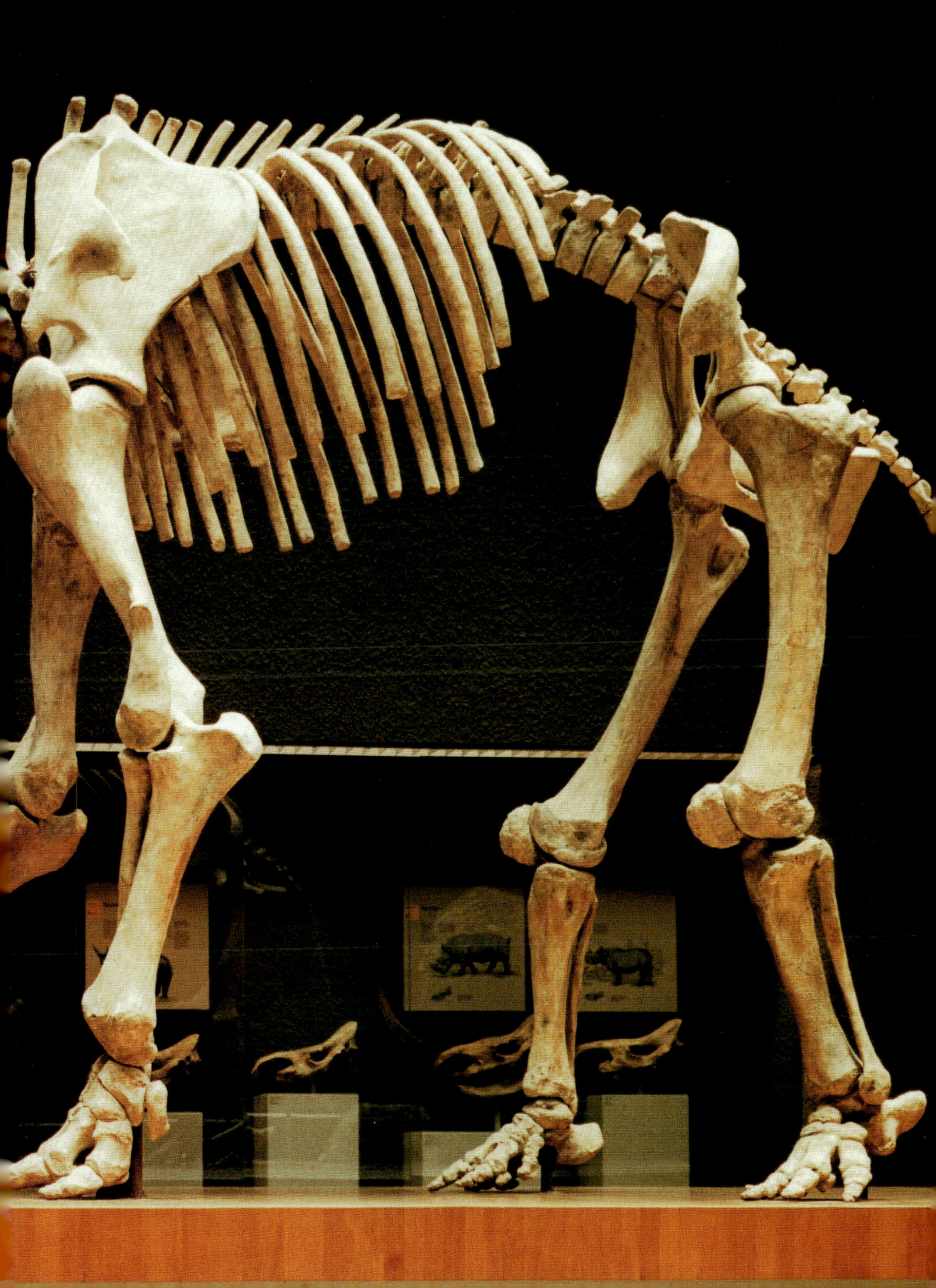

OBJEKT-MÖBLIERUNG GOLD

CONTRACT FURNITURE GOLD

HERSTELLER . MANUFACTURER
Wilkhahn
Wilkening + Hahne GmbH + Co
D-31848 Bad Münder

VERTRIEB . DISTRIBUTOR
Wilkhahn
Wilkening + Hahne GmbH + Co
D-31848 Bad Münder

DESIGN . DESIGNER
Wiege
D-31848 Bad Münder

Solis. Programm 190 : Bürostuhl . office chair

„Einfach sitzen" war die Leitidee bei der Entwicklung von „Solis". Der Bürostuhl ist konsequent einfach gestaltet. Die Reduktion auf homogene Flächen, gerade Linien und klare Körper schafft visuelle Ruhe und Ordnung – und damit Raum für Konzentration und Entspannung. Wo sonst zahlreiche Hebel, Knöpfe und Mechaniken die komplexen Einstellungen einer Sitzmaschine fordern, hat „Solis" gerade mal zwei Hebel, die ebenso griffgünstig wie unauffällig in einer zentralen Achse angeordnet sind.
Hinsetzen – Höhe und Gewicht einstellen. Alles Übrige macht „Solis" quasi von selbst: Die flexible Rückenschale stützt, wo sie stützen soll, und gibt nach, wo Anpassung gefordert ist. Das Zusammenspiel von flexiblen, ergonomisch geformten Schalen mit einer neuartigen Synchronautomatik erlaubt die Abstützung vom aufrechten Sitzen bis zum entspannten Relaxen. Der Sitz passt sich vollautomatisch den Bewegungsabläufen an. Dabei verstärkt sich mit zunehmender Neigung der Gegendruck – was dem natürlichen Bedürfnis beim Zurücklehnen entspricht –, synchron dazu senkt sich die Sitzvorderkante ab, damit der große Öffnungswinkel nicht zu verstärktem Druck auf die Oberschenkelunterseite führt.

.

"Simply sit" was the guiding idea behind the development of "Solis". The design of this office chair is simplicity itself. By reducing it down to homogenous surfaces, straight lines and clear elements, visual peace and order is created – and with it room for concentration and relaxation. Instead of the usual complex array of levers, buttons and mechanisms needed to adjust such a chair, "Solis" has just two levers arranged within easy reach yet unobtrusively around a central axis. Sit down – set height and weight. "Solis" does the rest more or less on its own. The flexible back rest provides support where it is needed and gives way wherever adjustment is needed. This interplay between flexible, ergonomically designed shells and an innovative synchro-adjustment mechanism allows support to be provided both when sitting upright and relaxing. The chair adjusts automatically to shifts in movement. Counter-pressure increases the more the chair is reclined – satisfying a natural need when reclining – and at the same time the forward edge of the seat is lowered, the large angle of aperture ensuring that this position does not result in increased pressure on the underside of the thighs.

Jury . Judges panel

Sehr gute Balance zwischen den neuen Anforderungen der modernen datenverarbeitenden Arbeitswelt und der klassischen Büromöblierung. Mit sehr schlanken Proportionen wurde eine optimale Funktion erreicht.

.

Very good balance between the new demands of the modern, data-processing world of work and classical office furniture. Using very slim proportions, the best possible function has been achieved.

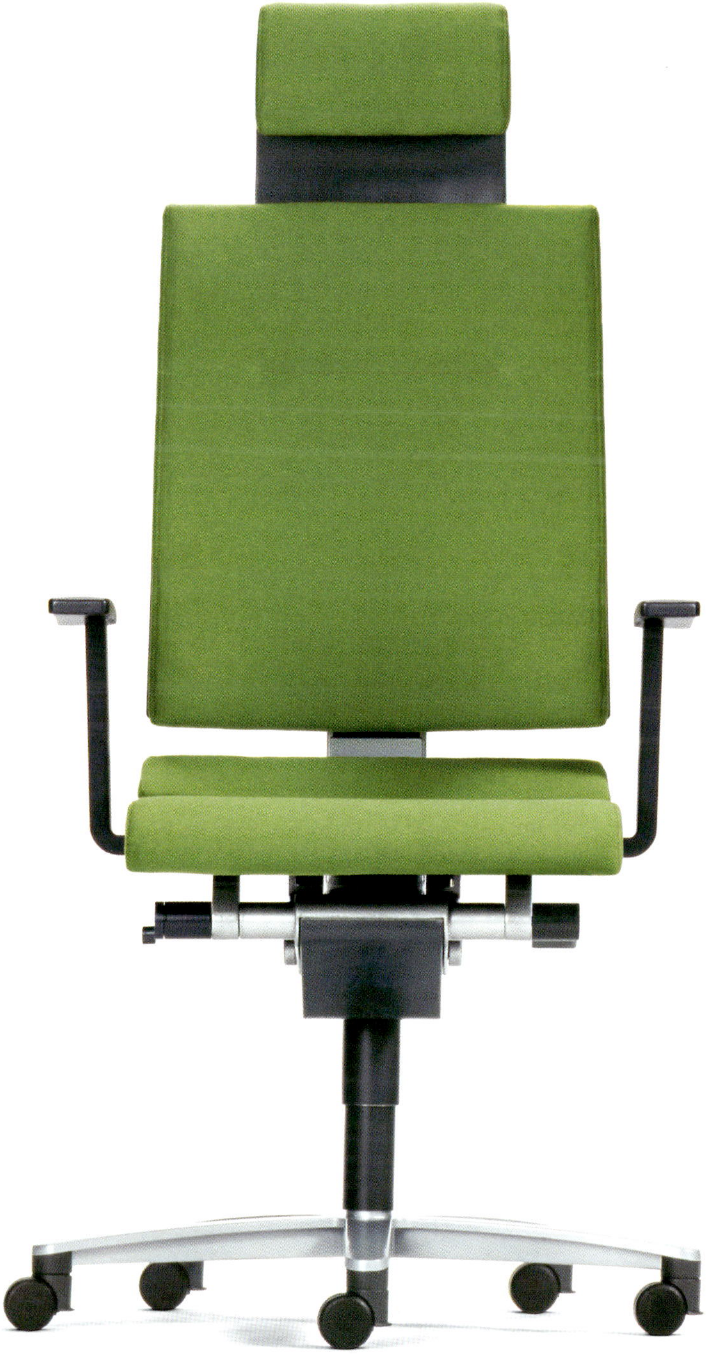

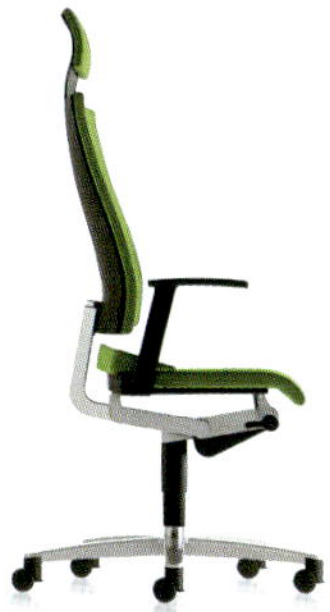

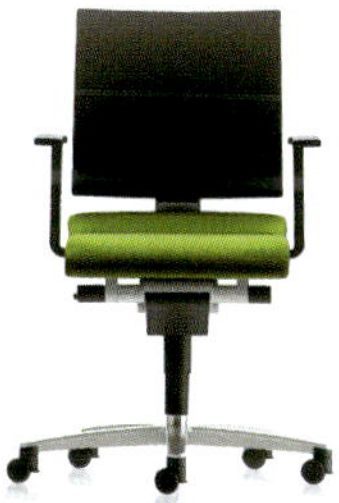

OBJEKT-MÖBLIERUNG SILBER

CONTRACT FURNITURE SILVER

HERSTELLER . MANUFACTURER
Wilkhahn
Wilkening + Hahne GmbH + Co
D-31848 Bad Münder

VERTRIEB . DISTRIBUTOR
Wilkhahn
Wilkening + Hahne GmbH + Co
D-31848 Bad Münder

DESIGN . DESIGNER
Wiege
D-31848 Bad Münder

Solis F. Programm 190 : Bürostuhl . office chair

„Solis F" zeichnet sich durch kompromisslos puristische Form und verblüffend hohen Sitzkomfort aus. Die weit öffnende Synchronautomatik verfügt über eine automatische Absenkung der Sitzvorderkante. Hier verbindet sich technische Ästhetik mit gesundem Sitzen. „F" steht für Frame und bezeichnet damit die prägnanten Sitz- und Rückenkonstruktionen: Zwei Längsprofile werden durch gewölbte Querstreben aus Aluminiumdruckguss gespannt.
Das schwarze, transparente Hochleistungsgewebe ist seitlich gespannt und bekommt dadurch oben und unten hohe Elastizität. So passt sich der Stuhlrücken individuellen Körperformen automatisch an. Gleichzeitig sorgen die ergonomisch ausgebildeten Rahmenprofile im Sitz für eine aktive Beckenstütze und im Rücken für die richtige Unterstützung im Lordosebereich.
Es gibt zwei Rückenlehnenhöhen, wobei der Hochlehner wahlweise mit einer Kopfstütze ausgestattet ist. Ein Modell hat mit Leder bezogene Polsterauflagen.

"Solis F" is uncompromisingly purist in form and an astonishingly comfortable seat. The wide opening synchro-adjustment mechanism automatically lowers the front edge of the seat. This seat merges a technical aesthetic and healthy posture. The "F" stands for frame, and alludes to the striking design of the seat and back. Two longitudinal sections are braced by arched cross-struts of die cast aluminium.
The black, transparent, hard-wearing fabric is tensed at the sides, making it very elastic at the top and bottom. The result is that the seat back automatically adjusts to individual body shape. At the same time, the ergonomically shaped frame sections mean that the seat surface actively supports the pelvis and the back surface provides correct support for the lumbar region.
The seat back is available in two heights, the higher version having an optional head rest. One model has leather-covered upholstered surfaces.

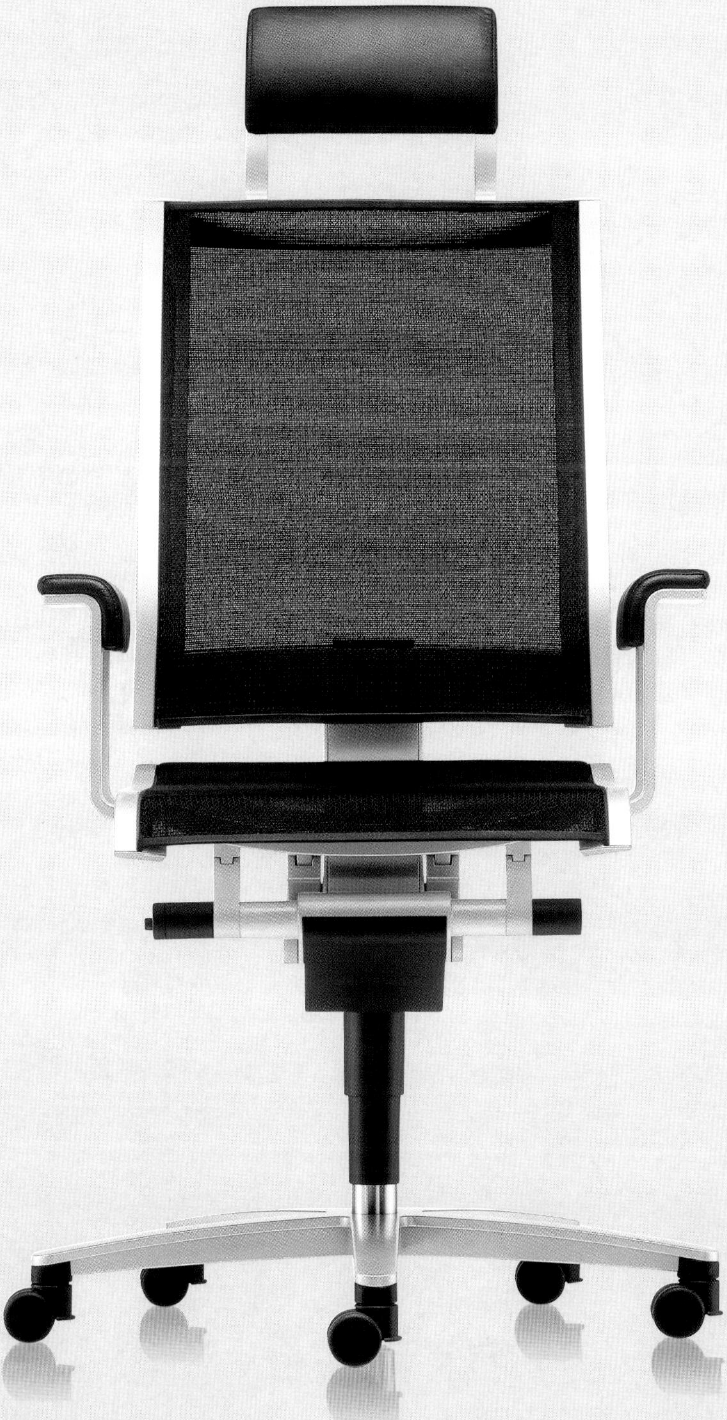

Jury . Judges panel

Ein elegantes Produkt mit gutem Materialeinsatz und Komfort. Es hat viele Funktionen, aber es ist trotzdem in der Erscheinung sehr einfach. Das ist daran die gute Balance.

.

An elegant product, comfortable, makes good use of materials. Despite its many functions, it is very simple in appearance. That is what makes it well-balanced.

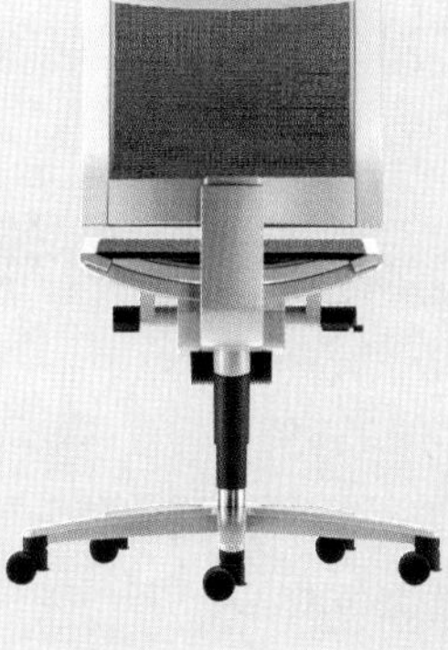

HERSTELLER . MANUFACTURER
Sedus Stoll AG
D-79761 Waldshut

VERTRIEB . DISTRIBUTOR
Sedus Stoll AG
D-79761 Waldshut

DESIGN . DESIGNER
Mathias Seiler
D-79539 Lörrach

Sedus open up workassistant : Arbeitstablar . writing tablet

Die aufrechte Sitzhaltung am Tisch wird durch die Entwicklung der modernen Informationstechnologien zunehmend in Frage gestellt. Besonders die zukünftige Verbreitung mobiler Notebooks mit LAN-Funkverbindungen werden größere Freiheiten beim Sitzen bringen.
„Sedus open up workassistant" ermöglicht abwechselndes Arbeiten im Sitzen, Stehen und Liegen. Er dient dabei als Beinablage und als Tablarträger, auf dem beispielsweise ein Notebook bedient werden kann. Natürlich kann man auch wie herkömmlich in aufrechter Sitzposition arbeiten. Da der Verstellmechanismus gasfederunterstützt arbeitet, lässt sich das Arbeitstablar schnell und bequem an die aufrechte Position anpassen. Auf der Beinablage können dann Unterlagen abgelegt werden. Will man stehend daran arbeiten, lässt sich die Höhe per Knopfdruck auf diejenige eines Stehpultes verstellen. Damit kann ganz spontan von der Sitz- zur Liege- und Stehposition gewechselt werden.

.

The development of modern information technology is increasingly casting doubt on the idea of upright posture at the desk. In particular, the spread of mobile LAN-linked notebooks in the future will bring about greater freedom of working posture in the office.
"Sedus open up workassistant" allows work to be done when sitting, standing or lying down. It serves as a footrest and a shelf support, on which a notebook can be used, for example. Of course, work can also be done in the usual upright seating position. As adjustment is done by means of a gas lift mechanism, the writing tablet can be adjusted quickly and conveniently to an upright seating position. In this case, the footrest can be used as a shelf for papers. If a standing position is desired, a push of a button is all that is needed to adjust its height to that of a lectern. In an instant, working position can be changed to a sitting, lying or standing posture.

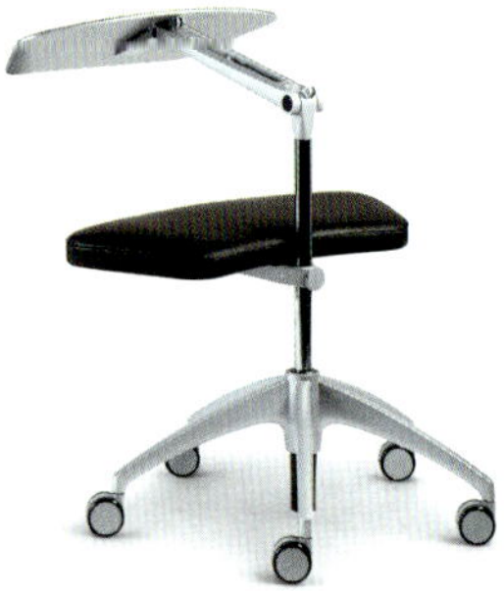

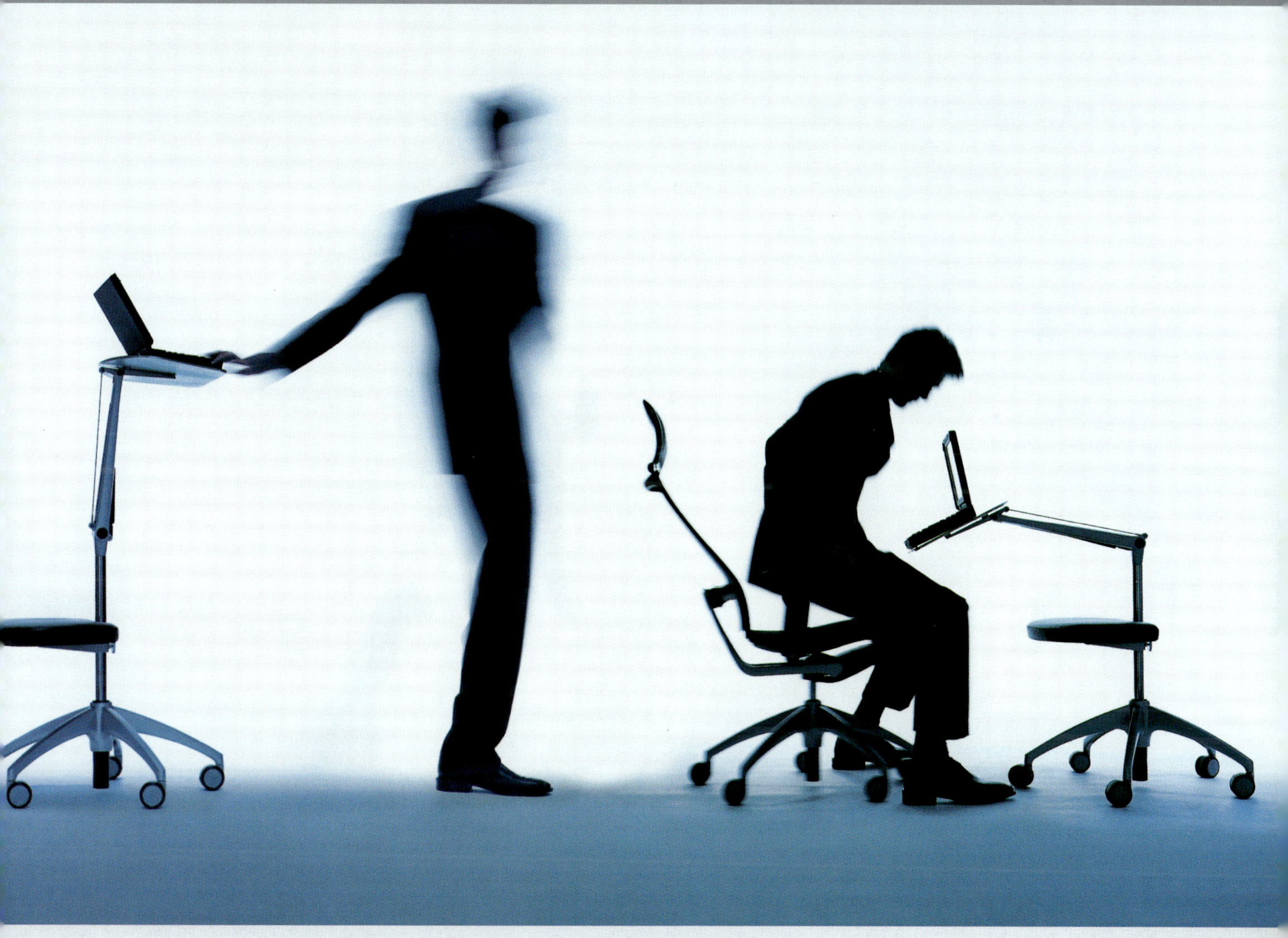

Jury . Judges panel

Das Produkt arbeitet mit der Frage, in welcher Position man arbeitet. Man kann sich hier in einer ganz anderen Art hinsetzen als an einen Tisch und an seinem Computer arbeiten oder eine Zeitschrift lesen. Das Produkt lässt Freiräume zu. Die technische Mechanik kann gestalterisch noch besser gelöst werden.

.

The product works with the issue of work posture. With this product, you can sit down quite differently than at a desk, and work at your computer or read a magazine. It allows room for movement. The design of its mechanical features could be improved.

OBJEKT-MÖBLIERUNG SILBER

CONTRACT FURNITURE SILVER

HERSTELLER . MANUFACTURER
Walter Knoll AG & Co. KG
D-71083 Herrenberg

VERTRIEB . DISTRIBUTOR
Walter Knoll AG & Co. KG
D-71083 Herrenberg

DESIGN . DESIGNER
Wolfgang C. R. Mezger
D-73054 Eislingen

Headoffice 3200 : Chefzimmerprogramm . range of executive furniture

Die moderne Reduktion auf das Wesentliche lässt Führungskräften Raum für Motivation und Konzentration. Das Chefzimmerprogramm zeichnet sich durch klare und rationale Strukturen aus, die kühl, sachlich und elegant wirken.
Auch die Auswahl der Materialien unterstreicht die funktionale und reduzierte Gestaltung: verschiedene hochwertige Furniere oder Glas für Arbeits- und Abdeckplatten, kühles Metall für die Konstruktion – strichmatt-verchromter Edelstahl für die Füße – und Strukturlack für die Kastenmöbel.
Das Chefzimmerprogramm besteht aus verschiedenen Schreib- und Konferenztischen, Containern, High- und Sideboards sowie einem Steharbeitsplatz.
Alle Teile lassen sich miteinander kombinieren. „Headoffice" bietet damit eine Vielfalt an Lösungen, individuelle Arbeitsplätze und Konferenzbereiche zu gestalten. Die Varianten reichen vom Einzeltisch über freie Kombination von Tischen mit Präsentationsmodul bis hin zur festen Konferenzanlage.

.

Modernist reduction to the bare essentials gives executives room to be motivated and focussed. This range of executive furniture stands out by virtue of its clear, rational structures, whose effect is cool, sober and elegant. The choice of materials also highlights the functional, reduced design, with various high-quality veneers or glass for desk tops and other surfaces, cool metal for the framework, matt chromium-plated stainless steel for the legs and structured paint for the containers.
The range of executive furniture comprises various desks and conference tables, containers, highboards and sideboards and a lectern. All the parts can be combined with each other. "Headoffice" therefore offers a variety of solutions for individually designed workplaces and conference areas. The possibilities range from a single desk via a free combination of tables with a presentation module to a fixed conference arrangement.

Jury . Judges panel

Ein eindrucksvolles Produkt, das aber etwas überzieht, was seine Proportionen, Materialien und das Gesamtdesign anbelangt. Alle Details sind wohldurchdacht und von herausragender Qualität. Ein luxuriöses, intelligentes Produkt.

.

An impressive product, but one which goes a bit too far in its proportions, materials and overall design. All details have been well thought through and are of excellent quality. A luxurious, intelligent product.

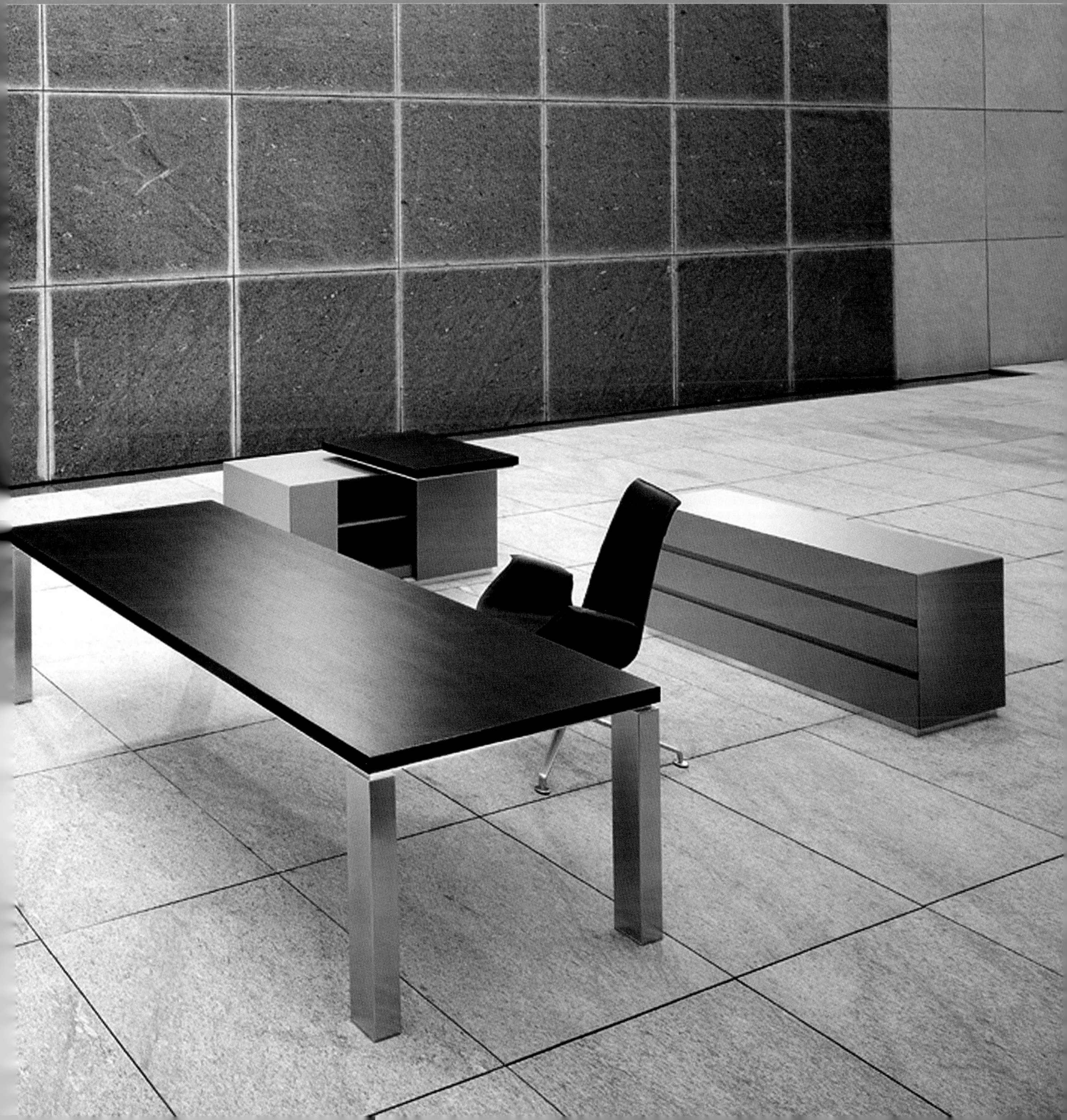

OBJEKT-MÖBLIERUNG | SILBER

CONTRACT FURNITURE | SILVER

HERSTELLER . MANUFACTURER
Helit Innovative Büroprodukte GmbH
D-58566 Kierspe

VERTRIEB . DISTRIBUTOR
Helit Innovative Büroprodukte GmbH
D-58566 Kierspe

DESIGN . DESIGNER
Porsche Design GmbH
Ferdinand A. Porsche
A-5700 Zell am See

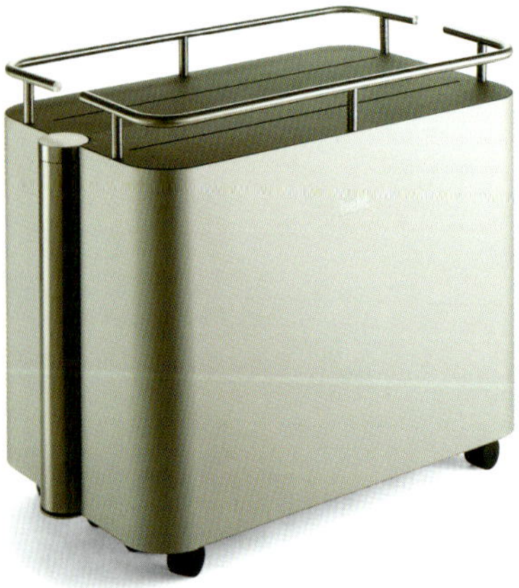

Büro- und Servicewagen . office and serving trolley

Der Büro- und Servicewagen für gehobene Ansprüche sorgt für Mobilität im Büro, aber auch im privaten Bereich. Persönliche Dokumente und vertrauliche Unterlagen stehen dank der leichtgängigen, stufenlosen Öffnung bis zu einem Winkel von 180 Grad gleich bereit und können ebenso einfach wieder unter Verschluss genommen werden. Aber auch als eleganter Servicewagen, in dem Geschirr oder Gläser griffbereit aufbewahrt werden, ist er vielseitig einsetzbar.
Die multifunktionale Inneneinrichtung mit variabel verstellbaren Zwischenböden und flexibel steckbaren Unterteilungen bietet große Flexibilität und Variationsmöglichkeiten. Hervorragende Verarbeitung und hochwertige Materialien sind weitere Merkmale dieses eleganten und nützlichen Produktes: gebürsteter Edelstahl, eloxiertes Aluminium und Perdur-Oberflächen.
Dank der bewusst sachlich und edel gewählten Formgebung fügt sich der Büro- und Servicewagen optisch in jede Umgebung ein.

.

This office and serving trolley for the discerning customer provides mobility in the office and the home. Because it can be opened easily by up to 180 degrees, personal papers and confidential documents are immediately accessible, and can be shut away again just as easily. It also has a variety of uses as an elegant serving trolley, however, in which crockery or glasses can be kept for immediate use.
Its multifunctional interior features, with variably adjustable shelves and flexibly positionable partitions, make it very flexible and versatile. Excellent workmanship and high-quality materials are further features of this elegant, useful product. It has brushed stainless steel, anodized aluminium and Perdur surfaces.
The deliberately sober, classic design means that the office and serving trolley harmonizes well in any environment.

Jury . Judges panel

Das Produkt ist sehr einfach, nüchtern und wendig, an unterschiedlichen Orten vielseitig einsetzbar. Auch für Messen gut, da der Wagen Lagerplatz bietet. Man hat immer etwas zu verstecken und etwas, das man schnell holen muss. Ansprechende Materialien und sehr elegante Detaillösungen.

The product is very simple, sober and easy to manoeuvre, and can be used in a variety of places. Good for trade fairs, too, because of the need for storage space there. There is always something that needs hiding away and something that has to be got quickly. Appealing materials and very elegant details.

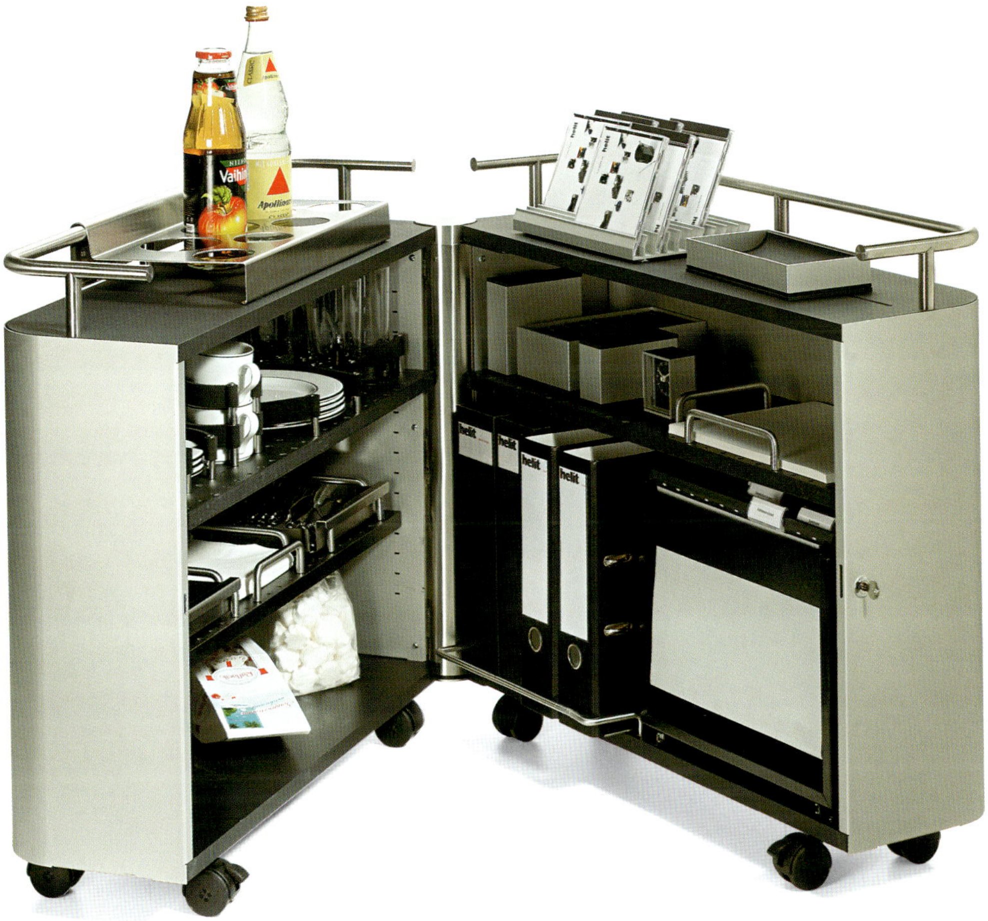

OBJEKT-MÖBLIERUNG | SILBER
CONTRACT FURNITURE | SILVER

HERSTELLER . MANUFACTURER
VS Vereinigte Spezialmöbelfabriken GmbH + Co
D-97941 Tauberbischofsheim

VERTRIEB . DISTRIBUTOR
VS Vereinigte Spezialmöbelfabriken GmbH + Co
D-97941 Tauberbischofsheim

DESIGN . DESIGNER
Hubert Eilers
D-14974 Gröben

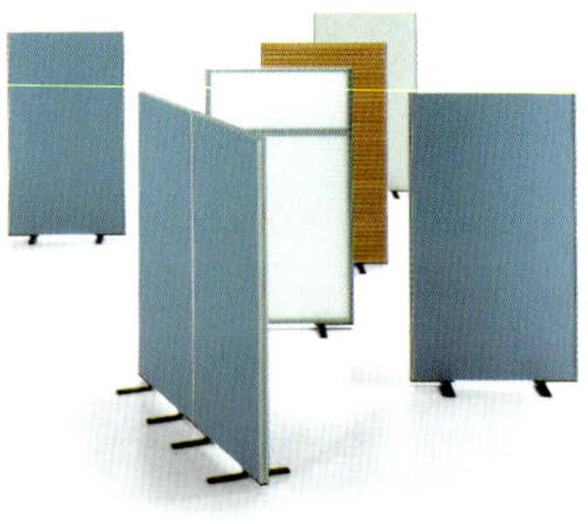

Jury . Judges panel

Raumsystem mit hoher Flexibilität zur Gliederung von Arbeitssituationen im Bürobereich. Verschiedenste Ausführungen möglich, von Akustikwänden bis zu transluzenten Wänden. Dadurch besonders vielseitig einsetzbar. Solide im Detail und in der Ausstattung der Elemente. Verschiedene Funktionen lassen sich dazu addieren.

.

Highly flexible room system for structuring work situations in the office. Very wide range of possible models, from soundproof walls to translucent walls. This makes the system extremely versatile. Sound in its details and the way the elements have been equipped. Various functions can still be added.

Serie 2000 : Stellwandsystem . partition wall system

„Serie 2000" ist ein leichtes, elegantes und flexibles Stellwandsystem, das vielfältige Anforderungen erfüllt: Akustik- und Sichtschutz, Abgrenzung von Büroarbeitsplätzen oder Besprechungsgruppen, pinfähige oder beschreibbare Wand, Ausstellungs- und Präsentationssystem, mobile oder stationäre Einheit. Diskretionsstellwände schaffen schnell und problemlos abgetrennte, geschützte Zonen. Funktionsstellwände dienen dem Aufbau einer flexiblen Raumgliederung. Die Wände sind mit Boards und Ablagen frei organisierbar. Adaptionsstellwände lassen sich mit Tisch- und Korpusmöbelsystemen von VS kombinieren. Sie sind leicht am Tisch zu befestigen und sorgen auf einfache Art für ein geschütztes Umfeld. Sie können Organisationsmittel aufnehmen, bei Doppelarbeitsplätzen auch wechselseitig. Vielfältige Materialien, Farben und Zubehörteile ermöglichen einen großen Gestaltungsspielraum. So kann man unter verschiedenen Füllelementen für den Aluminiumrahmen wählen: mit Stoff bespannte oder mit Holz beziehungsweise Schichtstoff beplankte Leichtbaupaneele sowie transluzente Kunststoff-Doppelstegplatten.

.

"Serie 2000" is a light, elegant and flexible partition wall system that fulfils a wide range of requirements. It can be a sight screen or keep noise out, partition office workplaces or meeting groups, function as a wall on which notices are pinned or written, be used for exhibitions or presentations, and be mobile or stationary.
"Discretion" partitions create separate, protected zones quickly and effortlessly. "Function" partitions help divide up a room flexibly. Storage units and shelves can be added to the walls as desired. "Adaption" partitions can be combined with VS desk and container furniture systems. They are simply fastened to a desk and provide a sheltered environment with a minimum of fuss. Organizational aids can be attached to them, on both sides in the case of double workplaces. A wide range of materials, colours and accessories allow ample room for design ideas. For the aluminium frame, for example, there is a choice of different panel elements: lightweight panels covered with fabric, or planked with wood or laminate, or translucent plastic double-ribbed panels.

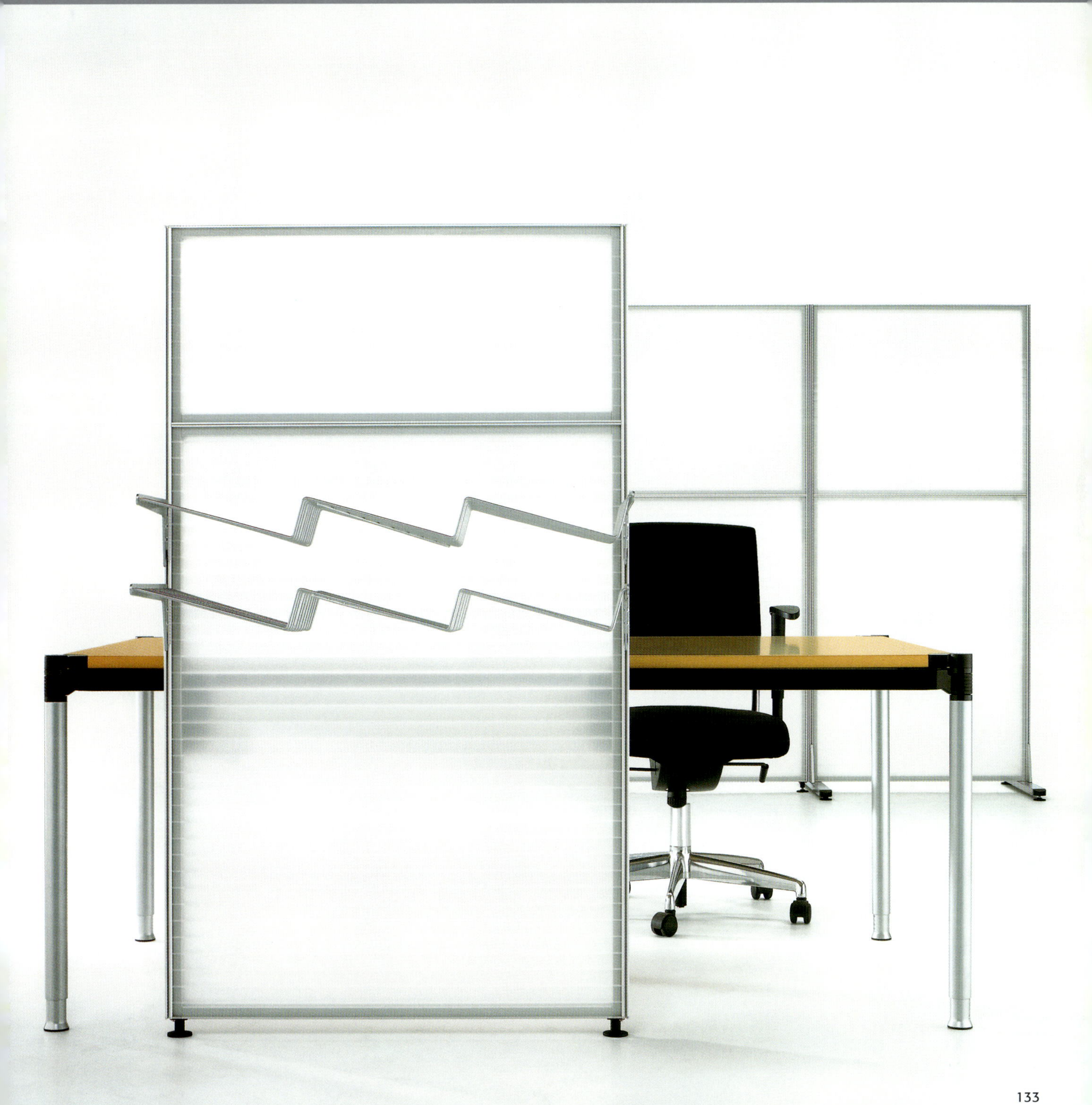

OBJEKT-MÖBLIERUNG	SILBER
CONTRACT FURNITURE	SILVER

HERSTELLER . MANUFACTURER
Sedus Stoll AG
D-79761 Waldshut

VERTRIEB . DISTRIBUTOR
Sedus Stoll AG
D-79761 Waldshut

DESIGN . DESIGNER
Werksdesign
Michael Kläsener

Mr. Charm : Bürodrehstuhl . office swivel chair

„Mr. Charm" spiegelt die Formensprache von moderner Architektur und Design wider. Die Reduktion auf einen eingeschränkten Formenkanon wird dazu genutzt, eine prägnante Struktur zu bilden. Die Hauptelemente des Stuhles sind ursprünglich aus dem Quader abgeleitet, er wurde modifiziert, verformt und verschieden proportioniert. Auffallend ist seine charakteristische Lehnenform, er hat großzügige Abmessungen. Der Stuhl wirkt dadurch sehr geschlossen und strahlt Geborgenheit und Komfort aus.
Zu den Drehstühlen gibt es auch passende Besucherstühle.

.

"Mr. Charm" reflects the formal language of modern architecture and design. Its reduction to a limited formal canon is used to make a striking structure. The chair's main elements are derived originally from a cuboid, which has been modified, distorted and given different proportions. One striking feature is the unusually generous size of its back rest. This gives the swivel chair a very closed effect, and it appears to emanate security and comfort.
There are matching visitor's chairs to go with the swivel chair.

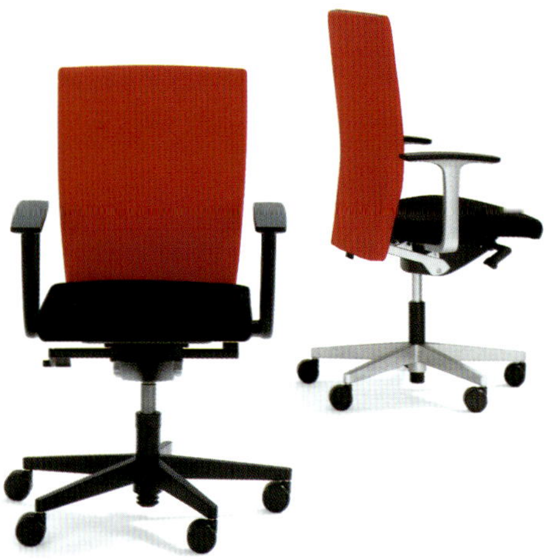

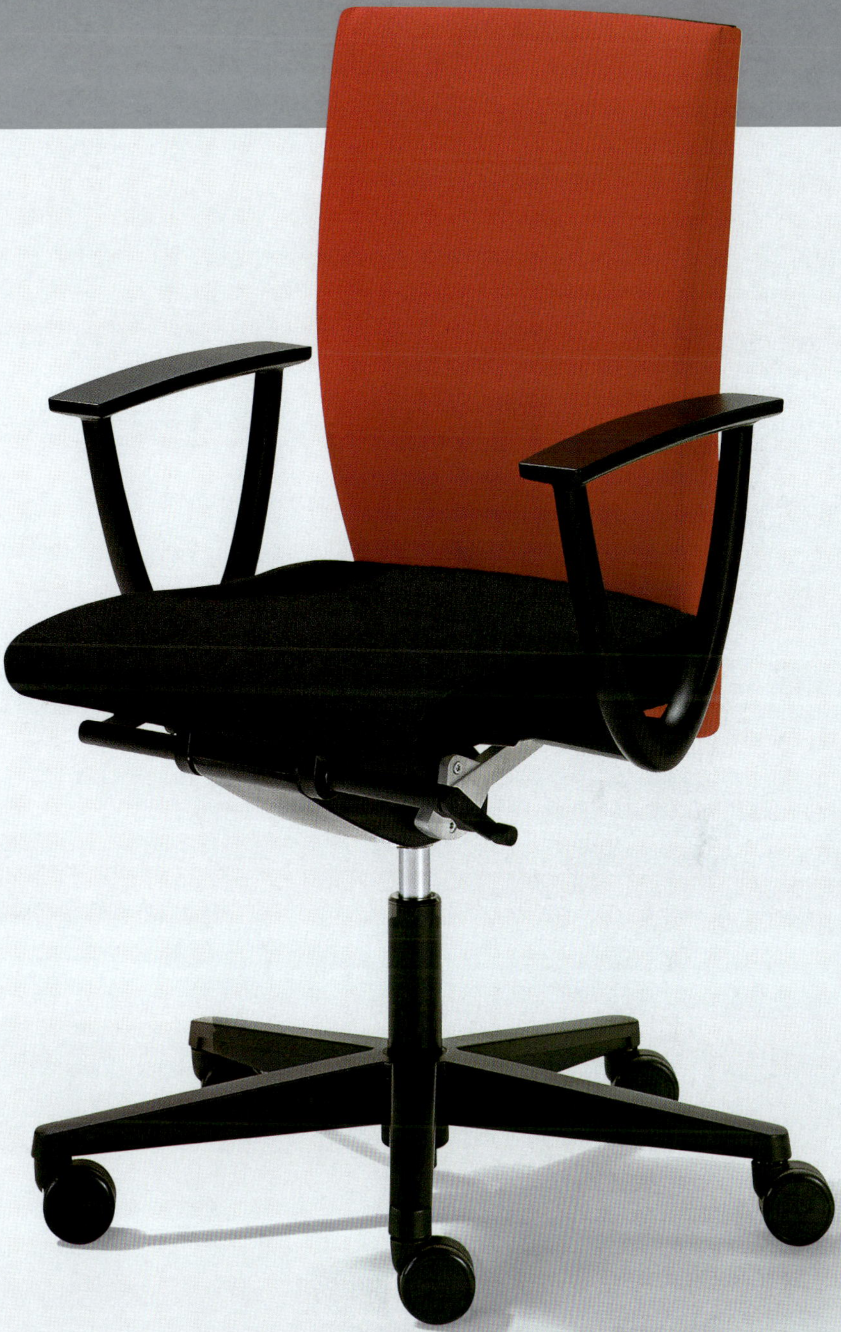

Jury . Judges panel

Komfortables Produkt für den Officebereich, das durch seine schlichte und auffällige Form überzeugt.

.

Comfortable product for the office, made convincing by its very simple yet striking form.

HERSTELLER . MANUFACTURER
König + Neurath AG
D-61184 Karben

VERTRIEB . DISTRIBUTOR
König + Neurath AG
D-61184 Karben

DESIGN . DESIGNER
Burkhard Vogtherr & Jonathan Prestwich
F-68100 Mulhouse

Re.Lounge : Loungestuhl . lounge chair

„Re.Lounge" gehört zu einem neuartigen Konferenz- und Lounge-Programm für gehobene Ansprüche, das sich edel und zugleich zwanglos zeigt. Dieser Loungestuhl lädt zum Verweilen, Relaxen, Arbeiten oder Konferieren ein. Mit ihm lassen sich ungewöhnliche Arbeits- und Gesprächsatmosphären schaffen, beispielsweise in fortschrittlichen Unternehmen mit flexibel gestalteten Arbeitssituationen. Er lädt Projektteams zum entspannten Wissensaustausch ein oder sorgt in halböffentlichen Bereichen für komfortable Meeting-Situationen. Seine niedrige Sitzhöhe erlaubt eine sehr viel entspanntere Sitz- beziehungsweise Arbeitsposition als herkömmliche Besprechungsstühle. Durch sein leichtes, zurückhaltendes und zeitloses Design trägt „Re.Lounge" zu einem ansprechenden komfortablen Ambiente bei, das sich von jeder konventionellen Einrichtungssituation abhebt.
Den Sessel gibt es wahlweise auf Rollen oder Gleitern und mit einem optionalen Tablar mit Drehgelenk. Außerdem wird er noch in einer niedrigeren Variante angeboten.

.

"Re.Lounge" is part of a new conference and lounge chair range for the demanding customer, and is both high-quality and casual in appearance. The lounge chair invites the user to sit a while, to relax, to work or to discuss. Unusual working and meeting atmospheres can be created with this chair – for example, in progressive companies with flexible working arrangements. It invites project teams to meet for a relaxed exchange of knowledge or provides a comfortable meeting setting in semi-public spheres. Its low seat height allows a much more relaxed seating and working position than conventional conference chairs. With its light, unassuming, timeless design, "Re.Lounge" contributes to an attractive, comfortable environment that sets itself apart from run-of-the-mill office furnishing.
The chair is available on either casters or grommets, with an optional hinged side table. It is also available in a lower seat height.

Jury . Judges panel

Ein Sessel, der Bequemlichkeit und interessante Optik miteinander verbindet. Ein außergewöhnlich bequemes Produkt mit interessanten Materialkombinationen.

.

A chair that combines comfort and an interesting appearance. An extraordinarily comfortable product with interesting combinations of materials.

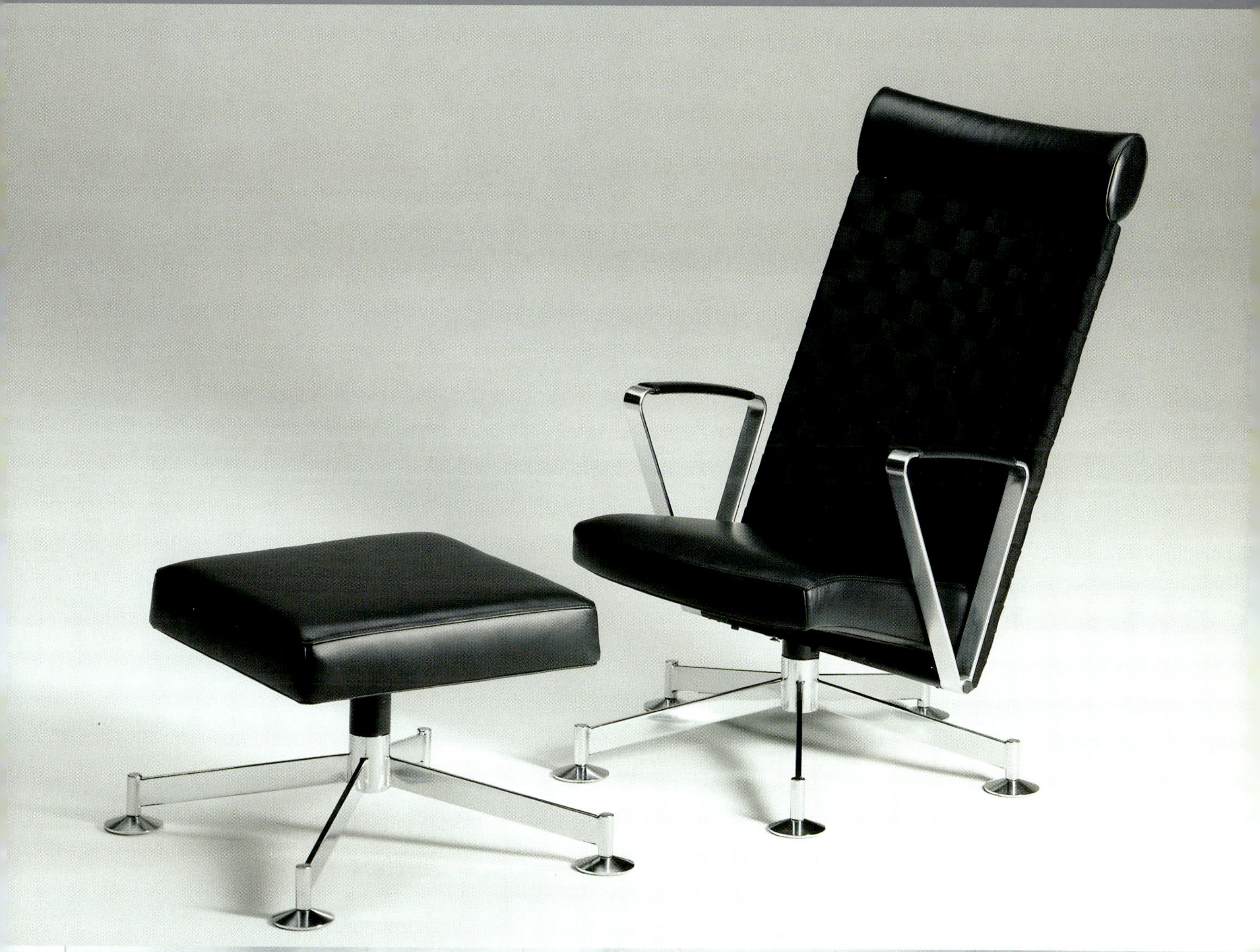

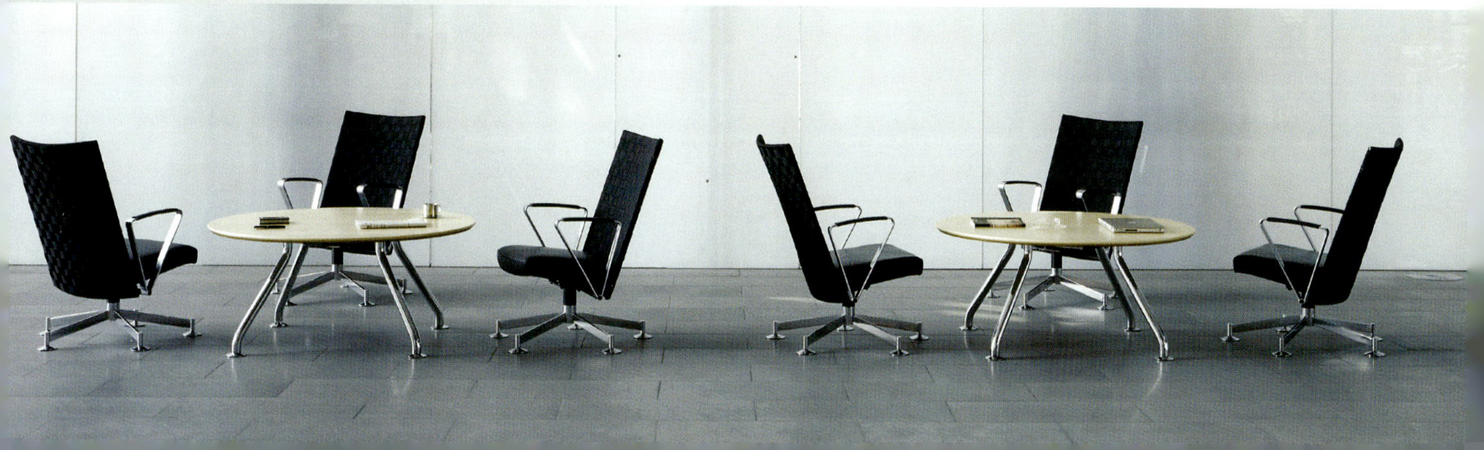

OBJEKT-MÖBLIERUNG SILBER

CONTRACT FURNITURE SILVER

HERSTELLER . MANUFACTURER
Linie K.
D-74613 Ohrnberg

VERTRIEB . DISTRIBUTOR
Lancaster Group/Jil Sander Colour
D-55116 Mainz

DESIGN . DESIGNER
Werksdesign

Jil Sander Open Sell

Das Verkaufs- und Präsentationssystem „Open Sell" ist sehr klar und übersichtlich aufgebaut. Es besteht aus einheitlichen Modulen, die sowohl in der Breite als auch in der Tiefe 20 Zentimeter messen. Diese werden für die einzelnen Systembreiten innerhalb eines 20-Zentimeter-Rasters kombiniert.
Durch ein speziell entwickeltes ausgeklügeltes Auszugsverfahren kann der Nutzer jedes Modul wie eine Schublade herausziehen und auf die hinter den Testern platzierte Originalware optimal zugreifen. Dadurch ist eine sehr gute Handhabung gewährleistet.
Der Korpus des Systems besteht aus silbermetallic beschichtetem Aluminium, die Testermodule sind aus silber hinterdrucktem Acryl. Jeder Boden ist durch eine separate Lichtschiene beleuchtet.

.

The "Open Sell" sales and presentation system is very clear and easy to understand. It is made up of standard modules 20 centimetres wide and deep. These are put together on a 20 centimetre grid pattern to make up the widths required.
A specially developed, sophisticated pull-out technique allows the user to pull out each module like a drawer and to reach the original goods placed behind the samples. This makes it very convenient and easy to use.
The carcass of the system is made of silver metallic coated aluminium, while the sample modules are made of silver-backed acrylic glass. The bottom of each module has its own separate track lighting.

Jury . Judges panel

Auf engstem Raum werden in dieser Schachteltechnik sehr viele Produkte anschaulich präsentiert. Übersichtlich, klar. Das Grundprinzip ist auf verschiedene Displayvarianten ausweitbar, verschiedene Produkte der Kosmetikindustrie können sinnfällig untergebracht werden. Der Lagerbedarf im hinteren Segment ist klar und simpel gelöst.

.

In the smallest of spaces, the box technique allows very many products to be presented attractively. Lucid and clear. The basic principle can be extended to form different display variants, and different cosmetic products can be accommodated logically. The need for storage space in the rear segment has been solved clearly and simply.

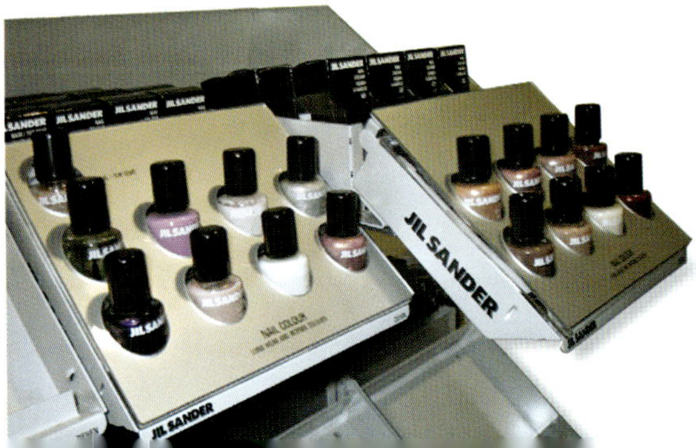

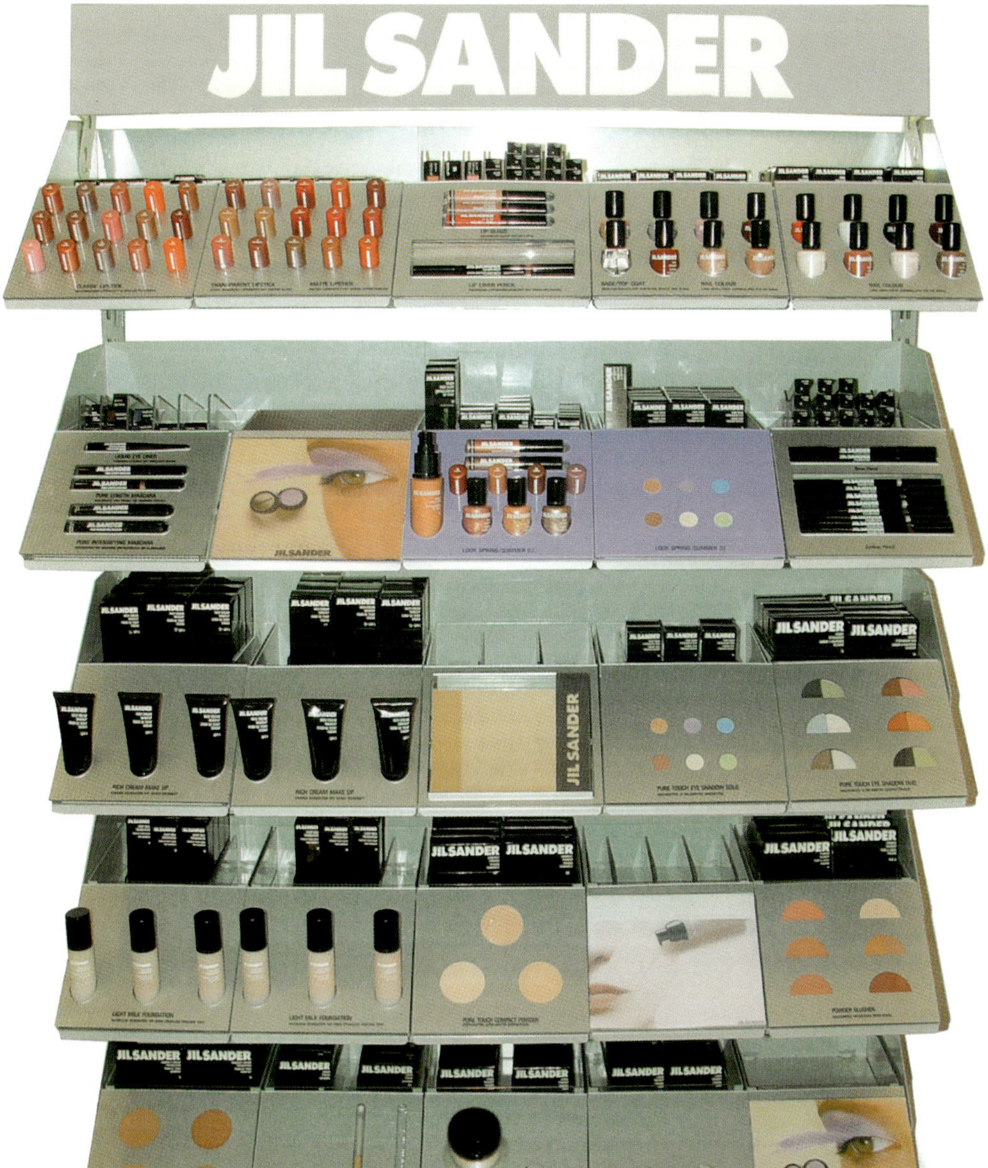
JIL SANDER

BÜRO-AUSSTATTUNG

OFFICE EQUIPMENT

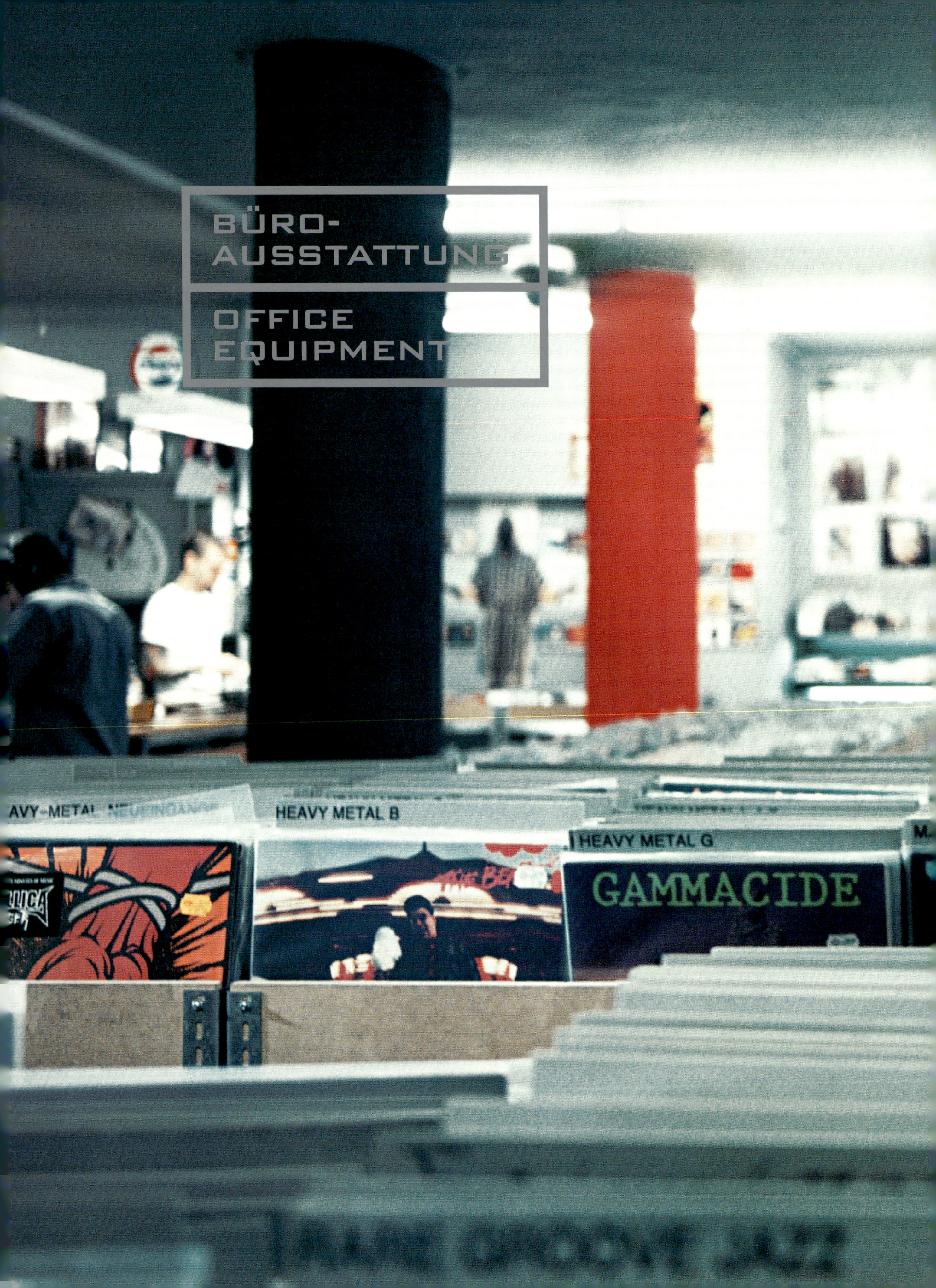

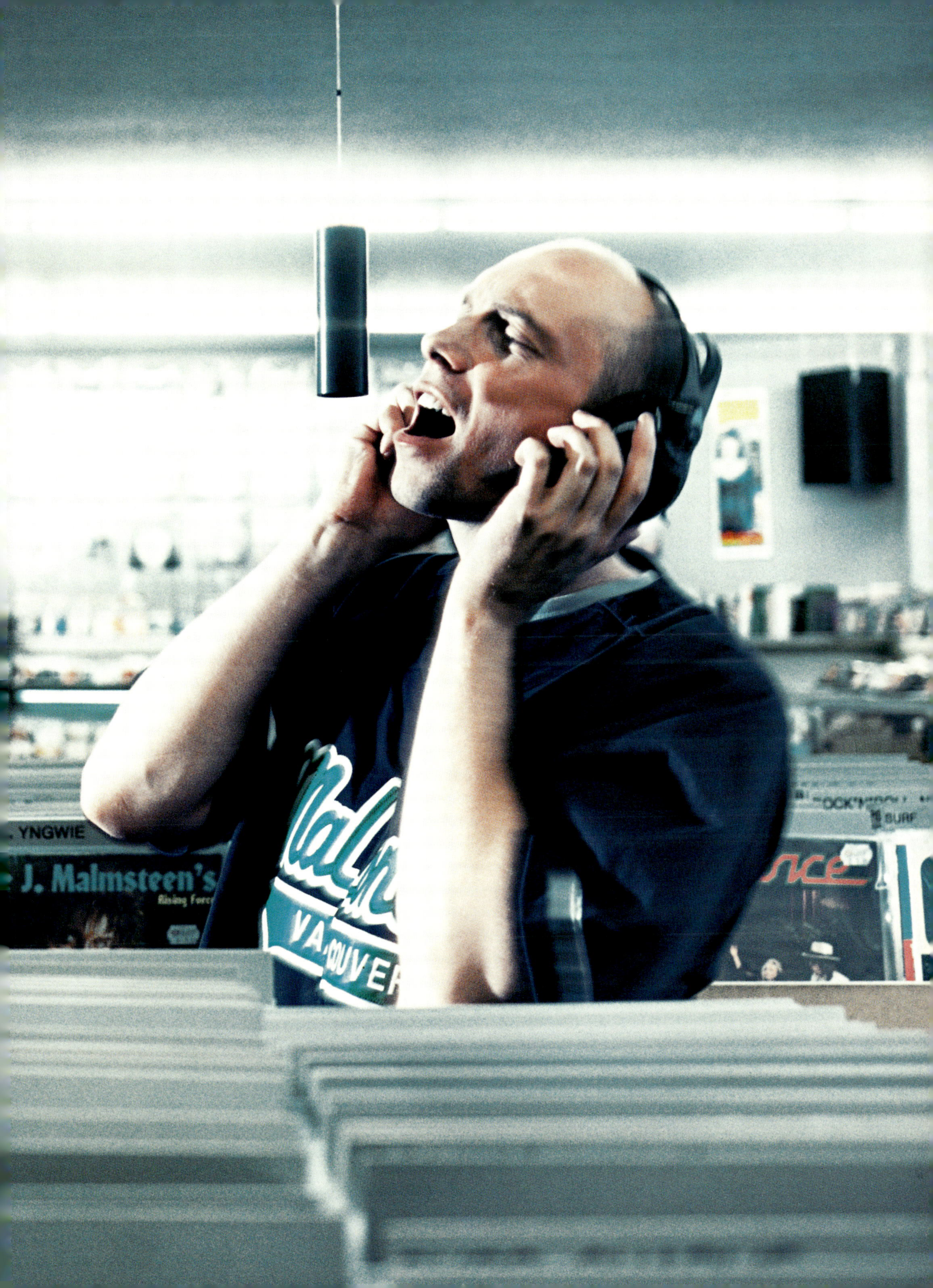
YNGWIE
J. Malmsteen's
SURF

BÜRO-AUSSTATTUNG | GOLD

OFFICE EQUIPMENT | GOLD

HERSTELLER . MANUFACTURER
KAAD
Karl-Axel Andersson Design
S-21121 Malmö

VERTRIEB . DISTRIBUTOR
Station Nord
D-10107 Berlin

DESIGN . DESIGNER
Werksdesign
Karl-Axel Andersson

Andersson Flag : Karten-Hänger . card hanger

„Flag" ist ein Präsentationssystem für unterschiedliche Sorten von Karten und Prospekten. Es lässt sich sowohl zu Hause als auch im Büro sinnvoll einsetzen. Die geniale und einfache Grundidee ist ein Pendel, das an einem langen durchsichtigen Doppelfaden von der Decke hängt. Zwischen die beiden Fäden werden die Karten und Prospekte eingesteckt und oben wie unten durch Gummikügelchen fest positioniert. Das Wechselspiel mit dem Gewicht macht das Übrige, und die ausgestellten Karten hängen und bewegen sich wie kleine Flaggen im Raum.
„Flag" ist in zwei Pendelgewicht-Versionen zu haben: eine volle zylindrische oder eine Kurvenform. Das Material ist Aluminium natur oder schwarz eloxiert.

.

"Flag" is a presentation system for different kinds of cards and brochures. It can be used equally well at home and in the office. The brilliant yet simple basic idea is a pendulum suspended from the ceiling on a long, transparent double thread. Maps and brochures are inserted between the two threads and held in place by rubber beads above and below. The downward pull of the weight takes care of the rest, and the cards on display hang and move in the room like small flags.
"Flag" is available in two versions: its pendulum weight can either be completely cylindrical or curved. It is made of aluminium, either untreated or with a black anodized coating.

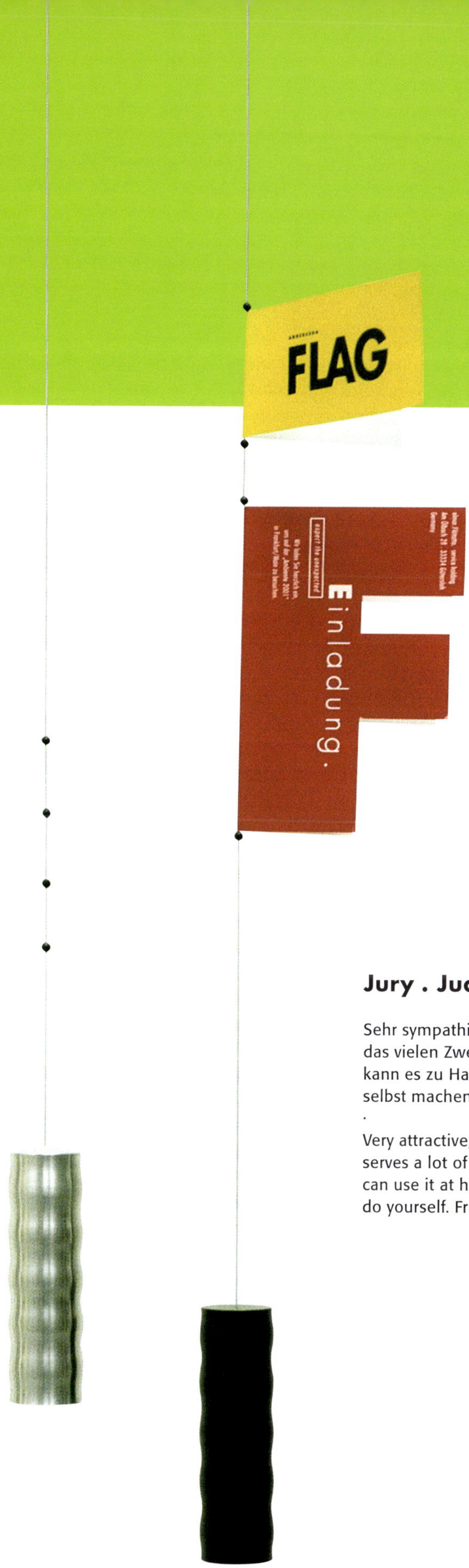

Jury . Judges panel

Sehr sympathisch, einfach, nützlich und schön. Die Qualität eines objet trouvé, das vielen Zwecken dient: eine Art Spielzeug, dann ist es sehr funktional, man kann es zu Hause oder im Büro nutzen. Eine gute Idee, die man im Grunde selbst machen könnte. In diesem Sinne ein sehr gutes Design.

.

Very attractive, simple, useful and beautiful. The quality of an objet trouvé that serves a lot of purposes: it is a kind of toy, then it is very functional, and you can use it at home or in the office. A good idea, something you could basically do yourself. From that point of view, excellent design.

HERSTELLER . MANUFACTURER
Eicher Siebdruck
D-71394 Kernen

VERTRIEB . DISTRIBUTOR
Eicher Siebdruck
D-71394 Kernen

DESIGN . DESIGNER
Büro Uebele Visuelle Kommunikation
Michaela Burger, Andreas Uebele,
Diane Ziegler
D-70180 Stuttgart

Sero Flex : Raumkennzeichnung . room designation

„Sero Flex" ist ein klassisches, analoges Türschild, das sehr flexibel einzusetzen, einfach zu montieren und leicht zu handhaben ist. Dank der reduzierten, sachlichen Gestaltung fügt es sich leicht in verschiedene architektonische Umgebungen ein. Als Informationsträger wird schlicht ein weißes Blatt Papier verwendet, das mit handelsüblichen Bürodruckern bedruckt wird. Dieses legt man zwischen dem Wandträger aus silber eloxiertem Aluminium und einer polierten Acrylglasplatte ein. Beide werden über einen Schwalbenschwanz miteinander verbunden und durch eine Innensechskantschraube gesichert. Der untere Teil der Schwalbenschwanzverbindung dient gleichzeitig als Montagehilfe des Einlegeblattes, das so – auch ohne Acrylglasabdeckung – gehalten wird. Das Türschild kann sowohl horizontal als auch vertikal montiert werden. „Sero Flex" wird zusammen mit einem Computer-Programm geliefert, mit dem die Einlegeblätter selbst typografisch gestaltet werden können. Zusätzlich gibt es eine Konstruktion mit einem Schieber, der bei Besprechungsräumen „frei" (Grün) oder „besetzt" (Rot) anzeigt.

.

"Sero Flex" is a classical, analogue door sign, which can be used very flexibly and is easy to assemble and mount. Its reduced, sober design means that it harmonizes easily with different architectural environments. All that is needed for the information to be displayed is a white sheet of paper, which can be printed using any commercial office printer. The paper is then placed between the wall bracket, made of silver anodized aluminium, and a polished pane of acrylic glass. The two are connected by dovetailing, and secured with an Allan key. The lower half of the dovetail joint also serves as an aid when positioning the sheet to be inserted, holding it in place even without the acrylic glass cover. The door sign can be mounted horizontally and vertically. "Sero Flex" comes together with software for the typographic design of the inserts. The range also includes a model with a slide for conference rooms, which displays "not in use" (green) or "in use" (red).

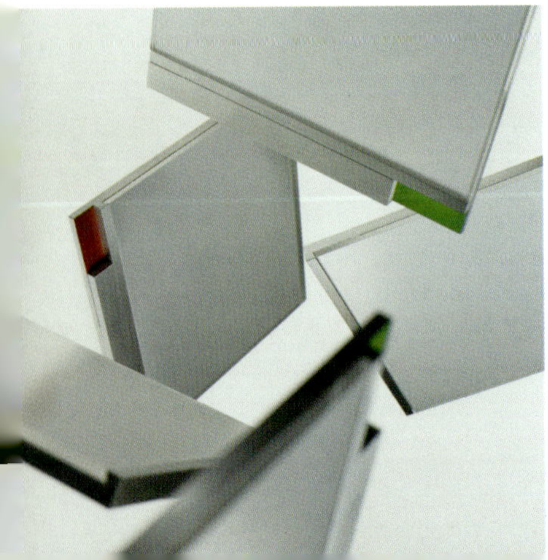

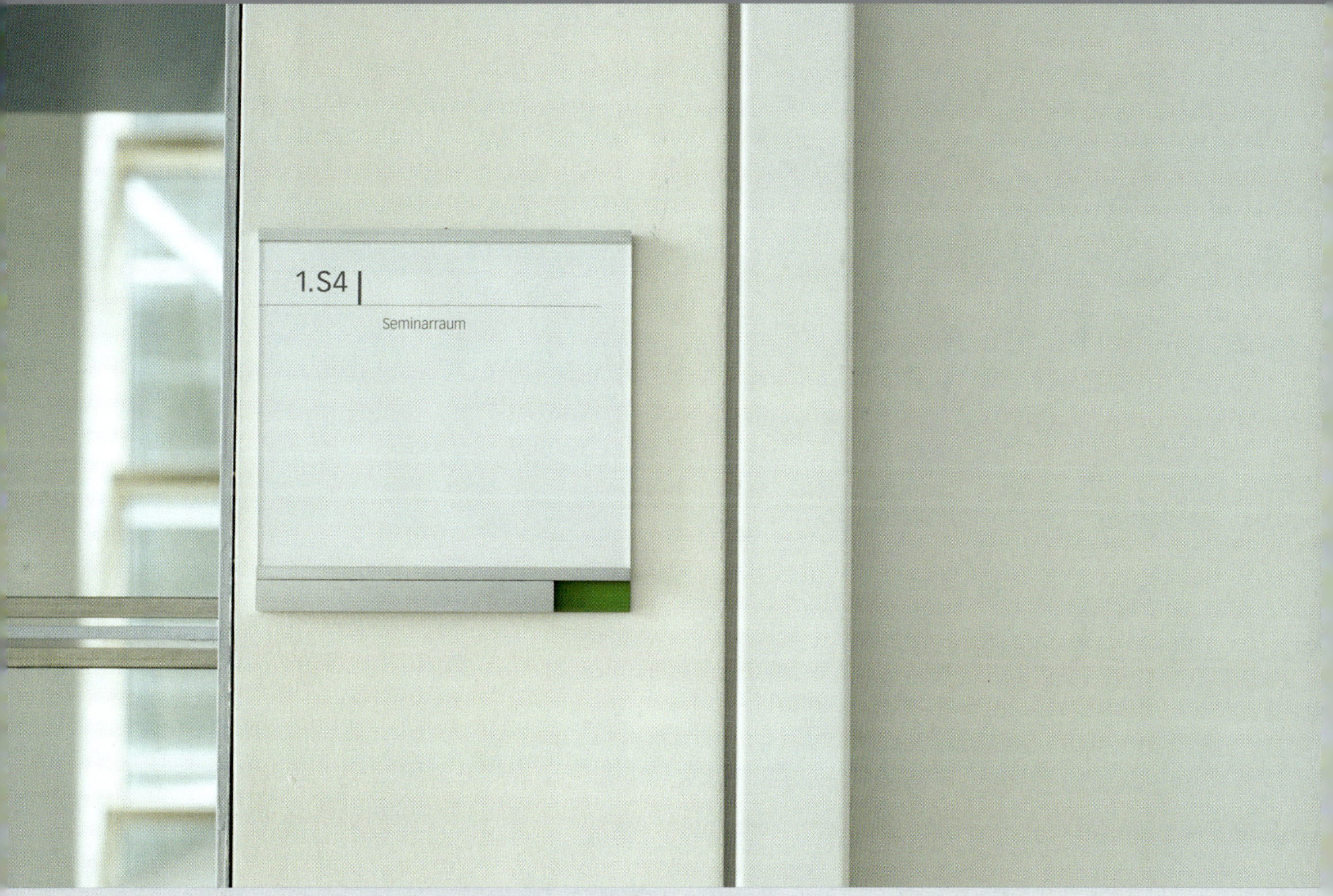

Jury . Judges panel

Eine Lösung, die in ihrer Einfachheit nicht übertroffen werden kann. Mit minimalen Elementen wird die Funktion der Raumkennzeichung erreicht.

·

A solution whose simplicity cannot be beaten. With a minimum of parts it fulfils a function that is needed for almost every door.

HERSTELLER . MANUFACTURER
Staedtler Mars GmbH & Co. KG
D-90427 Nürnberg

VERTRIEB . DISTRIBUTOR
Staedtler Deutschland Vertriebs-GmbH
D-90427 Nürnberg

DESIGN . DESIGNER
Staedtler Design Team
D-90427 Nürnberg

Triplus : Fineliner . fineliner

Die charakteristische Dreikantform des Fineliners in hochwertigem, edel anmutendem Design sorgt für leichtes und ermüdungsfreies Schreiben, da sich der Stift ergonomisch in die Form der Finger einfügt. Ein Drehen des Stiftes wird auf diese Weise beim Schreiben verhindert, wodurch der Griff locker und entspannt bleibt. Selbst häufiges Schreiben, Zeichnen und Malen lässt die Hand nicht ermüden. Die superfeine, metallgefasste Spitze gibt ein besonders angenehmes Schreibgefühl und eignet sich perfekt für den täglichen Einsatz. Dank der Dry-Safe-Eigenschaft der Tinte kann der Stift – hat man vergessen die Kappe aufzusetzen – einige Tage offen liegen bleiben ohne einzutrocknen. Zusätzlich garantiert der Polypropylen-Schaft eine lange Lebensdauer. Bis zu 20 brillante Schreibfarben stehen zur Auswahl. Für die Fineliner gibt es eine patentierte Box, die nicht nur die Stifte beim Transport schützt, sondern auch mit einem Handgriff zu einem attraktiven und praktischen Stiftständer aufgestellt werden kann.
Zur Triplus-Serie gehören noch Tintenroller, Kugelschreiber, Druckbleistift und Fasermaler.

.

The fineliner's characteristic triangular form, with its high-quality, classic design makes writing easy and fatigue-free, because the pen's ergonomic shape fits snugly into the fingers. This prevents the pen turning during writing, and the writer's grip thus remains light and relaxed. Even frequent writing, drawing and painting does not tire the hand. The superfine tip in its metal casing provides for a particularly pleasant writing sensation, and is perfect for daily use. The ink's "dry-safe" properties mean that, should one forget to replace the cap, the pen can be left open for some days without drying up. In addition, the polypropylene barrel ensures long life. There is a choice of up to 20 brilliant ink colours. The fineliner comes with its own patented box, which not only protects the pens during transport, but can also be converted in a thrice into an attractive, practical pen stand.
The triplus series also includes a rollerball pen, ballpoint pen, propelling pencil and felt-tip pen.

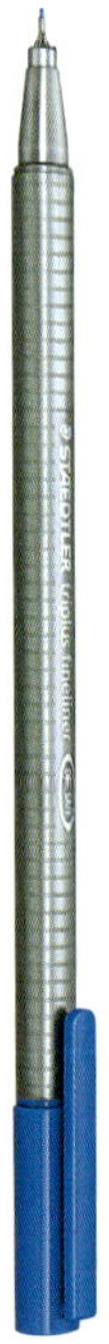

Jury . Judges panel

Der Stift hat einen guten Griff, er ist schön und leicht zu handhaben. Die Box ist sehr praktisch und gut zu nutzen, da man sie offen auf dem Tisch stehen lassen kann und ständig Zugriff auf die Stifte hat. Obwohl der Fineliner kein ungewöhnliches Produkt ist, ist die Präsentation und die Art, wie er gemacht ist, sehr gut.

.

The pen gives a good grip, it is attractive and easy to use. The box is very practical and easy to use, because it can be left open on a desk while using the pen that is needed. Although the fineliner is not an unusual product, the way it is presented and made is very good.

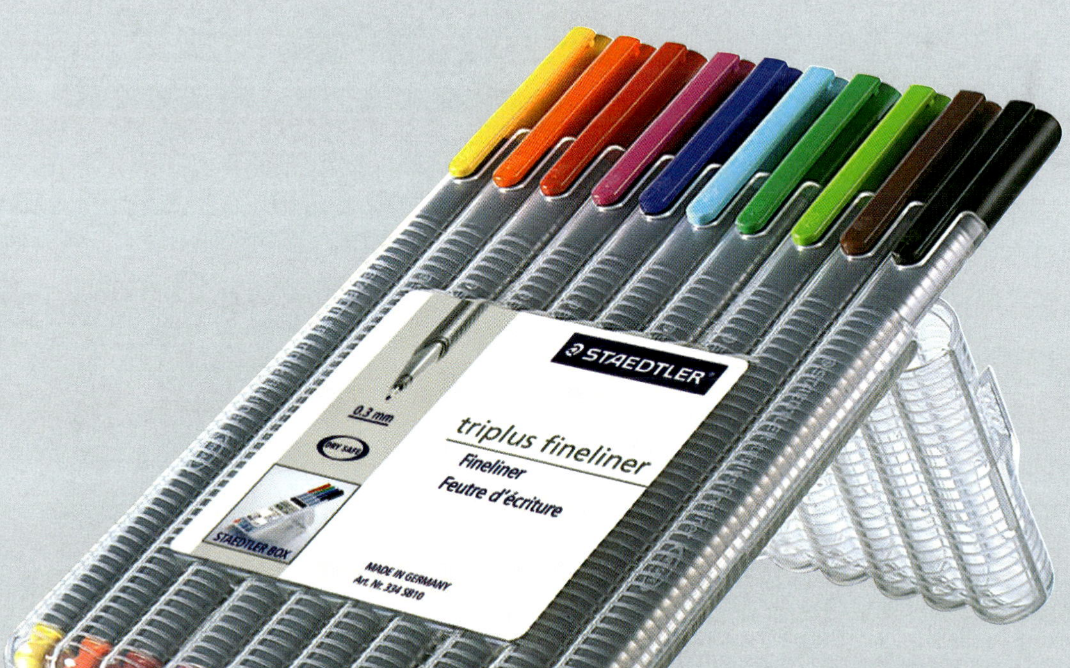

BÜRO-AUSSTATTUNG | SILBER

OFFICE EQUIPMENT | SILVER

HERSTELLER . MANUFACTURER
Humanscale
New York, NY 10010

DESIGN . DESIGNER
Werksdesign
Manuel Saez

M7 : Flachbildschirm-Monitorarm . flat screen monitor arm

Flachbildschirme verändern die Bürolandschaft. Sie brauchen weniger Platz, bieten bessere Auflösung und blenden weniger als herkömmliche Monitore. Ihr größter Vorteil aber ist, dass sie leicht auf einen verstellbaren Arm montiert werden können. Damit können sie individuell für jeden Nutzer positioniert und, wenn man sie gerade nicht braucht, einfach weggeschoben werden. Die Beweglichkeit des Arms ermutigt den jeweiligen Nutzer, Winkel, Tiefe und Höhe des Monitors nach seinem Maß einzustellen, um so das für ihn komfortabelste Blickfeld zu bekommen.
„M7" bietet mehr als andere Monitorarme. Er besteht aus einer Aluminiumrahmenstruktur, lässt sich in alle Richtungen bewegen und in zwei Minuten installieren. Er bietet extreme Stabilität, eingebautes Kabelmanagement und elegantes, funktionales Design. Außerdem ist er sogar noch umweltfreundlich, da er aus 90 Prozent recycelten Materialien besteht.

.

Flat screens are changing the office environment. They take up less room, provide better resolution and are kinder to the eyes than traditional monitors. But their greatest benefit is that they can easily be attached to an adjustable monitor arm. This allows them to be positioned individually for each user and simply pushed to one side again when they are not in use. The mobility of the monitor arm also encourages the individual user to adjust the angle, depth and height of the monitor to suit his requirements, so that he has the best possible view.
"M7" has more to offer than other monitor arms. It has an aluminium framework structure, can be moved in any direction, and can be set up in just two minutes. Extremely robust and with an elegant, functional design, it comes with an integrated cable management system. Moreover, it is environmentally friendly, as it is made of materials that are 90 percent recyclable.

Jury . Judges panel

Für unsere wechselnde Arbeitswelt braucht man immer überzeugende Produkte. Dieses ist ein Beispiel für ein Produkt, das unser Arbeitsleben leichter und weniger stressreich macht. Eine sehr gute und interessante Lösung für die Arbeitswelt mit Datenarbeitsplätzen.

.

For our changing world of work, convincing products are always needed. This is an example of a product that makes our working life easier and less stressful. A very good, interesting solution for the IT workplace.

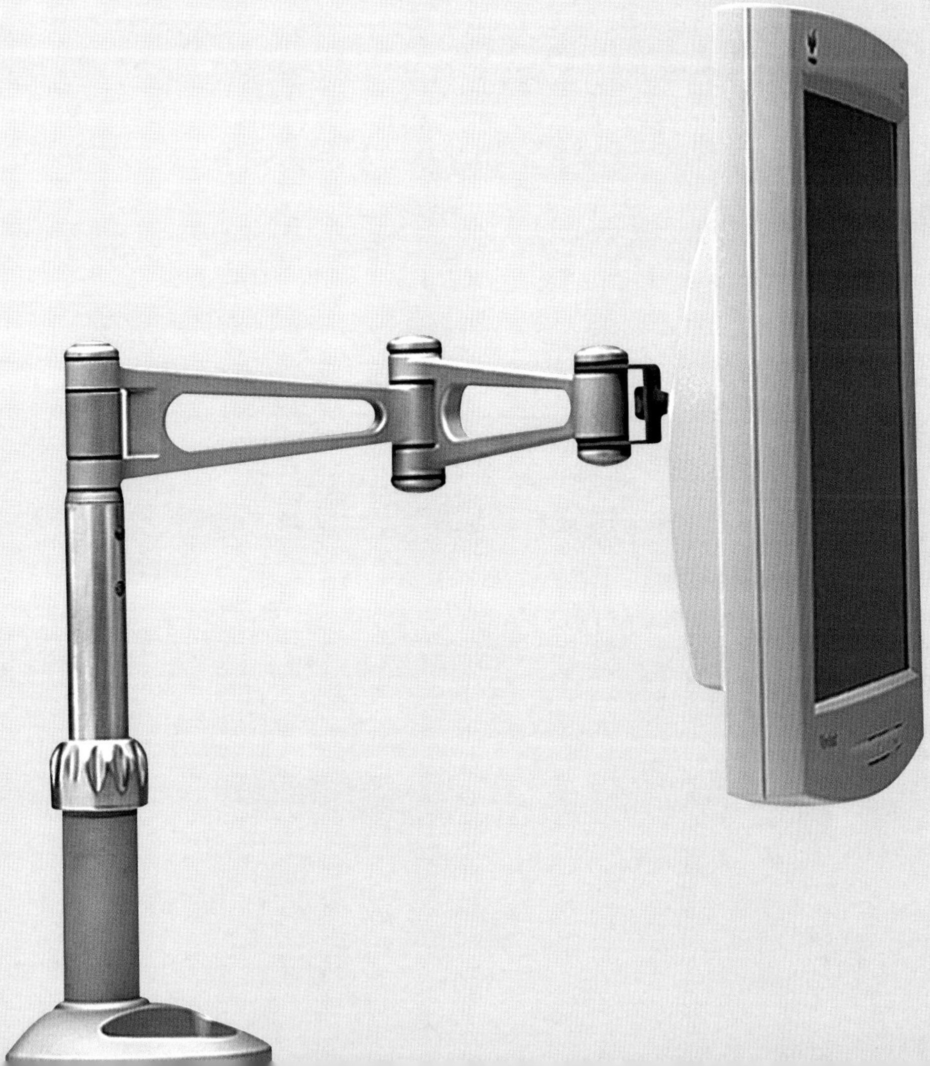

BELEUCHTUNG

LIGHTING

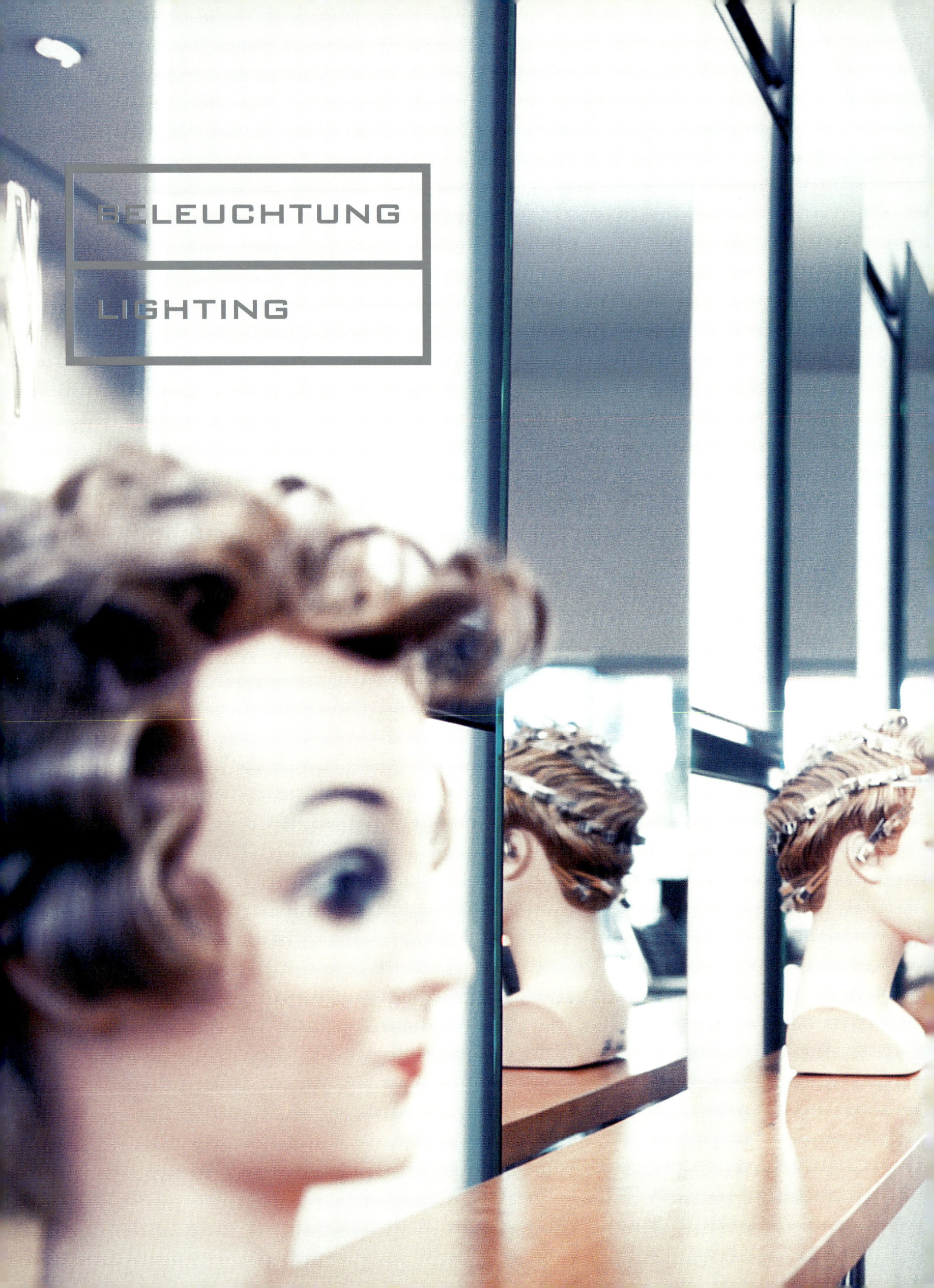

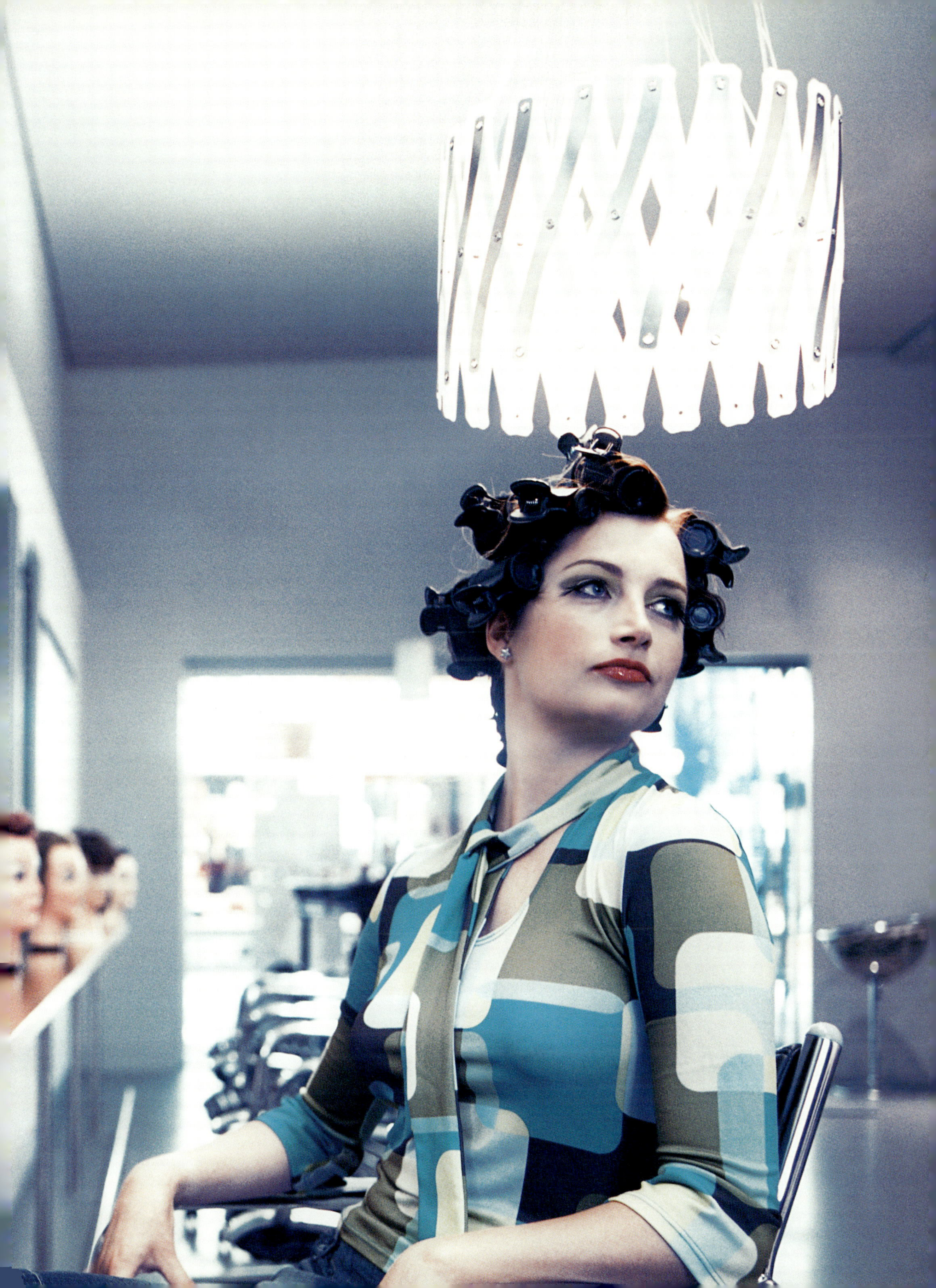

HERSTELLER . MANUFACTURER
Serien Raumleuchten GmbH
D-63110 Rodgau

VERTRIEB . DISTRIBUTOR
Serien Raumleuchten GmbH
D-63110 Rodgau

DESIGN . DESIGNER
Floyd Paxton
c/o D & W Designstudio
D-60318 Frankfurt

Zoom : Pendelleuchte . pendant lamp

„Zoom" – die Pendelleuchte, die sich verwandelt. Sie kann nach Belieben klein oder groß werden. Sie besteht aus einer Scherengitterkonstruktion aus flexiblem Edelstahl-Federband, die zu einem Kreis geschlossen ist. Dies erlaubt ein stufenloses Verstellen des Leuchtrings von 20 bis 130 Zentimeter, wodurch die Leuchte nicht nur verschiedene Möglichkeiten im Design, sondern auch zahlreiche Variationen der Lichtgebung bietet. Denn auch das Licht, das durch eine transluzente Folie nach außen tritt, wandelt sich, wenn die Pendelleuchte verstellt wird. Schiebt man „Zoom" eng zusammen, wird sie zu einer geschlossenen, nach oben und unten abstrahlenden zylindrischen Schirmleuchte. Wird das Scherengitter gedehnt, verändert sie sich zunächst in eine offene Leuchtskulptur und wird, ganz auseinandergezogen, schließlich zu einem modernen Kronleuchter mit 20 frei abstrahlenden Niedervolt-Halogenlämpchen.
Für sehr große Räume wie Foyers und Festsäle gibt es „Zoom" auch in einer größeren Version „Zoom XL", die von 40 bis 260 Zentimeter verstellt werden kann.

.

"Zoom" – the pendant lamp that changes its form. As desired, it can become small or large. It comprises a lattice grate design made of flexible stainless spring band steel, the ends of which have been joined to form a circle. This allows the light ring to be adjusted to any size between 20 and 130 centimetres. The result of this is not only various design possibilities, but also a wide variety of lighting effects. If the shape of the pendant lamp is changed, then the light – which emanates through a translucent film – changes too. If "Zoom" is squeezed up tight, it becomes a closed, cylindrical shaded lamp, emitting light upwards and downwards. If the lattice grate is opened out it changes first of all into an open light sculpture. Finally, when fully opened out, it becomes a modern chandelier from which 20 low-voltage halogen lamps shine.
For very large rooms such as entrance halls and banqueting rooms, "Zoom" is available in a larger version, "Zoom XL", which can be adjusted to any size between 40 and 260 centimetres.

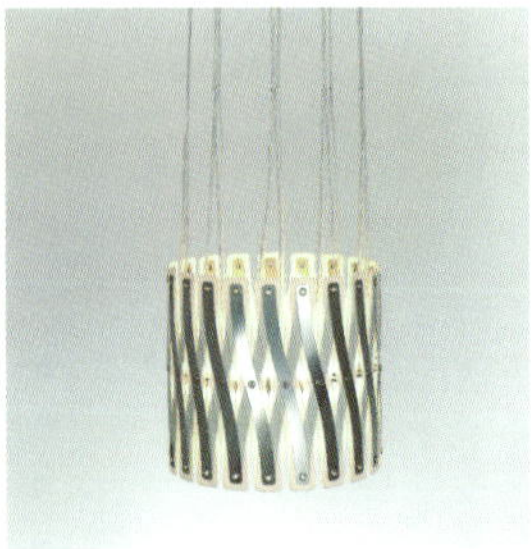

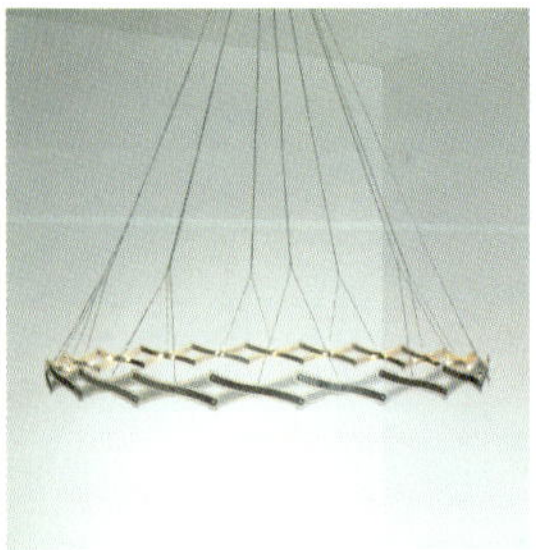

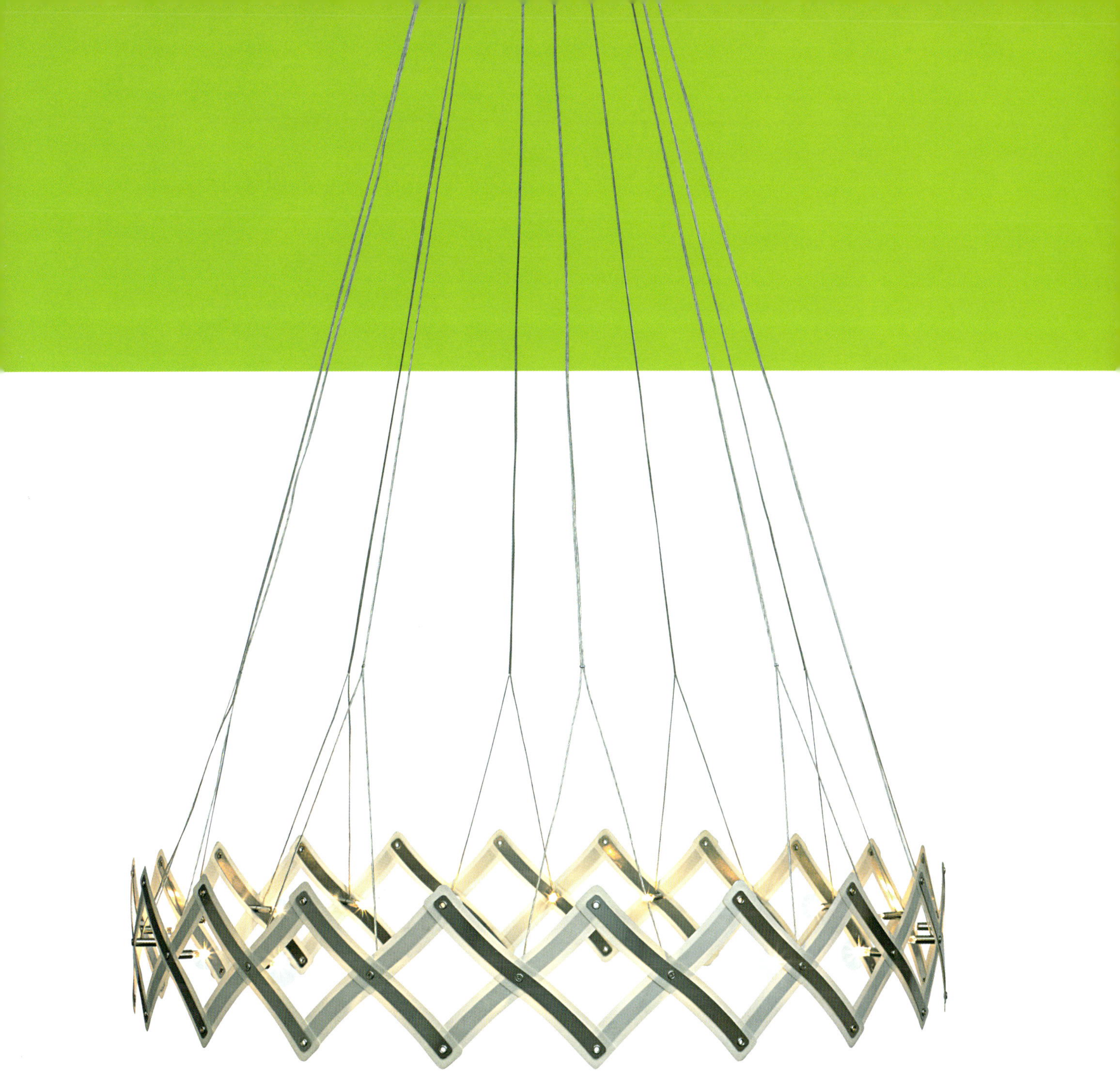

Jury . Judges panel

Sehr innovative Interpretation eines klassischen Kronleuchters. Überzeugende Lösung. Die Leuchte kann wachsen, sie kann wieder zusammenschrumpfen. Sehr schönes Design. Eine gute Balance zwischen Stimmung und Innovation.

.

Very innovative interpretation of a classical chandelier. Convincing solution. The lamp can grow and can shrink again. Very elegant design. A good balance between mood and innovation.

BELEUCHTUNG SILBER

LIGHTING SILVER

HERSTELLER . MANUFACTURER
Serien Raumleuchten GmbH
D-63110 Rodgau

VERTRIEB . DISTRIBUTOR
Serien Raumleuchten GmbH
D-63110 Rodgau

DESIGN . DESIGNER
Nextspace
Claus Zimmermann, Roland Knieg
D-34119 Kassel

Reef Wall : Wandleuchte . wall light

Die Wandleuchte bekommt durch ein innovatives Material atmosphärisches Licht. Hier wird zum ersten Mal der Hightechwerkstoff Keramikschaum im Leuchtendesign verwendet. Dieses Material zeichnet sich durch eine leicht poröse Beschaffenheit aus, die an die Struktur von Korallen erinnert. Die zerbrechliche Anmutung täuscht: Der Leuchtenschirm ist sogar spülmaschinenfest und dadurch besonders pflegeleicht. Um ihn abzunehmen, muss nur eine Schraube gelöst werden.
Aber der Keramikschaum hat nicht nur eine ungewöhnliche Beschaffenheit sondern verfügt auch über eine außergewöhnliche gestalterische Qualität und besondere Lichteigenschaften. Gehäuse und Schirm können in zwei Arten montiert werden: entweder nach oben oder unten abstrahlend. Das Licht kommt durch den Schirm diffus, aber nach oben oder unten jedoch direkt, je nachdem wie die Leuchte montiert ist. Das Gehäuse ist entweder aus gebürstetem oder hochglanzpoliertem Aluminium.

.

Thanks to an innovative material, this wall light emanates an atmospheric light. This is the first time that the high-tech material ceramic foam has been used in luminaire design. The characteristic property of this material is its slight porosity, reminiscent of the structure of corals. But this fragile appearance is deceptive, because the light shade is even dishwasher-safe, making it particularly easy to clean. All that has to be done to take it down is loosen a screw. Ceramic foam does not only have this unusual property, however. It also has extraordinary design qualities and special lighting properties. The housing and the shade can be mounted in two ways, so that light is projected upwards or downwards. The light that comes through the shade is diffuse, but direct light is also projected upwards or downwards depending on how the lamp has been mounted. The housing is made of either brushed or mirror-finish aluminium.

Jury . Judges panel

Das Material, das hier bei dem Leuchtschirm eingesetzt wird, hat die Lebendigkeit, die wir sonst eigentlich nur von Holz, Leder und anderen natürlichen Materialien kennen.

.

The material used for the light shade here has a vitality that we otherwise only find with wood, leather or other natural materials.

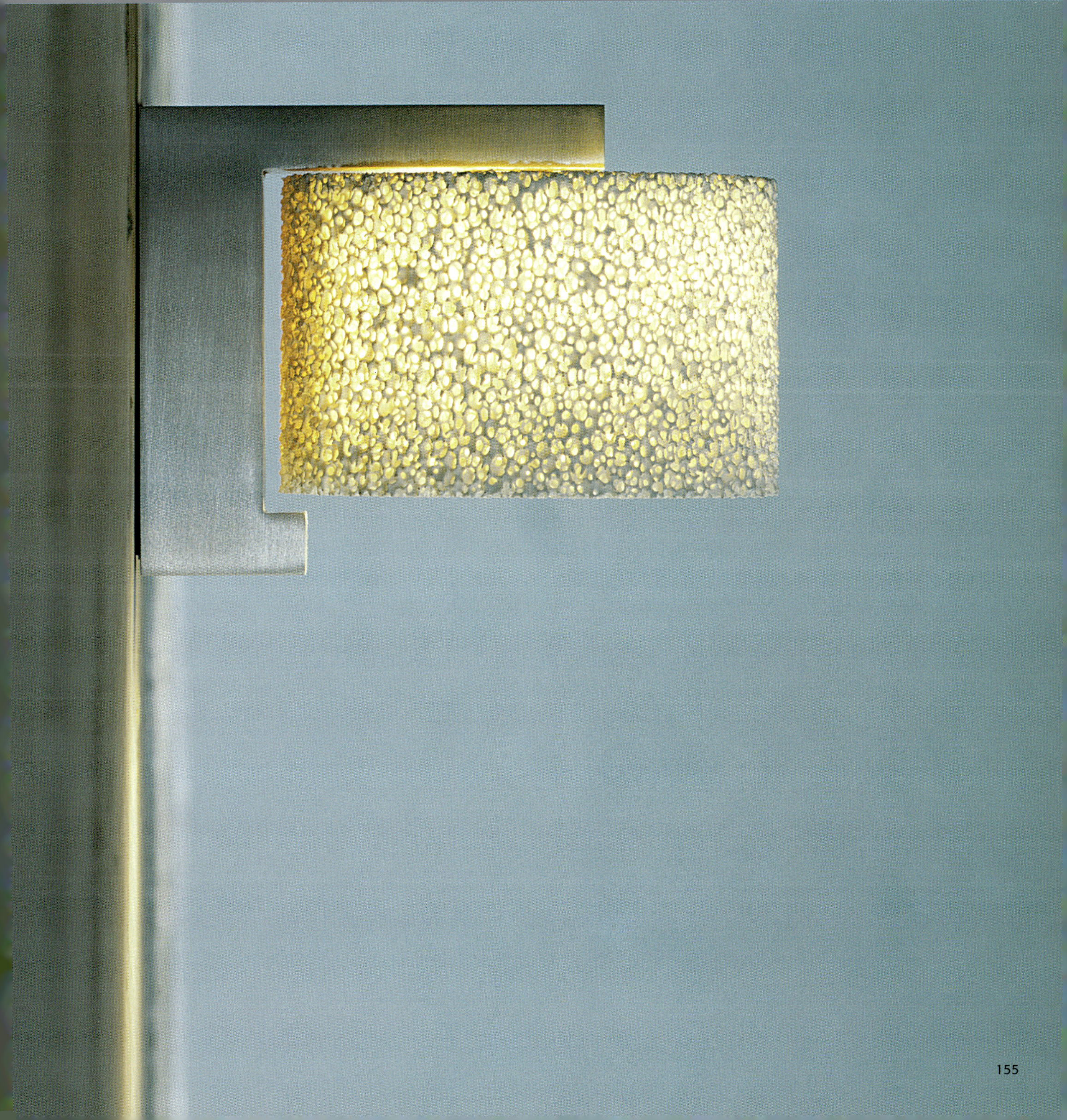

HERSTELLER . MANUFACTURER
Serien Raumleuchten GmbH
D-63110 Rodgau

VERTRIEB . DISTRIBUTOR
Serien Raumleuchten GmbH
D-63110 Rodgau

DESIGN . DESIGNER
Jean-Marc da Costa
c/o Serien Raumleuchten GmbH
D-63110 Rodgau

SML Wall : Wandleuchte . wall light

Das Leuchtenprogramm „SML" bietet extrem große Variationsmöglichkeiten. Die formal minimalistischen, nach oben und unten abstrahlenden Halogenleuchten kann sich jeder wie bei einem Baukastensystem individuell und je nach Einsatzbereich zusammenstellen.
Man wählt aus drei verschiedenen Längen – S, M oder L – und aus verschiedenen Abdeckungen.
Es gibt drei unterschiedliche Keramikglas-Abdeckungen, die mit hochtemperierten Spezialfarben bedruckt sind. Je nach ausgesuchter Abdeckung bekommt man eine andere Lichtwirkung: direktes, leicht oder stark gedämpftes Licht.
Es können auch unterschiedliche Abdeckungen für den oberen und unteren Lichtaustritt kombiniert und damit verschiedenste Lichtwirkungen erzielt werden.
Die Leuchtkörper des Baukastens verbinden sich durch ihren klaren architektonischen Charakter und die minimalen Abmessungen unaufdringlich mit der Gebäudesubstanz zu einer geschlossenen harmonischen Einheit, was das Gehäuse aus glasperlgestrahltem und silber eloxiertem Aluminium unterstreicht.

.

The "SML" luminaire range can be varied in any number of ways. Like a modular system, the formally minimalist halogen luminaires, which project light up or down, can be put together individually and as needed.
There is a choice of three different lengths – S, M, or L – and different covers.
There are three different ceramic glass covers, printed with high temperature special paints. Depending on the cover selected, a different light effect is achieved: direct, soft or very subdued light. Different covers can also be combined for the upward and downward light rays, allowing extremely varied light effects to be achieved.
The clear, architectural character and extremely small size of the luminous elements in the set merge in discreetly with the building to form a closed, harmonious unit, a feature which is underlined by the housing of pearl-finish and silver anodized aluminium.

Jury . Judges panel

Ein Produkt, das in seiner Erscheinung vollkommen in den Hintergrund tritt und die Funktion, nämlich das Leuchten, die Hauptsache sein lässt.

.

A product which merges completely into the background and allows its lighting function to take centre stage.

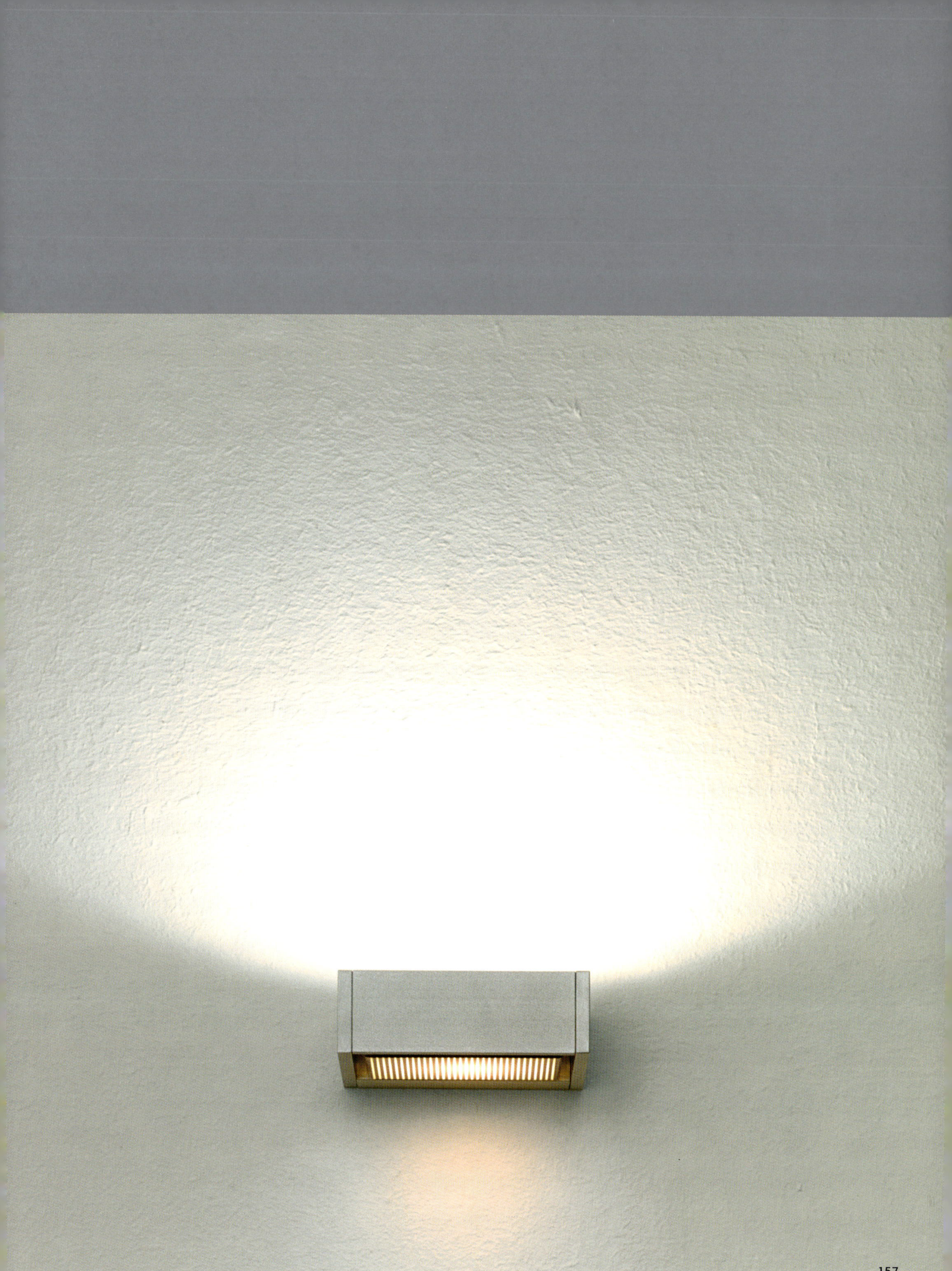

BELEUCHTUNG SILBER

LIGHTING SILVER

HERSTELLER . MANUFACTURER
Serien Raumleuchten GmbH
D-63110 Rodgau

VERTRIEB . DISTRIBUTOR
Serien Raumleuchten GmbH
D-63110 Rodgau

DESIGN . DESIGNER
Jean-Marc da Costa
c/o Serien Raumleuchten GmbH
D-63110 Rodgau

SML Wall T5 : Wandleuchte . wall light

„SML Wall T5" steht äußerlich in der Tradition des SML-Baukastensystems: minimalistisches Design bei maximaler Gestaltungsmöglichkeit. Wie die Halogenversion ist sie extrem variabel einsetzbar und auch in den Größen S, M und L erhältlich. Das Gehäuse ist ebenfalls aus glasperlgestrahltem und silber eloxiertem Aluminium und unterstreicht den klaren architektonischen Charakter. Technisch ist „SML Wall T5" allerdings völlig anders aufgebaut und setzt neue Maßstäbe in der Leuchtstofftechnik. Mit einer Bauhöhe von 25 Millimetern ist sie die flachste Wandleuchte mit energiesparendem Leuchtmittel am Markt. Das gesamte elektronische Betriebsgerät ist – aufs Äußerste minimiert – im Wandprofil untergebracht.
Die Abdeckung der Leuchtmittel ist aus speziell extrudiertem Acrylglas in zwei Versionen mit unterschiedlicher Lichtwirkung. Das klare Glas mit Prismenstruktur entblendet und lenkt das Licht von der Wand in den Raum hinein. Das satinierte Glas ohne Prismenstruktur strahlt sanftes, mildes Licht ab.

.

"SML Wall T5" is ostensibly part of the tradition set by the SML modular system, combining minimalist design with a maximum of design possibilities. Like the halogen version, it can be used in any number of ways, and is available in small, medium and large sizes. The housing is also made of pearl-finish and silver anodized aluminium and underlines its clear, architectural character. Technically, however, "SML Wall T5" is constructed completely differently, and sets new standards in the technology of luminous elements. Only 25 millimetres thick, it is the flattest wall light using energy-saving bulbs on the market. All the electronics needed to operate it have been reduced to a minimum and are concealed in the wall.
The cover is made of specially extruded acrylic glass, available in two versions, each with different lighting effects. First there is clear glass with a prism structure and a non-dazzle effect, which guides the light from the wall into the room. Then there is satin-finish glass without a prism structure, which emanates a soft, gentle light.

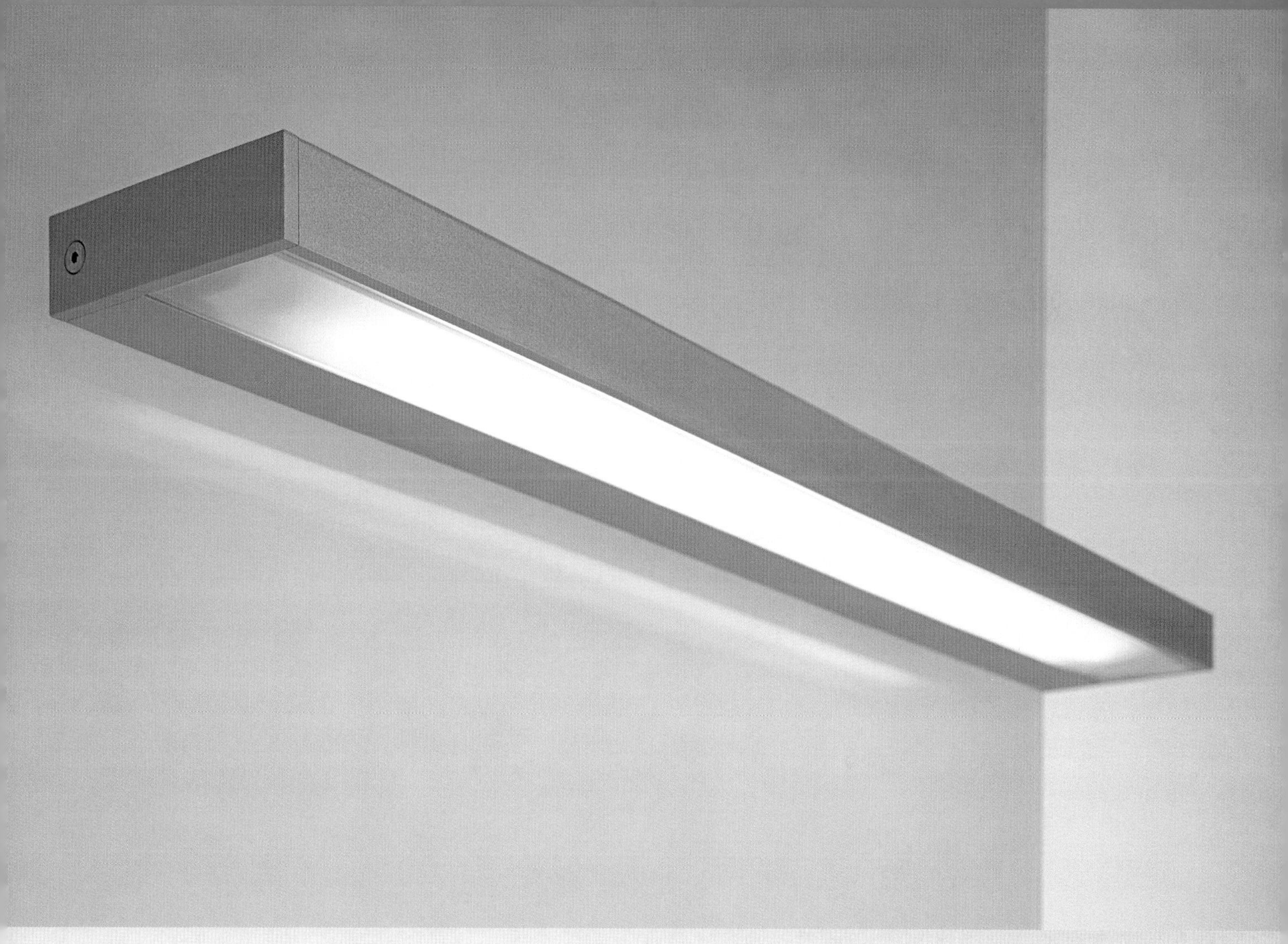

Jury . Judges panel

Eine Leuchte, reduziert auf das absolute Minimum, mit sehr klaren Details und sehr guten Materialübergängen. Ein Produkt, das an Einfachheit nicht zu überbieten ist.

.

A luminaire reduced to the absolute minimum, with excellent details and very good transitions between materials. A product whose simplicity cannot be beaten.

BELEUCHTUNG SILBER

LIGHTING SILVER

HERSTELLER . MANUFACTURER
Hess Form + Licht GmbH
D-78050 Villingen-Schwenningen

VERTRIEB . DISTRIBUTOR
Hess Form + Licht GmbH
D-78050 Villingen-Schwenningen

DESIGN . DESIGNER
Karsten Winkels
D-44147 Dortmund

LED-Lichtlinien . LED strip lights

Die 91 Zentimeter langen Lichtlinien für den Außenbereich geben je nach eingesetzten LEDs ein unterschiedlich farbiges, aber immer gleichmäßiges Licht ab. Sie markieren nicht nur Wege, sondern sorgen auch für eine stimmungsvolle Gestaltung und Atmosphäre. Sie sind absolut dicht und witterungsbeständig. Das garantieren das Gehäuse aus V4A-Edelstahl und das Einscheiben-Sicherheitsglas in Verbindung mit der hochwertigen Verarbeitung. Außerdem können die Lichtlinien mit bis zu eineinhalb Tonnen überrollt werden, ohne Schaden zu nehmen.
Die Installation ist sehr einfach, die Kabel können problemlos in der Unterkonstruktion verlegt werden.

.

These 91 centimetre long strip lights for external use emanate a light whose colour varies depending on the LEDs used, but which is always constant. They can be used not only to mark paths, but also to provide atmospheric design and moods. They are absolutely waterproof and weather-resistant. This is guaranteed by the V4A stainless steel housing and the single-pane safety glass, as well as by excellent workmanship. Moreover, the strip lights can be run over by a weight of up to one and a half tons without being damaged.
Installation is very simple, and cables can easily be concealed in the base construction.

Jury . Judges panel

Design auf das Minimalste reduziert, aber trotzdem mit großem Effekt. Mit der Anwendung von verschiedenen Farben werden unterschiedliche Atmosphären geschaffen, die im Außenbereich sehr schön sind. Man sieht so etwas sehr gern, wenn man draußen ist.

.

Minimalist design, yet to great effect. The use of different colours creates different moods which are very beautiful in an outdoor setting. This is something that is a pleasure to see when you are outside.

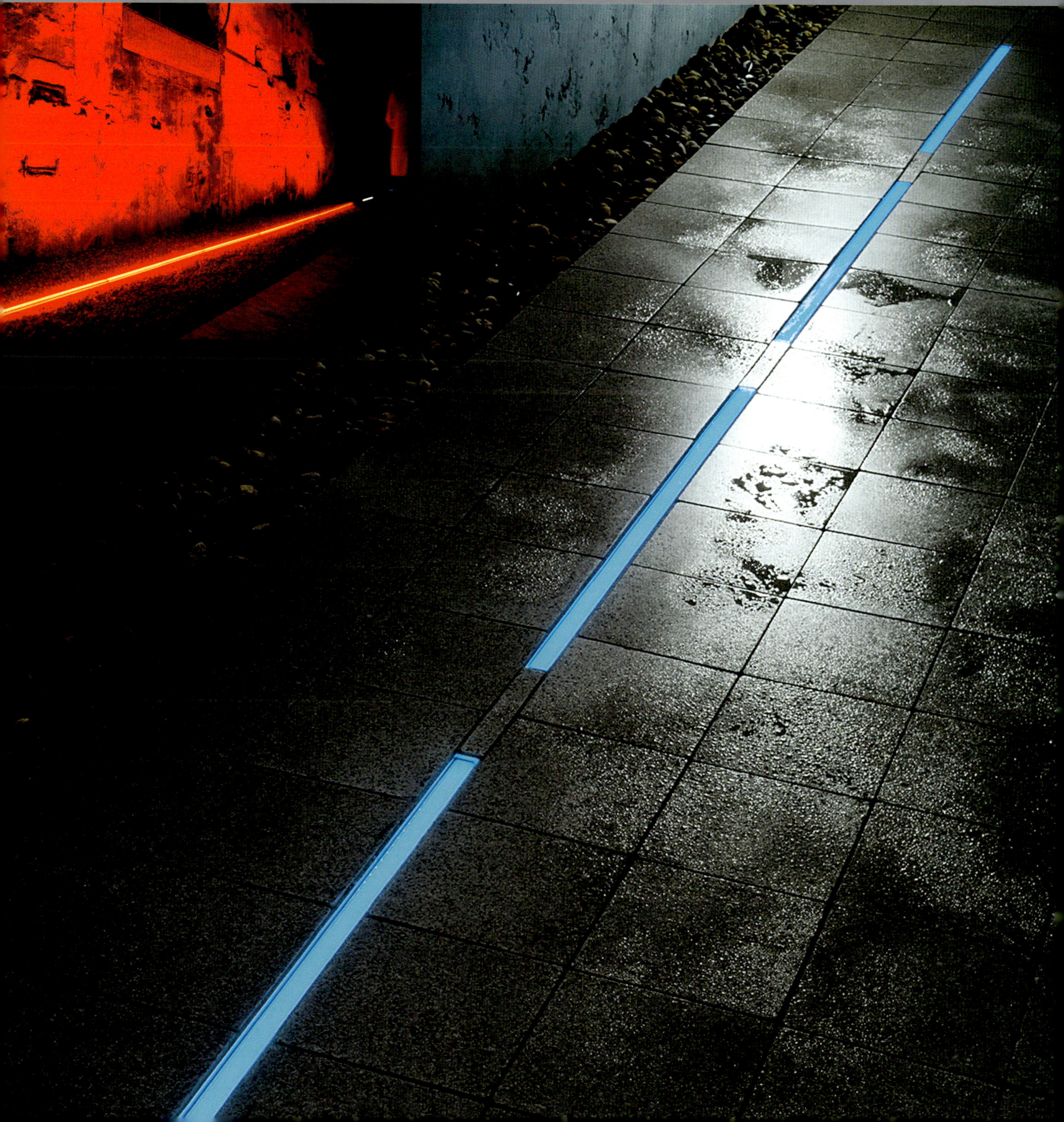

BELEUCHTUNG SILBER

LIGHTING SILVER

HERSTELLER . MANUFACTURER
Tobias Grau GmbH
D-25462 Rellingen

VERTRIEB . DISTRIBUTOR
Tobias Grau GmbH
D-25462 Rellingen

DESIGN . DESIGNER
Werksdesign
Tobias Grau

Go Track + Go Ceiling : Lichtschiene . track lighting

Das Lichtschienensystem besteht aus zwei Teilen: der Stromschiene „Go Track" und den Leuchten „Go Ceiling", die miteinander kombiniert werden können. Die Stromschiene „Go Track" hat eine nach oben gerichtete Standard Euro-3-Phasenschiene und darüber hinaus eine zusätzliche Kabelführungsmöglichkeit. Dabei ist die Stromschiene gar nicht als solche zu erkennen, da sie – und das ist völlig neu – von unten geschlossen ist und so optisch nicht stört. Die Stromschiene wird an Stäben mit 12 beziehungsweise 40 Zentimetern Länge von der Decke abgehängt. „Go Track" kann mit den Leuchten „Go Ceiling", einem Programm filigraner Deckenleuchten etwa für Büroräume, verknüpft und mit verschiedenen Hängeleuchten und Spots bestückt werden. So ergibt sich die Möglichkeit, Grundbeleuchtung, Spotbeleuchtung und abgependeltes Licht in einem System miteinander zu kombinieren.
Mit der Verbindung der Leuchten „Go Ceiling" und der Stromschiene „Go Track" lassen sich beispielsweise für Flure lange Strukturen von Lichtschienen bilden. Die Leuchten sind formal stark minimiert und gleichzeitig bis ins Detail hochwertig gefertigt.

.

This track lighting system comprises two parts: the "Go Track" electrified track and the "Go Ceiling" luminaires, which can be combined with each other. The "Go Track" electrified track has a standard 3-phase Euro-track, directed upwards, and an additional cable conduit. However, the electrified track is not recognizable as such, because – and this is a completely new feature – its lower surface is closed and it is thus not visually distracting. The track lighting is suspended from the ceiling by 12 or 40 centimetre long rods. "Go Track" can be combined with "Go Ceiling", a range of filigree ceiling-mounted luminaires for rooms such as offices, and can also be used in combination with various pendant lamps and spotlights. This allows background lighting, highlighting and pendant lighting to be combined with each other in one and the same system. Combining the "Go Ceiling" luminaires with the "Go Track" electrified track allows long structures of track lighting to be put together for long corridors, for example. Formally speaking, the luminaires are very minimalist, yet manufactured to a high quality, right down to the last detail.

Jury . Judges panel

Dieses Lichtschienensystem versteckt beziehungsweise lässt das technische Aufhängungsprinzip im Hintergrund und hat dadurch einen sehr ruhigen Charakter, der in jeder Bürolandschaft eingesetzt werden kann.

.

This track lighting system conceals or pushes the technical suspension principle into the background, which gives it a very restful property which can be used in any office environment.

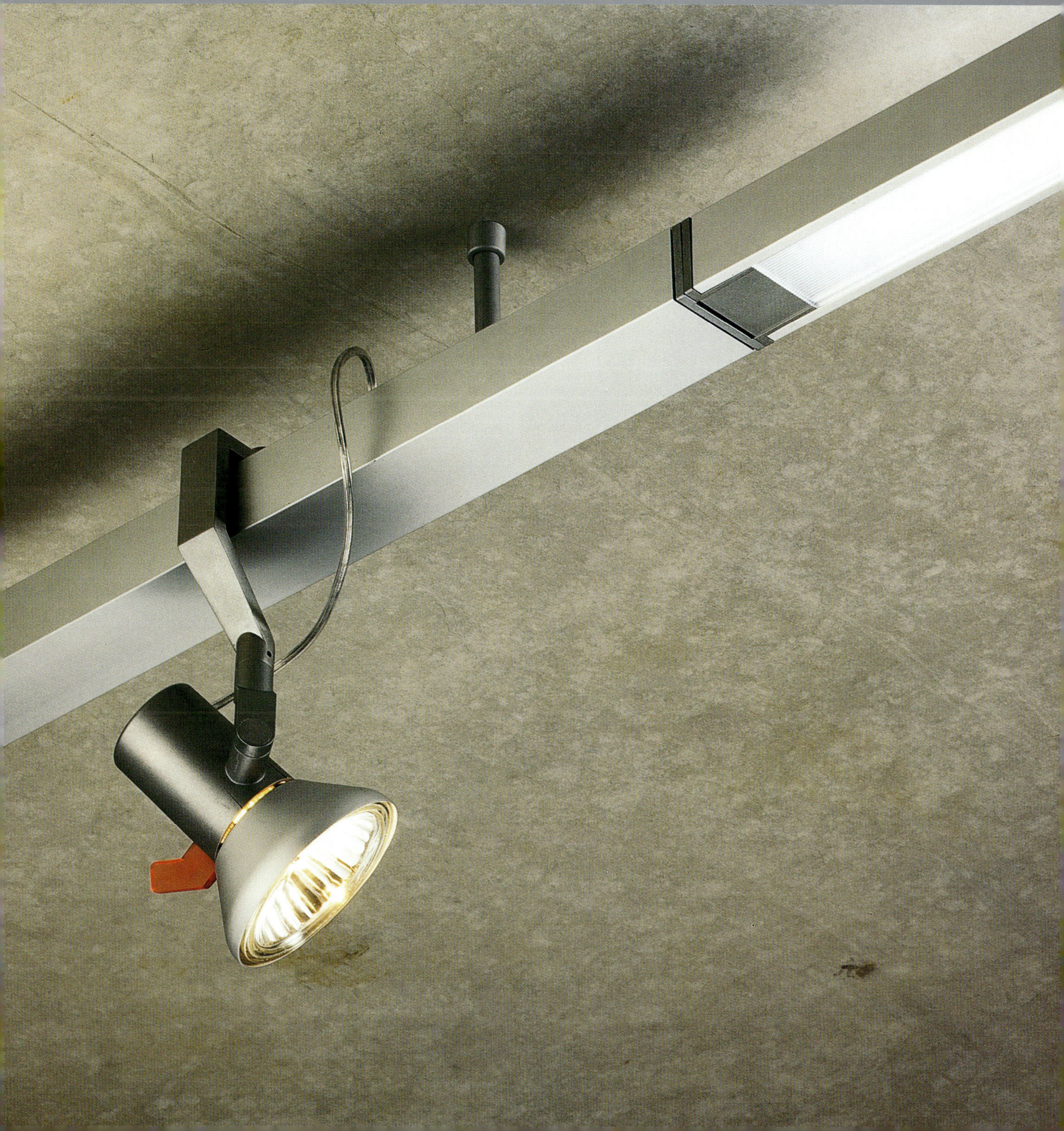

HERSTELLER . MANUFACTURER
Tobias Grau GmbH
D-25462 Rellingen

VERTRIEB . DISTRIBUTOR
Tobias Grau GmbH
D-25462 Rellingen

DESIGN . DESIGNER
Werksdesign
Tobias Grau

Bill : Tischleuchte . table lamp

Das schlanke Aluminiumgestell und die beweglichen Elemente aus Zinkdruckguss geben der Tischleuchte einen besonders klaren, ästhetischen Ausdruck. Die minimierte Formensprache und das zurückhaltende Design unterstreichen den zeitlosen Charakter.
„Bill" lässt sich als Tisch-, Steh-, Klemm- oder als Wandleuchte verwenden. Der im Aluminiumgehäuse versteckte Federmechanismus bewirkt eine sehr leichtgängige Verstellbarkeit. Der orangefarbene oder transparente Ein- und Ausschalter am Griff der Leuchte ist ein markantes Erkennungszeichen von „Bill". Das neue 60-Watt-Halogenhochvolt-Leuchtmittel bringt eine hohe Lichtintensität und brillante Farbwiedergabe.

.

The slim aluminium stand and the movable parts of die cast zinc give this table lamp a particularly clear, aesthetic expression. Its minimalist formal expression and discreet design underline its timeless character.
"Bill" can be used as a table lamp, standard lamp, clamp light or wall light. The spring mechanism concealed in the aluminium housing makes it extremely easy to adjust. The orange or transparent on/off switch on the lamp grip is "Bill's" conspicuous hallmark. The new 60 watt halogen high-voltage bulb provides an extremely intense light and brilliant colour reproduction.

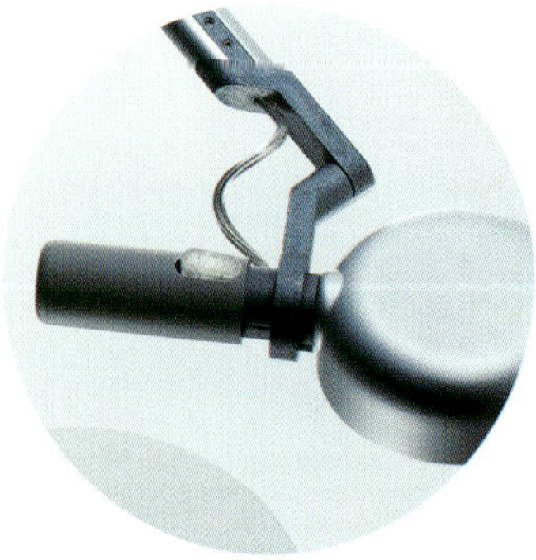

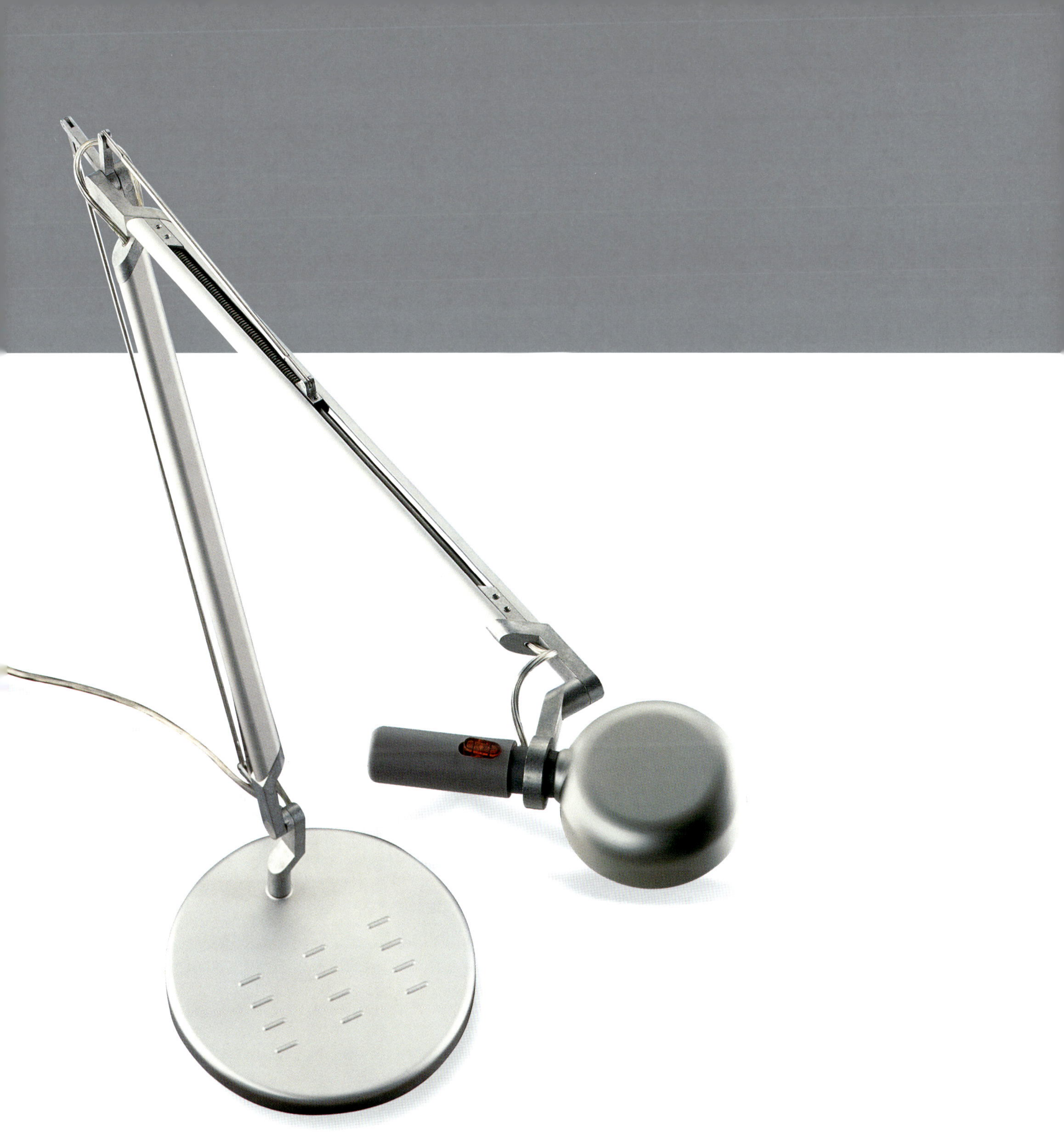

Jury . Judges panel

Eine Leuchte, die neue Wege in der Materialqualität und in technischen Detaillösungen geht. Die Leuchte besticht auf der einen Seite durch Einfachheit, auf der anderen Seite durch wirklich intelligente Verarbeitung und Detailarbeit.

.

A lamp that explores new areas of material quality and technical detail. The lamp is captivating on the one hand because of its simplicity, and on the other because of the truly intelligent workmanship and details.

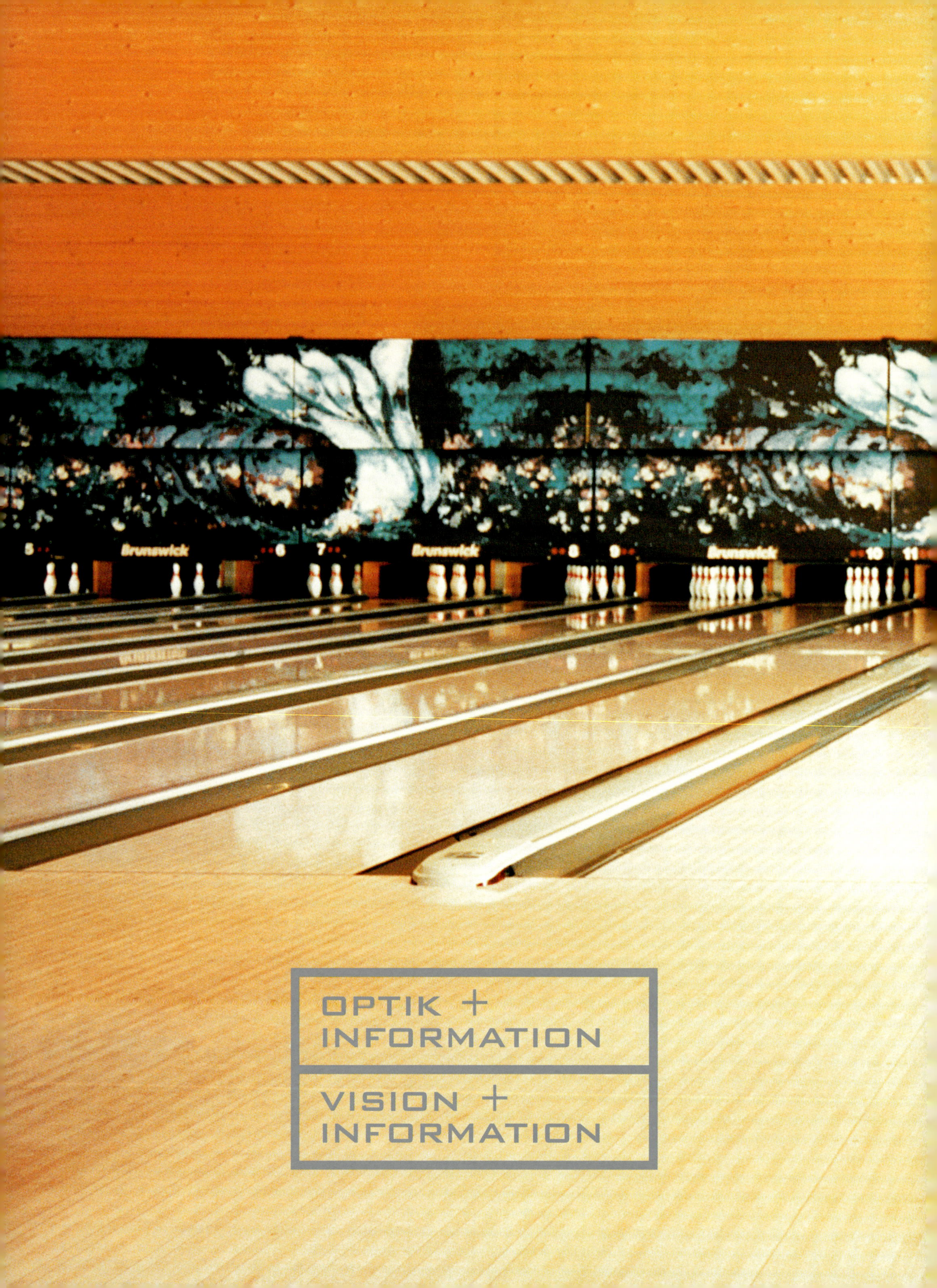

OPTIK + INFORMATION

VISION + INFORMATION

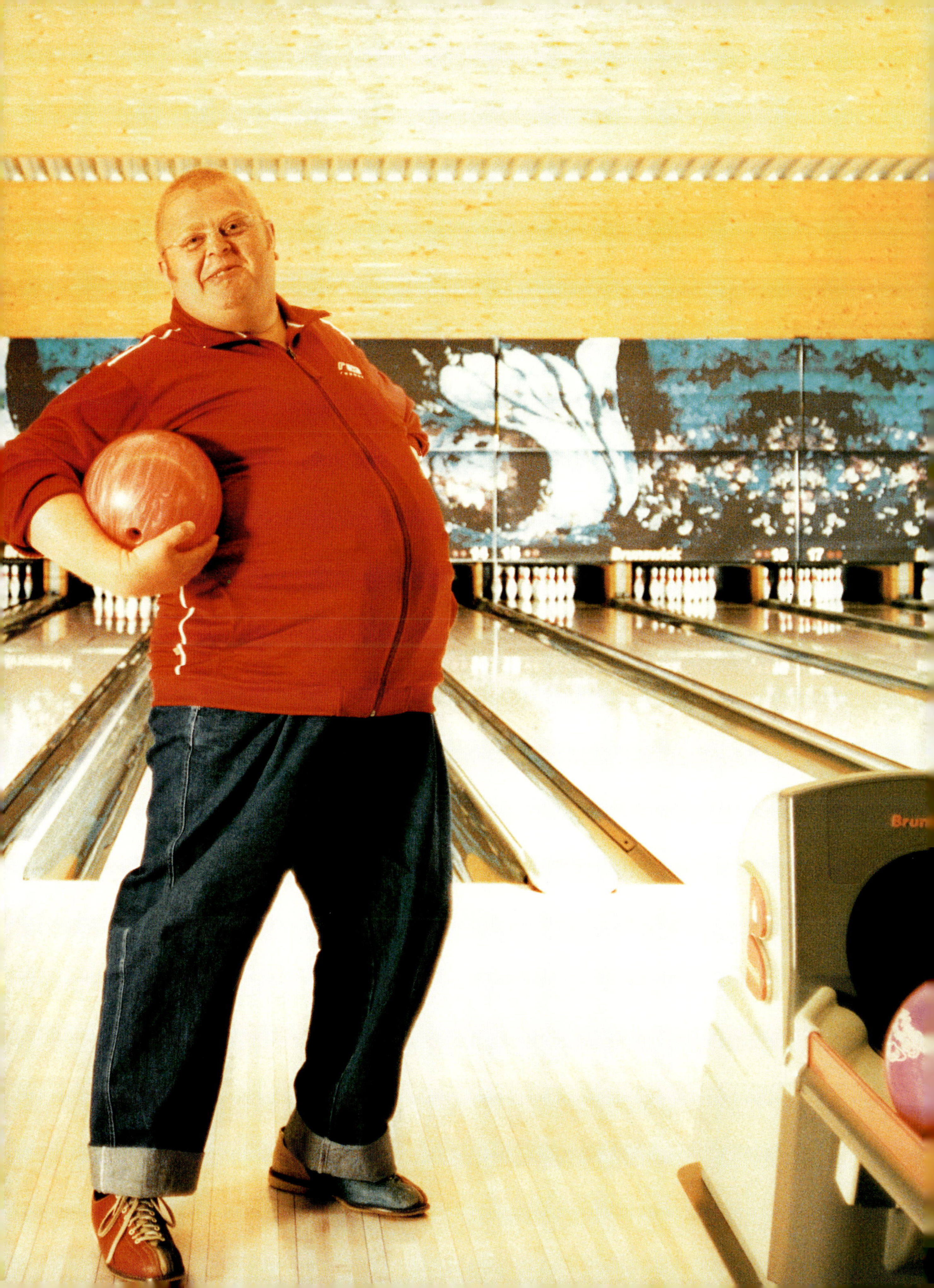

OPTIK + INFORMATION GOLD

VISION + INFORMATION GOLD

HERSTELLER . MANUFACTURER
EBM
D-94036 Passau

VERTRIEB . DISTRIBUTOR
EBM
D-94036 Passau

DESIGN . DESIGNER
Thomas Trauth Design GmbH
D-73547 Lorch

EBM Balgrip System : Brille . spectacles

„Balgrip“ ist ein neues System, Brillen ohne Schrauben zu verglasen. Das Glas der Fassung hängt in einer vorgespannten, federnden Klammer und wird mit zwei Kunststoffhülsen verankert. Diese Kunststoffhülsen werden auf die Stärke des Glases abgestimmt. Auf das Glas wird dabei kein Druck ausgeübt.
Dank „Balgrip“ können die Gläser in Sekundenschnelle von Hand gewechselt werden, beispielsweise vom Optiker bei der Kundenberatung um verschiedene Formen und Farben vorzuführen, oder vom Kunden selbst um auf Sonnengläser zu wechseln.
Das Brillengestell wurde aus dem neuen Hightechmaterial Genium® gefertigt, einem diamantgezogenen Metall, das sich durch höchste Oberflächendichte und -härte auszeichnet. Genium® ist eine korrosionsbeständige Speziallegierung, die absolut nickelfrei und antiallergen ist. Es ist nicht nur flexibel und superelastisch – dabei aber extrem formstabil durch Memory-Effekt –, sondern auch enorm belastbar und langlebig. Da das Material federleicht ist, ist „Balgrip“ eine der leichtesten Brillen der Welt.

.

"Balgrip" is a new system for fitting lenses into spectacles without screws. The lens is suspended in a pre-stressed, spring-mounted clip and is held in place by two plastic sleeves. These plastic sleeves match the thickness of the lens. No pressure is exerted on the lens.
Thanks to "Balgrip", lenses can be changed manually in a few seconds, by the optician advising his customer, for example, and wanting to demonstrate different forms and colours. Or by the customer himself, if he wants to change to sunglasses.
The frame has been made of Genium®, a new high-tech material which is diamond polished, and which is characterized by extreme surface density and hardness. Genium® is a corrosion-resistant special alloy, which is absolutely nickel-free and hypo-allergenic. It is not only flexible and highly elastic (yet with a memory effect which means that it always preserves its shape) but also extremely tough and long-lasting. As the material is extremely light, these are among the lightest spectacles in the world.

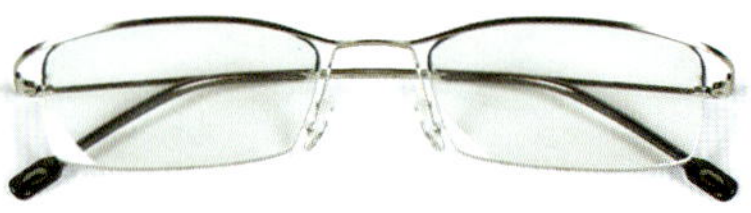

Jury . Judges panel

Hervorragende Harmonie zwischen Benutzung eines Materials mit guten Eigenschaften und einer Funktion, die die Handhabung erleichtert. Das Produkt ist auch dank der Anwendung des Materials sehr widerstandsfähig. Im Design überzeugend.

.

Excellent harmony between use of material with good properties and a function that makes it easier to use. Thanks to the material used, the product is also very tough. Convincing design.

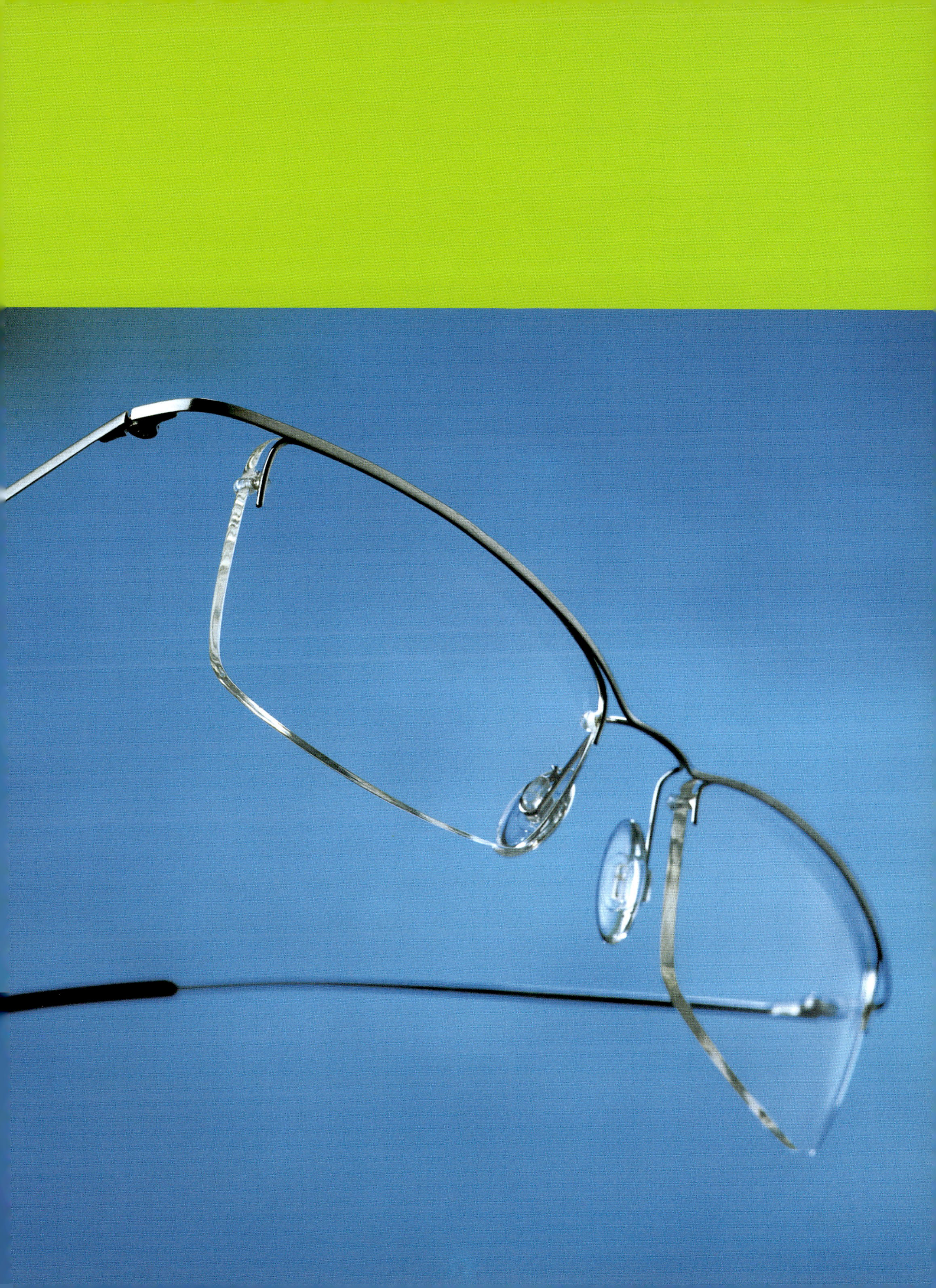

OPTIK + INFORMATION | SILBER

VISION + INFORMATION | SILVER

HERSTELLER . MANUFACTURER
Troika Böll GmbH & Co. KG
D-57629 Müschenbach

VERTRIEB . DISTRIBUTOR
Troika Böll GmbH & Co. KG
D-57629 Müschenbach

DESIGN . DESIGNER
Alexander Schenk
Produktgestaltung & Designberatung
D-73614 Schorndorf
und
Roy Schäfer
D-75181 Pforzheim

Balance : Tischuhr . desk-top clock

Diese ungewöhnliche Schaukel-Tischuhr in klassischem, funktionalem Design findet immer wieder ihre Balance. Sie regt den Spieltrieb an. Obwohl sie rund ist, rollt die Uhr nicht weg. Sie kehrt immer zu ihrer Mitte zurück. Ist man im Stress, tippt man „Balance" an und schaut zu, wie sie wieder ins Gleichgewicht kommt – und findet dadurch möglicherweise seine eigene Balance wieder. Die Tischuhr aus Metallguss kann auch als Wecker benutzt werden, denn sie hat eine eingebaute Weckfunktion.

This unusual, swinging desk-top clock with its functional, classic design always regains its balance. It appeals to people's playful side. Although it is round, the clock will not roll away. It always returns to its centre. Anyone under pressure can give "Balance" a tap and watch how it regains its balance again, and may even regain his own inner poise in the process. The cast metal desk-top clock can also be used as an alarm clock, as it comes with an alarm function.

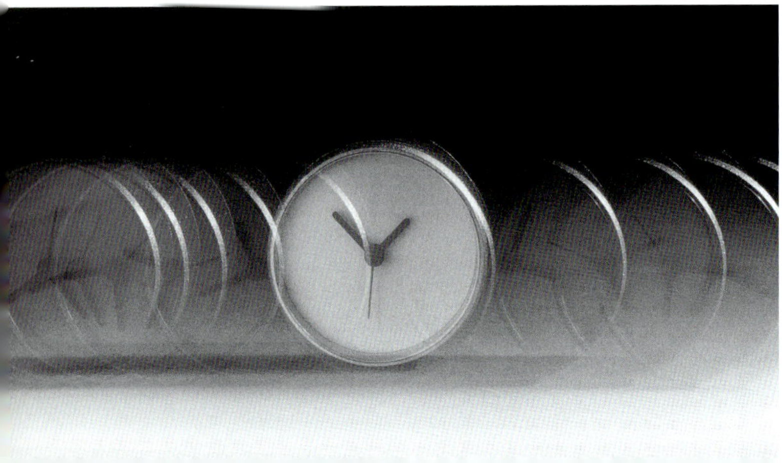

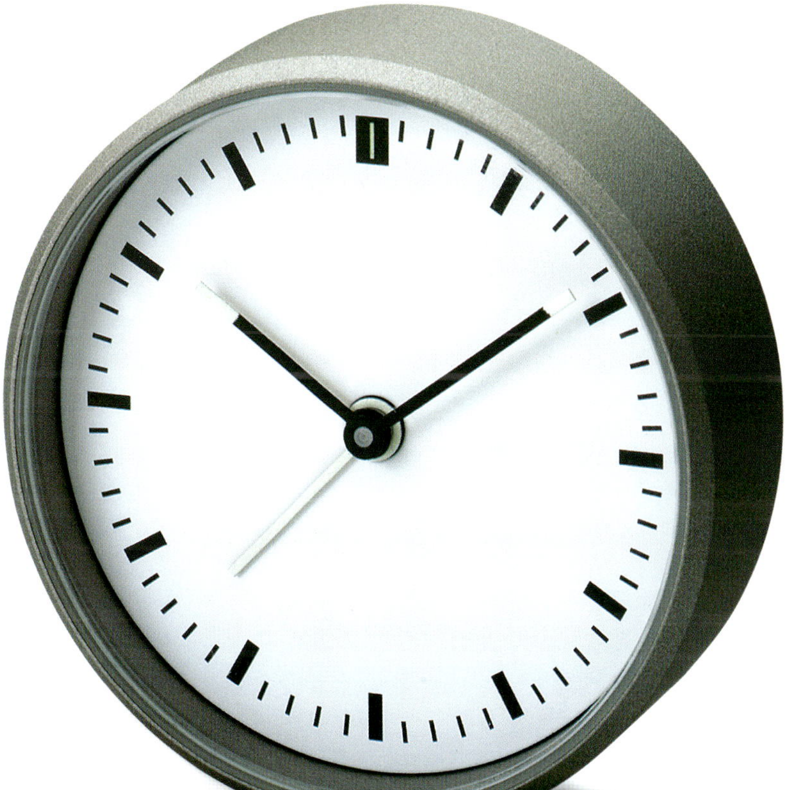

Jury . Judges panel

Diese Tischuhr ist das einzige Produkt, das das Thema völlig getroffen hat. Sehr einfaches Produkt, ein Basic. Die Rückansicht des Produktes kann noch deutlich verbessert werden.

.

This desk-top clock is the only product to have captured the competition's theme one hundred percent. Very simple product, a basic. The rear view of the product can be improved considerably.

OPTIK + INFORMATION | SILBER
VISION + INFORMATION | SILVER

HERSTELLER . MANUFACTURER
Loewe AG
D-96317 Kronach

DESIGN . DESIGNER
Phoenix Product Design
D-70376 Stuttgart

RC 1 : Fernbedienung . remote control

„RC 1" ist mehr als eine Fernbedienung. Sie ist die zentrale Steuereinheit, hier laufen alle Informationen zusammen. So steuert sie jedes einzelne Modul, Fernsehen, DVD/Video, Audio. Mit entsprechenden Aufrüstsätzen reguliert sie aber auch Raumklima, Lichtstimmung, Markise und Jalousie, simuliert Anwesenheit oder dirigiert HomeMultiMedia-Funktionen: Online-Anwendungen und im Internet surfen.
Grundlage ist die intelligente Systemverlinkung, über die alle Komponenten miteinander kommunizieren konnen. „RC 1" ist die erste bidirektionale Fernbedienung von Loewe. Da sie rund ist, sieht sie auch nicht aus wie eine Fernbedienung.
Klartextmenüs auf dem runden, beleuchteten Display dienen der Kontrolle und Steuerung und machen die Bedienung einfach und bequem. Die einzelnen Bedienelemente sind sehr übersichtlich angeordnet.

.

"RC 1" is more than a remote control. It is a central control panel where all information converges. It controls individual modules of all kinds – television, DVD/video, hi-fi. With supplementary equipment, it can be upgraded to function as a control for air conditioning/heating, light moods, awnings and blinds, can simulate the presence of people in the house or control home multi-media functions such as online applications or surfing the net.
It is based on intelligent system linking, via which all components can communicate with each other. "RC 1" is the first bi-directional remote control by Loewe. As it is round, it does not look like a remote control.
Plain-text menus on the round, illuminated display help to direct and control, and make operation easy and convenient. The individual control elements are very clearly arranged.

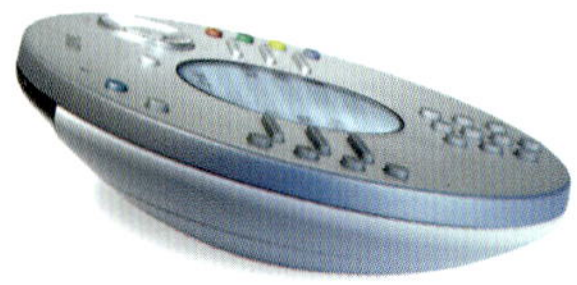

Jury . Judges panel

Ein Beispiel für eine progressive Interface-Gestaltung, die Einteilung in unterschiedliche Zonen, besonders die Zweihandbedienung und Übersichtlichkeit ist sehr gut gelöst.
Bei der Einhandbedienung auf dem Tisch ist die Stabilität des Produktes noch nicht zufriedenstellend und sollte noch verbessert werden.

.

An example of progressive design for an interface, divided into different zones. The two-handed operation and clear arrangement have been particularly well done. One-handed operation at the table is not yet satisfactory, as the product still tends to be wobbly, and this shortcoming needs to be solved.

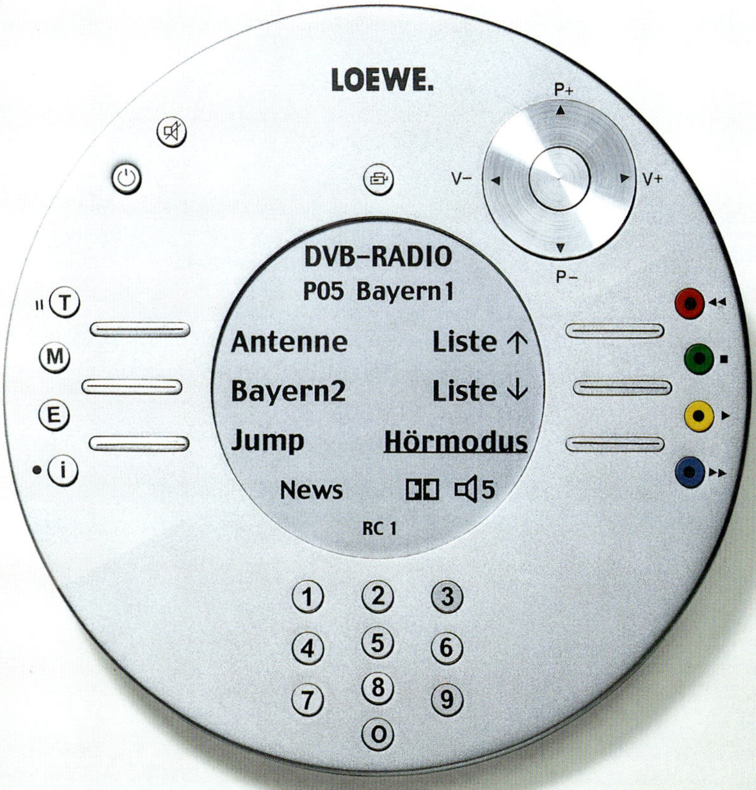
LOEWE.
P+
V–
V+
P–
DVB-RADIO
P05 Bayern1
Antenne
Liste ↑
Bayern2
Liste ↓
Jump
Hörmodus
News
RC 1
T
M
E
i
1
2
3
4
5
6
7
8
9
0

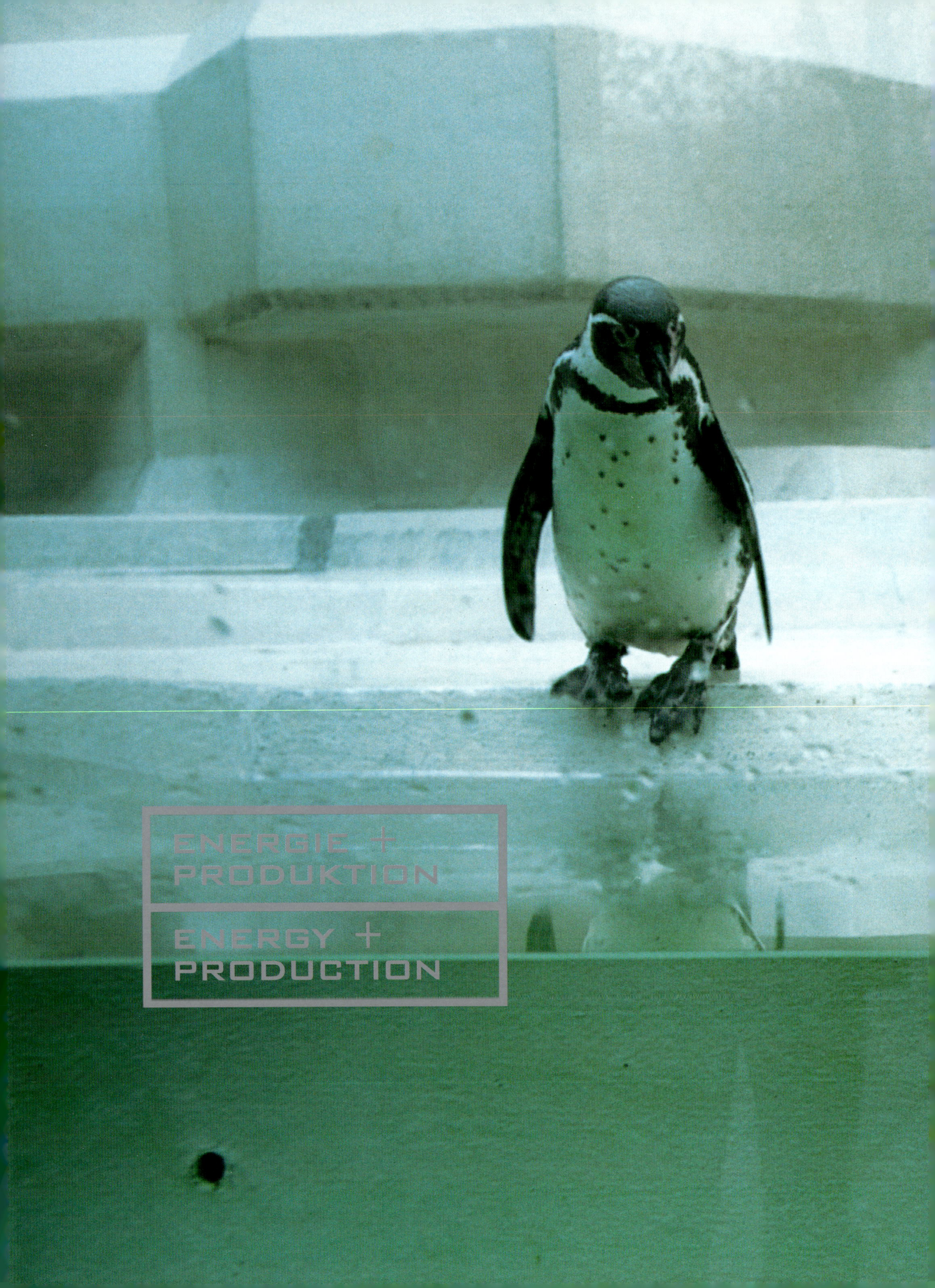

ENERGIE + PRODUKTION

ENERGY + PRODUCTION

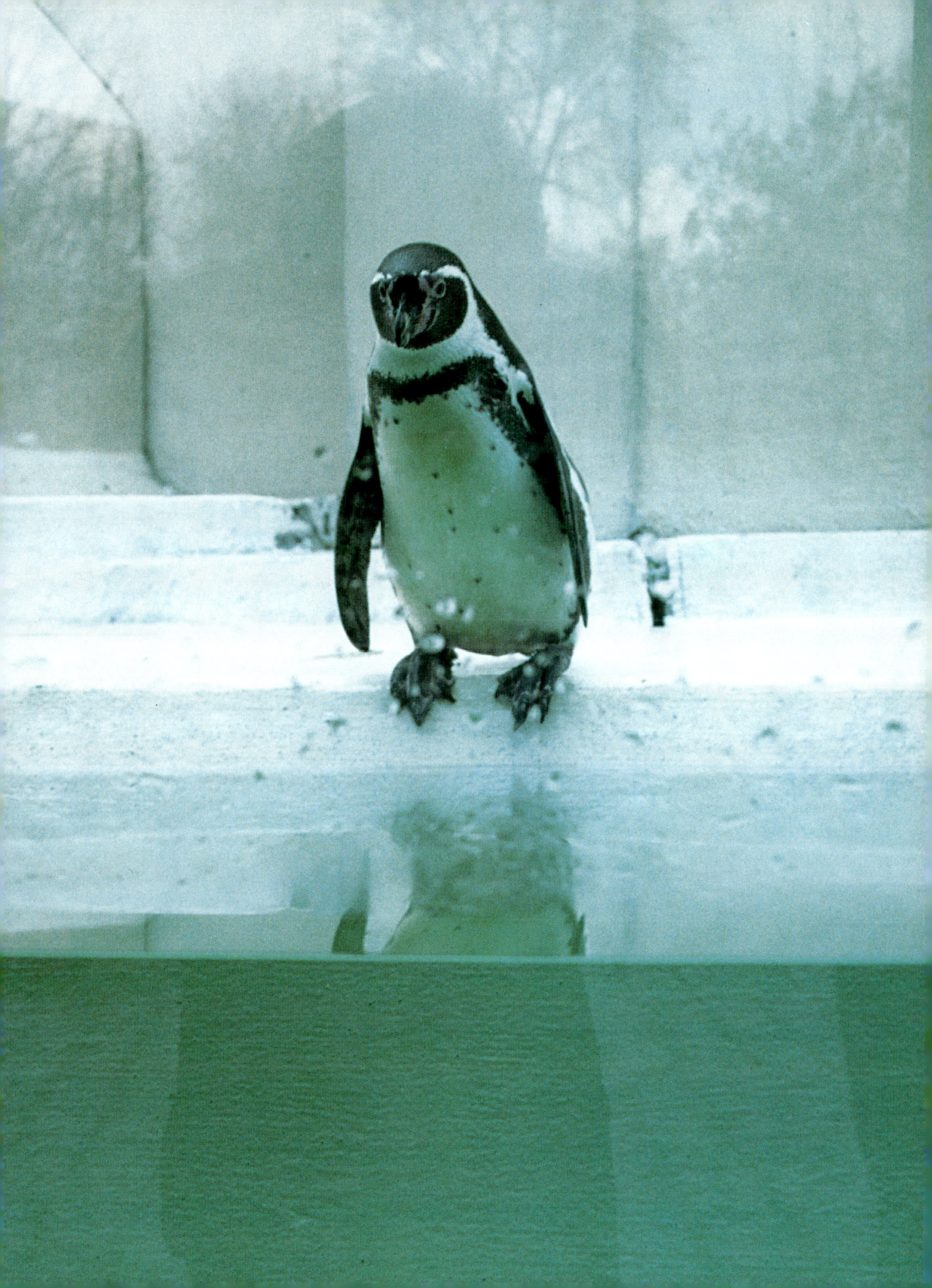

ENERGIE + PRODUKTION | SILBER

ENERGY + PRODUCTION | SILVER

HERSTELLER . MANUFACTURER

Power Jack OEG
A-8010 Graz

DESIGN . DESIGNER

WIGL-Design
A-8010 Graz

PowerJack : Heizleiste . strip heater

„PowerJack" ist die kleinste Heizleiste der Welt: nur 10 Zentimeter hoch und 2,8 Zentimeter tief, Länge nach Bedarf. Sie kann vielseitig eingesetzt werden und ist eine ideale Komponente zur Raumgestaltung. Durch die extrem niedrige Wassermenge kann sie sehr viel schneller als normale Heizkörper auf Temperaturschwankungen reagieren. Bei herkömmlichen Heizkörpern entstehen Wärmepolster unter der Decke und eine starke örtliche Luftzirkulation, wobei Staub aufgewirbelt wird, die Wandfläche neben den Heizkörpern aber kalt bleibt. „PowerJack" dagegen sorgt für angenehme Strahlungswarme an der gesamten Wand und auf dem Fußboden, da die Heizleiste entlang der Außenwand angebracht wird. Durch die bessere Wärmenutzung ist eine Energieeinsparung bis zu 15 Prozent möglich. „PowerJack" vermittelt Behaglichkeit wie ein Kachelofen. Da durch die geringere Luftzirkulation weniger Staub aufgewirbelt wird, werden Reizungen der Atmungsorgane verhindert und das Raumklima verbessert. So ist die Heizleiste auch besonders für Allergiker geeignet. Dank einer einfachen Klippmontage ist sie leicht und problemlos in allen Bereichen zu montieren. Sie ist in allen RAL-Farben zu haben. Für hohe Qualität sorgt die korrosionsbeständige Aluminiumlegierung.

.

"PowerJack" is the world's smallest strip heater: only 10 centimetres high and 2.8 centimetres deep, and available in any length. It can be used in very many ways, and is an ideal interior design element. Because it only holds a very small amount of water, it can react to temperature fluctuations very much faster than normal radiators. When conventional radiators are used, pockets of heat are created under the ceiling together with strong local circulation of air, which stirs up dust, while the wall surface next to the radiators remains cold. "PowerJack", by contrast, ensures that there is pleasant radiated heat along the entire wall and on the floor, as the strip heater is attached along the outside-facing wall. The improved use of heat means that energy-saving can be as high as 15 percent. "PowerJack" gives off the cosy heat of a tiled oven. The reduction in air circulation means that less dust is stirred up, and so irritation of the respiratory tract is prevented and the room climate is improved. This also means that the strip heater is particularly suitable for allergy-sufferers. A simple clip-on mount means that it can be attached easily and effortlessly to any wall. It is available in every RAL colour. High-quality is also ensured by the corrosion-resistant aluminium alloy used.

Jury . Judges panel

Sehr schönes Heizungssystem, das nicht wie ein Heizungssystem sondern wie eine Wandverkleidung aussieht. So beherrscht es nicht den Raum, sondern wird zu einem schönen Teil des Raumes.

.

Very attractive heating system that looks more like a wall panel than a heating system. It does not dominate the room, but becomes an attractive part of it.

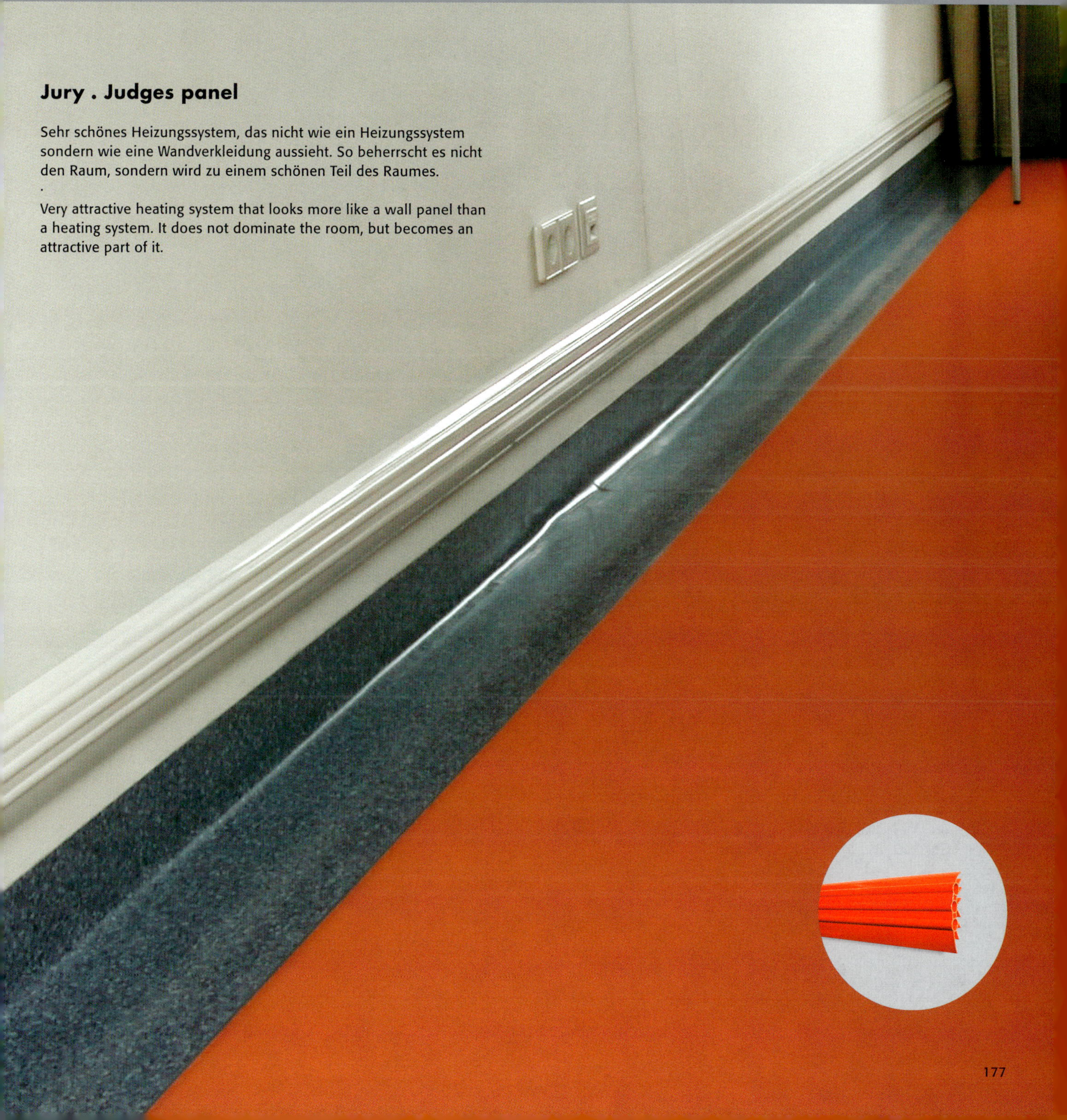

ENERGIE + PRODUKTION — SILBER

ENERGY + PRODUCTION — SILVER

HERSTELLER . MANUFACTURER
Solarpower GmbH
D-08529 Plauen

VERTRIEB . DISTRIBUTOR
Solarpower GmbH
D-08529 Plauen

DESIGN . DESIGNER
Thomas Ellerstorfer
D-93413 Cham

Solartrap 300 : Solaranlage/Speicherkollektor . solar heating system/collector

„Solartrap" ist ein besonders wirtschaftlicher und einfach zu installierender Speicherkollektor, der mit sehr viel weniger technischem Aufwand betrieben wird als konventionelle Solaranlagen. So erzeugt er ohne zusätzlichen Pufferspeicher, Wärmetauscher oder Umwälzpumpen täglich bis zu 300 Liter 50 bis 60 Grad warmes Wasser, in südlicheren Regionen noch mehr. „Solartrap" besitzt knapp drei Quadratmeter Kollektorfläche und zusätzliche Reflexionsspiegel, die vor allem bei tiefstehender Sonne im Frühjahr und Herbst den Ertrag um bis zu 40 Prozent erhöhen. Diese Ausstattung genügt, um das Brauchwasser eines Drei- bis Vier-Personenhaushaltes an Sonnentagen zu erwärmen. Eine patentierte transparente Wärmedämmung verringert den Wärmeverlust über Nacht, denn sie isoliert das Heißwasser von der kälteren Umgebungsluft, lässt jedoch das Sonnenlicht zur Wasseraufheizung hinein. Die Solaranlage ist an jedem Ort einsetzbar, egal ob Grünfläche, Flach- oder Schrägdach. Auch die Integration in bestehende Heizsysteme ist einfach realisierbar. Und „Solartrap" spart nicht nur Energie, die Solaranlage wird auch energiesparend hergestellt.

.

"Solartrap" is an especially economical, easy-to-install solar collector, which is technically far less trouble to operate than conventional solar heating systems. Without additional storage tanks, heat exchangers or circulation pumps, it generates up to 300 litres of 50 to 60-degree hot water per day, and even more in southern regions. "Solartrap" comprises just under three square metres of collector surface and additional reflecting mirrors, which can increase yield by up to 40 percent, especially when the sun is low in the sky in spring and autumn. When the sun shines, such a system is sufficient to heat the water used by a three to four-person household. A patented transparent heat insulating material reduces heat loss over night, insulating hot water against colder ambient air but letting sunlight through to heat the water. The solar heating system can be installed anywhere, in the open, on a flat roof or on a sloping roof. It is also simple to integrate with existing heating systems. "Solartrap" not only saves energy, but the heating system itself is produced in an energy-saving way.

Jury . Judges panel

Obwohl die Jury in Bezug auf die gestalterische Qualität ihre Einschränkungen für dieses Produkt zum Ausdruck gebracht hat, wurde es im Bereich der Ökologie und Ökonomie sehr positiv beurteilt und mit Silber ausgezeichnet.

.

Although the judges qualified their verdict with respect to the design quality of the product, its ecological and economical features were judged very positively, hence the silver award.

ENERGIE + PRODUKTION | SILBER

ENERGY + PRODUCTION | SILVER

HERSTELLER . MANUFACTURER
Steitz Secura GmbH
D-67292 Kirchheimbolanden

VERTRIEB . DISTRIBUTOR
Steitz Secura GmbH
D-67292 Kirchheimbolanden

DESIGN . DESIGNER
Werksdesign
Margit Ahnert

Alu Inside AL 90, AL 103 : Sicherheitsschuh . safety boot

Der Sicherheitsschuh vereint hohe Funktionalität und Sicherheit, optimalen Gesundheitsschutz, maximalen Komfort, ansprechendes Design und hohe Wirtschaftlichkeit. Bisher bestand der Zehenschutz eines Sicherheitsschuhs aus einer Stahlkappe, die meist nicht fußgerecht geformt war und deren hohes Eigengewicht zu einer gewissen „Kopflastigkeit" führte. Hier dagegen wird erstmalig eine Zehenschutzkappe aus Aluminium eingesetzt, die das Gewicht des Schuhs spürbar reduziert und den Schuhschwerpunkt zum Mittelpunkt verlagert. Dies hilft, Stolperunfälle zu vermeiden. An die Stelle der früheren Überkappe aus Polyurethan tritt ein Carbon-verstärktes Leder, das in Kombination mit hochwertigem Nubukleder und Textilien das Erscheinungsbild des Schuhs wesentlich prägt. In den reflektierenden Gurtbändern und Logoelementen verbindet sich Design und Sicherheit. Die Schuhe werden in vier Weiten gefertigt. Zusammen mit der vertikalen Fersenführung werden Unfälle durch Ausrutschen, Umknicken und Stolpern deutlich reduziert. Eine gewichtsangepasste Dämpfung durch ein Modul im Fersenbereich fördert die Gesunderhaltung von Rücken und Gelenken.

.

This safety boot combines excellent functionality and safety, optimum health protection, maximum comfort, appealing design and very good value for money. The reinforced toe of a safety boot used to be made of steel, which usually did not conform to the shape of the foot, and whose weight meant that the boot was generally "top-heavy". This boot, by contrast, uses an aluminium reinforced toe for the first time, which appreciably reduces the boot's weight and shifts its centre of gravity to the middle of the boot. This helps to prevent accidental tripping. What used to be a polyurethane toe cap has now been replaced by carbon-reinforced leather. In combination with the high-quality nubuk leather and textiles used here, this has a significant effect on the boot's appearance. The reflecting tapes and logos combine design and safety. The boots are produced in four widths. The vertical heel grip means that accidents as a result of slipping, twisting the ankle or stumbling are significantly reduced. Weight-adjusted damping provided by a module in the heel helps keep back and joints healthy.

Jury . Judges panel

Eine perfekte Balance zwischen Sicherheit, Komfort und Ästhetik. Es ist sehr gut gelungen, ein Produkt für die Arbeitswelt zu schaffen, das den strengen Anforderungen gerecht wird, dem Benutzer aber nicht den Eindruck vermittelt, dass er damit nur zur Arbeit gehen kann.

.

A perfect balance between safety, comfort and aesthetics. What has been successfully created here is a product for the working environment which does justice to its strict requirements but still does not give the wearer the impression that he can only go to work when he is wearing them.

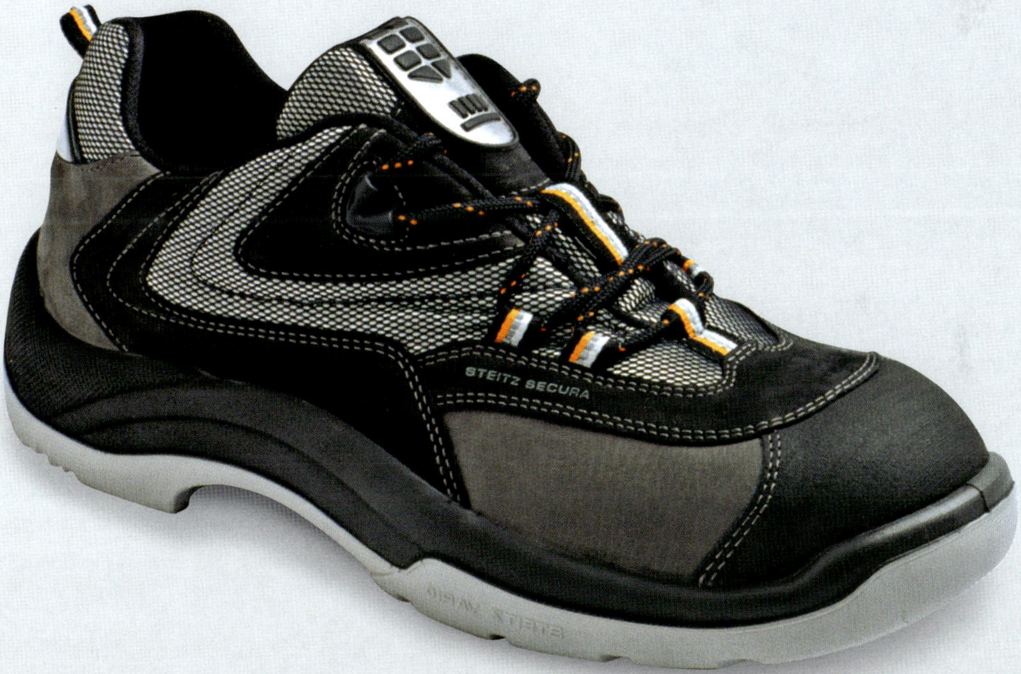

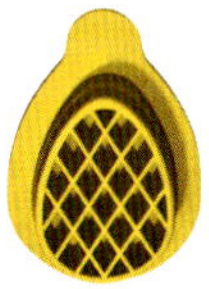
bis 57 kg

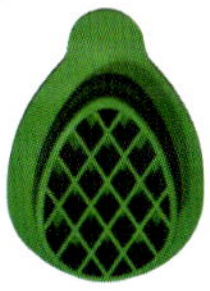
bis 79 kg

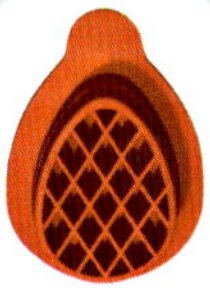
bis 91 kg

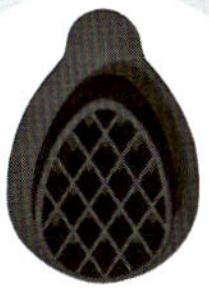
über 91 kg

ENERGIE + PRODUKTION | SILBER

ENERGY + PRODUCTION | SILVER

HERSTELLER . MANUFACTURER

Festool GmbH
D-73240 Wendlingen

VERTRIEB . DISTRIBUTOR

TTS Tooltechnic Systems Deutschland GmbH
D-73240 Wendlingen

DESIGN . DESIGNER

Studiowerk Design
D-82266 Inning

TS 55 EBQ und FS 1400 : Tauchsäge . plunge-cut saw

Die Arbeitssicherheit der Säge wird durch das Tauchsägenprinzip erreicht. Weitere Arbeitssicherheit bietet die Bedieneinheit im Führungs- und Tragegriff. Dort bedient man griffgünstig den Ein- und Ausschalter, die Eintauchverriegelung und den Fast-Fix-Hebel mit Spindelstop und Höhenarretierung. Dank dem Fast-Fix-Spindelstop lassen sich die Sägeblätter schnell und einfach wechseln. Das besonders geringe Eigengewicht der Tauchsäge schont die Kräfte beim Arbeiten. Der rutschfest gestaltete und weiche Soft Grip-Handgriff, die in das Motorgehäuse integrierte Elektronik und der ergonomische Zusatzhandgriff sorgen für Bedienfreundlichkeit.
Durch das Sägen auf der neuen Führungsschiene wird exakt und sicher gearbeitet, man erhält schnurgerade Arbeitsergebnisse. Der Splitterschutz sorgt dabei für saubere, ausrissfreie Schnitte. Mit der Absaugung wird umweltfreundlich und sauber gearbeitet.

.

The plunge-cutting principle used here makes the saw safe to work with. Other safety features include the operating controls in the guiding and carrying handle. While gripping the handle it is easy to operate the on/off switch, the plunge lock and the fast-fix lever with its spindle lock and blade-height locking device. The fast-fix spindle lock means that saw blades can be exchanged quickly and simply.
The unusually low weight of the plunge-cut saw reduces work load. The non-slip, soft handgrip, the electronics integrated in the motor housing and the ergonomic additional handle make it user-friendly.
By sawing along the new guide slot, work is done accurately and safely, and the results are perfectly straight. The chip deflector ensures that the cut is clean, and that there is no tearing out. The suction attachment makes work environmentally friendly and clean.

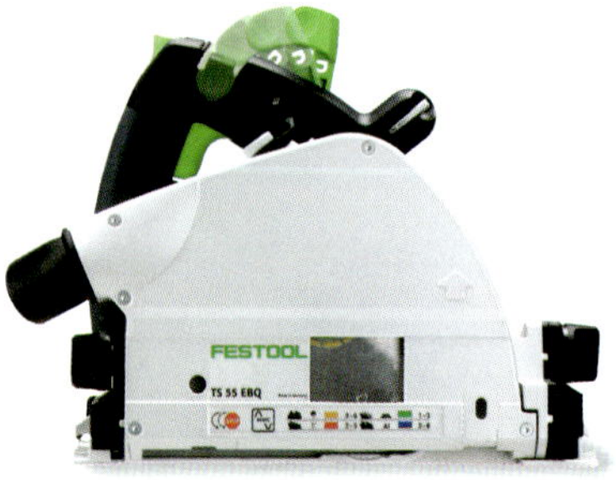

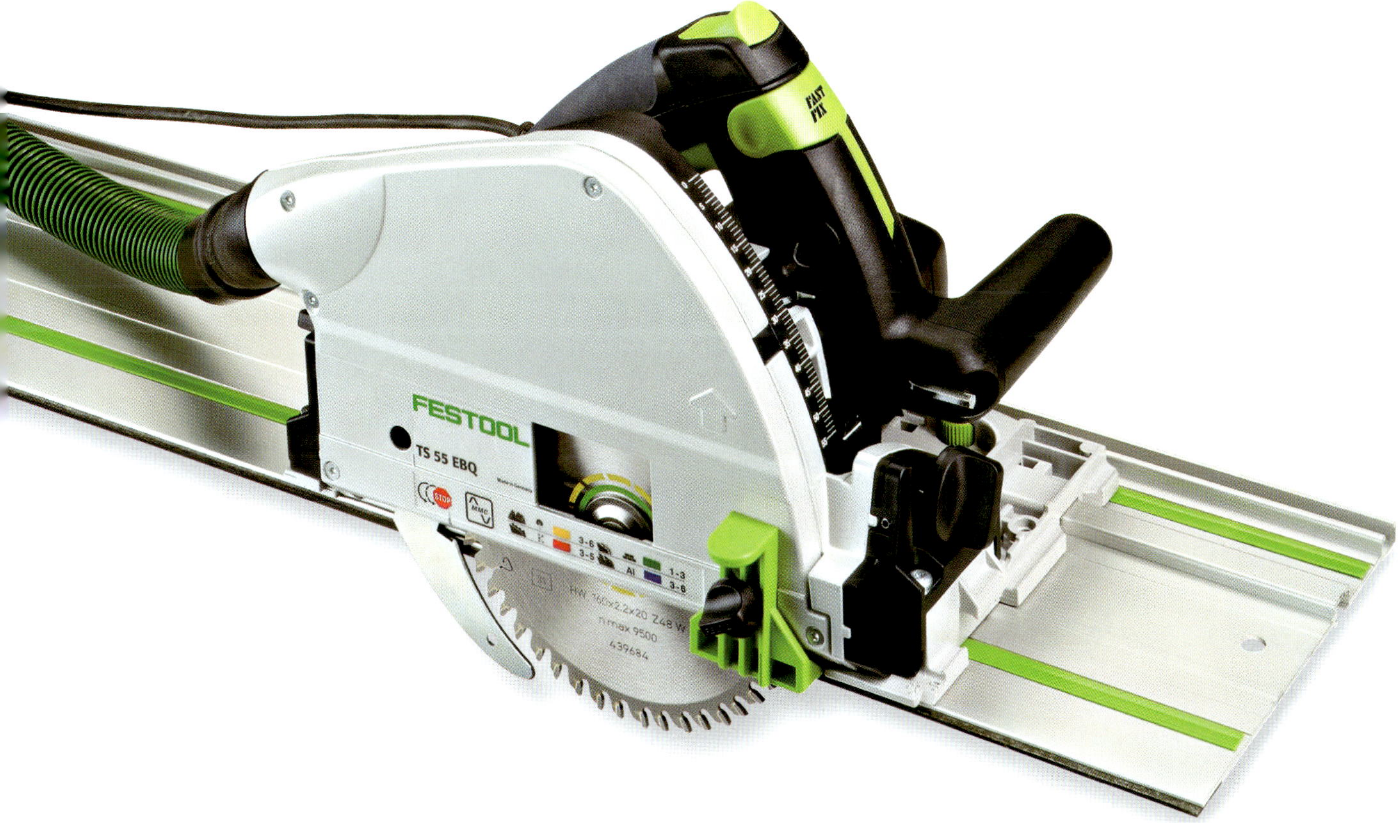

Jury . Judges panel

Eine solide Designarbeit, die die Komplexität dieses Handgeräts in eine Einheit bringt. Ein Produkt mit hoher Qualitätsanmutung, mit interessanten durchdachten Details, die dem Benutzer stets den Eindruck vermitteln, ein Qualitätswerkzeug mit hohen Sicherheitsmerkmalen zu haben, und das Ganze auch noch praktisch in einer Box.

.

Sound design work which makes a unity out of a complex piece of hand-held machinery. A product which transmits an outstanding sense of quality, with interesting, well thought-out details which constantly give the user the sense of having a quality tool with outstanding safety features. All of this comes in a practical box.

PROF. SIGMAR WILLNAUER
Mitglied der Jury, Leiter des Studiengangs Produktgestaltung an der Hochschule für Gestaltung in Schwäbisch Gmünd und Chefdesigner von Shelter and Roam in Mannheim

EIN KOMPLEXES HAUS MIT 100 RÄUMEN _ PRODUKTGESTALTUNG

Alle Versuche, einheitlich zu definieren, was Produktgestaltung ist und wie der Design-Prozess abläuft, scheitern an der Realität. Als ein Erklärungsmodell taugt aber, sich Produktgestaltung als ein komplexes Haus mit 100 Räumen vorzustellen. Jeder dieser Räume enthält einen bestimmten Inhalt des Design-Prozesses. Bei jeder Produktgruppe, Nutzergruppe, jedem Projekt und jedem Unternehmen sind jedoch eine andere Kombination und Anordnung der Räume entscheidend. Wenn beispielsweise Herman Miller und Vitra im selben Jahr einen Bürostuhl entwickeln, sind ihre Raum-Sets dennoch verschieden.

Bei jeder Organisation des Hauses bestehen unterschiedliche Wertigkeiten der Bedingungen, und ein bestimmtes Projekt benötigt beispielsweise jene siebzig Räume, aber nicht die anderen dreißig. Auch der Wandel der marktwirtschaftlichen und kulturellen Entwicklungen – sagen wir vom vergangenen Jahrzehnt hin zum ersten des neuen Millenniums – wird mit dem 100-Räume-Modell verständlich: In der Praxis, bei Designwettbewerben und in der Hochschulausbildung versinken einige Räume in die Bedeutungslosigkeit, neue zuvor irrelevante Räume mit neuen Inhalten kommen hinzu. Out sind unbenutzbare und unbezahlbare zweihundertprozentige Über-Tech-Designlösungen, tonangebend sind dagegen Light-Tech-Designlösungen.

Zum etablierten Raum „Nachhaltige Produktgestaltung" kommt das neue Zimmer „Drei-Liter-Produktentwicklung": die Fähigkeit, mit weniger von der Ressource „Entwicklungszeit" weiter zu kommen. Dies fordert aber auch vom Kunden, dass er eine erste gute Designlösung akzeptiert, statt erst der fünften Version Vertrauen zu schenken.

Designwettbewerben, Hochschulen und Designzeitschriften kommt für die gesamte industriell-kulturelle Entwicklung die Rolle von Barometern zu, um diese neuen Räume zu ertasten und verantwortlich mitzugestalten. In unseren Designstudios ist diese Barometerrolle auf Grund der rasanten, auch Budget-bedingten Geschwindigkeit nur selten einzulösen. Designentscheider in den Unternehmen können, wenn sie diese Barometer über einige Zeit genau beobachten, erkennen, ob ihre Designentwicklung den zukunftsfähigen 100 Räumen entspricht oder ob sie beladen mit Überholtem hinterher hinken. Designwettbewerbe und Hochschulen entwickeln das neu entstehende Set der hinzukommenden Räume mittels hunderter Pilotprojekte in Form von Wahrscheinlichkeits- und Möglichkeitsstudien. An der Hochschule für Gestaltung in Schwäbisch Gmünd forschen wir derzeit an der Entstehung der neuen Räume „Entgrenzung" und „Regionalität": Bei welchem Set ist es erforderlich, entgrenztes, internationales Produktdesign einzusetzen, und/oder bei welchem Set muss das Design eine regionale Sprache sprechen?

Die zu Focus Balance eingereichten Produkte zeigten dieses Jahr eine Wetterlage der Vorsicht: Vorsicht im Sinne von Risiko-Reduktion, nicht eine Vor-Sicht im Sinne von Vorhersehen und Voraussehen, was der derzeitigen Wirtschaftslage besser entspräche, da die Sets der Räume und Chancen gerade neu konfiguriert werden. Es erfordert natürlich unternehmerischen und gestalterischen Mut, in dieser Phase großzügig anstatt kleinmütig zu handeln.

Das Thema „Focus Balance" weist auf eine weitere Kernkompetenz der ProduktgestalterInnen hin: Wir sind Experten und Mittler im Umgang mit den Widersprüchen in der Produktentwicklung. Gegensätzliche Produktkriterien, wie zum Beispiel „schnell" und „langlebig", in einem Produkt gleichwertig zu vereinen, gehört zu unseren Kernkompetenzen. Wir sind auch jene Experten, die die Balance zwischen der Akzeptanz des Essenziell-Realen und der Entsorgung des Unwichtig-Realen erzeugen, den Schlüssel zum Raum der wertschöpfenden Innovation. Diese Leistung ist bei allen Produkten, die mit dem Focus in Gold ausgezeichnet wurden, auf höchstem Niveau erbracht.

Große wertschöpfende Neuerer der letzten Jahrzehnte wie Olivetti, Braun, Apple, Nike und Nokia haben schon zu Anfang, als sie noch Mikro-Marken waren, auf strategisches Design gesetzt, ihre Kunden und Nutzer nicht als Untertanen behandelt und auf die neuen Räume (statt der gesicherten) gesetzt. Ich bin sicher, dass das Design Center Stuttgart als Design-Schrittmacher mit seinem Internationalen Designpreis auch im nächsten Jahr einen weiteren neuen Raum öffnen wird.

PROF. SIGMAR WILLNAUER
Member of the judges panel, Head of the Industrial Design Department
Hochschule für Gestaltung Schwäbisch Gmünd and Chief Designer of Shelter and Roam, Mannheim

A COMPLEX BUILDING WITH 100 DIFFERENT ROOMS _ PRODUCT DESIGN

Any attempt to reach a standard definition of what product design is and how the design process functions is thwarted by multiple realities. However, as a way of explaining what it is, it is useful if we imagine product design as a complex building with 100 different rooms. Each of these rooms contains one particular aspect of the design process. However, for each product group, user group, project and company, a different combination and arrangement of rooms will be decisive. If Herman Miller and Vitra, for example, develop an office chair in the same year, the sets of rooms they use will nevertheless be different.

In each organization of the building, conditions are valued differently, and a certain project will require seventy of these rooms, for example, but not the other thirty. The 100-room model also explains the change in market-economic and cultural developments – from the past decade, say, up to the first decade of the new millennium: in everyday practice, in design competitions and in university education, certain rooms fade into insignificance, while other, previously irrelevant, rooms with new contents come to the fore. Out go unusable, unaffordable 200 percent hyper-tech products, and in come light-tech design solutions.

The established "sustainable product design" room is being joined by the new "90 mpg product development" room: the ability of progressing further using less of the "development time" resource. At the same time, however, this requires that the customer accepts the first good design solution instead of only placing his faith in the fifth version.

Design competitions, colleges and design journals are taking on the role of barometers for the whole complex of industrial-cultural development, sounding out these new rooms and playing a responsible part in their design. Breakneck, budget-driven speed means that this barometer role can only rarely be fulfilled in our design studios. Design decision-makers in companies can, if they observe these barometers carefully over a period of time, recognize whether their design development is aligned with the 100 rooms which have a future, or whether, loaded down with obsolescence, they are limping behind. Design competitions and colleges develop the newly emerging set of rooms to be admitted to the canon by means of hundreds of pilot projects, in the form of probability and feasibility studies. At the Schwäbisch Gmünd Hochschule für Gestaltung (University of Applied Sciences) we are currently researching the emergence of the new "globalism" and "regionality" rooms. Which set needs a boundaryless, international product design, and/or in which set does design have to speak a regional language?

The products submitted to the Focus Balance competition this year were evidence of a precautionary mood: precaution in the sense of risk reduction, but not in the sense of prediction and foresight, which would be more suited to the current economic situation, given that the sets of rooms and opportunities are currently being re-configured. Of course, in a phase like this, it takes entrepreneurial and design courage to act generously instead of parsimoniously.

The "Focus Balance" theme alludes to a further core competence of product designers: we are experts and mediators when dealing with contradictions in product development. Unifying contradictory product criteria such as "fast" and "long-lasting" on equal terms in one product is one of our core competences. We are also the experts who generate balance between acceptance of the essential and real and disposal of the insignificant and real, the key to the room of innovation that provides value added. In all the products awarded the Focus in Gold, this has been achieved to the highest standard.

The great value-added-generating innovators of the past decades, such as Olivetti, Braun, Apple, Nike and Nokia put their faith in strategic design right from the beginning, when they were still micro-brands. They did not treat their customers and users like mere subjects, and they staked their future on new rooms (rather than safe ones). I am sure that the design pacemaker that is the Design Center Stuttgart and its International Design Award will open up a further new room next year, too.

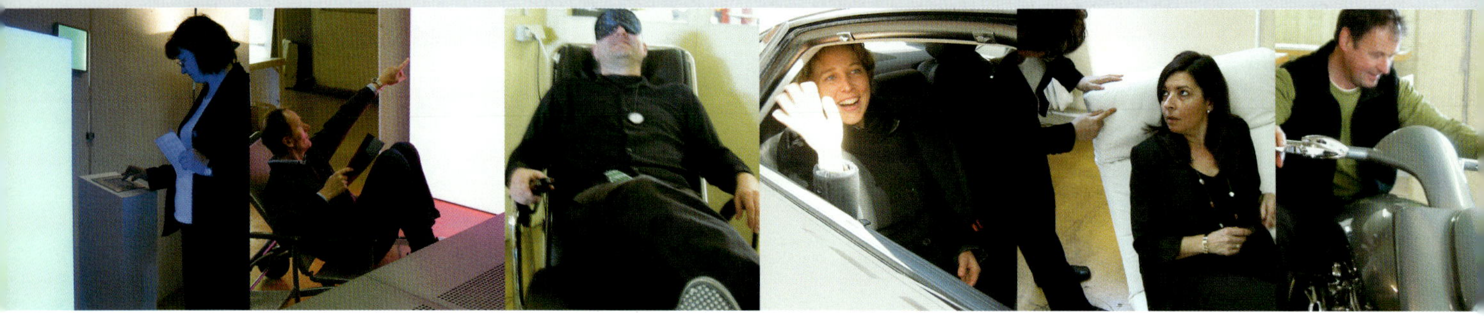

JURY . JUDGES PANEL

ĽUBICA FÁBRI
Direktorin des slowakischen Designzentrums in Bratislava, Chefredakteurin von „Designum"
·
Director, Slovak Design Centre, Bratislava
Editor-in-chief of "Designum"

Ľubica Fábri studierte an der Bauhaus-Universität Weimar, Fachrichtung Architektur. Nach ihrem Diplom in Innenraumgestaltung und Möbeldesign 1984 arbeitete sie in einem Forschungsinstitut der Möbelindustrie in Bratislava. Neben theoretischen Untersuchungen realisierte sie auch eigene Möbelentwürfe. Dies mündete schließlich in die Selbstständigkeit. Neben ihrer Designarbeit war sie auch als Journalistin und Publizistin für Interior- und Grafikdesign tätig. Seit 1999 ist sie nun Direktorin des Slowakischen Designzentrums SCD in Bratislava und seit 2000 zudem Chefredakteurin der Designfachzeitschrift „Designum". Auf internationaler Ebene wirkte Ľubica Fábri in zahlreichen Jurys mit und war als Kuratorin für Ausstellungen im Produkt- und Grafikdesign verantwortlich.
·
Ľubica Fábri studied Architecture at the Bauhaus University in Weimar. After graduating in Interior Design and Furniture Design in 1984, she worked in a research institute of the furniture industry in Bratislava. Apart from theoretical studies, she also produced some furniture designs of her own. This finally resulted in her becoming freelance. Apart from design work, she also worked as a journalist and publisher for interior and graphic design. Since 1999 she has been the director of SCD, the Slovak Design Centre in Bratislava, and since 2000 she has been editor-in-chief of the design journal "Designum". On the international stage, Ľubica Fábri has been involved in many judging panels and has acted as the organizer of exhibitions of product and graphic design.

THOMAS GERLACH
Designer, via 4 Design in Nagold

Thomas Gerlach studierte Design an der Universität Essen GHS (Folkwangschule). Nach seinem Studienabschluss begann er 1983 bei frogdesign in Altensteig, wo er 1990 bis zum President for Europe aufstieg. Später machte er sich selbstständig und gründete das Designbüro via 4 Design in Nagold im Schwarzwald. Parallel dazu arbeitete er von 1996 bis 1997 in der Rosenthal AG als Leiter von Marketing, Design, Kommunikation und Art Direction. Heute ist via 4 Design in den verschiedensten Designbereichen tätig, wie E-Commerce, Markenbildung, Grafik- und Produktdesign sowie Design-Strategien. Zu den Kunden zählen unter anderem die Deutsche Bank, Mercedes-Benz, Samsung und zahlreiche andere weltweit operierende Konzerne. Von Thomas Gerlach erschienen verschiedene Publikationen, er erhielt zahlreiche internationale Design- und Corporate-Identity-Preise.

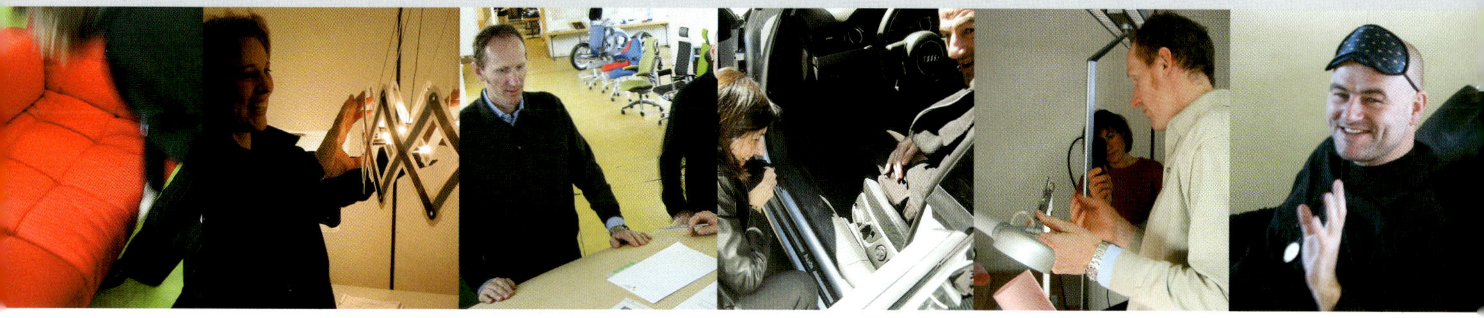

Thomas Gerlach studied Design at the University of Essen (Folkwangschule). After graduating, he joined frogdesign in Altensteig in 1983, progressing to become President for Europe in 1990. Later he became a freelancer, founding the via 4 Design bureau in Nagold (Black Forest). From 1996 to 1997 he also worked at Rosenthal AG as the Head of Marketing, Design, Communication and Art Direction. Today, via 4 Design is involved in a wide area of design fields, such as e-commerce, branding, graphic and product design, and design strategies. Its customers include Deutsche Bank, Mercedes-Benz, Samsung and many other globally operating groups. Thomas Gerlach is the author of many publications, he has received many international design and corporate identity awards.

HENRIK HOLBÆK

Architekt und Industrial Designer
Tools Design in Kopenhagen

.

Architect and industrial designer
Tools Design in Copenhagen

Henrik Holbæk studierte an der Royal Academy of Arts in Kopenhagen Industrial Design. 1989 gründete er zusammen mit Jens Clausen Tools Design in Kopenhagen. Das Designbüro gestaltet elektronische und medizinische Produkte und hat vor allem in der Haushaltsbranche und mit Accessoire-Entwürfen weltweiten Erfolg. Tools Design hat maßgeblich die Produkte der Unternehmen Tommy Larsen und Eva Denmark geprägt. Die Produkte von Tools Design werden oft als sehr einfach und innovativ charakterisiert. Henrik Holbæk und sein Partner erhielten für ihre Produkte bisher mehr als 40 dänische und internationale Preise, darunter 2002 bei Focus Lebensart – Internationaler Designpreis Baden-Württemberg zweimal den Focus in Silber, und repräsentierten dänisches Design auf mehreren internationalen Ausstellungen.

.

Henrik Holbæk studied Industrial Design at the Royal Academy of Arts in Copenhagen. In 1989, he and Jens Clausen founded Tools Design in Copenhagen. The design bureau designs electronic and medical products, and has especially experienced international success in the household appliance industry and with designs for accessories. Tools Design has had a formative influence on the products manufactured by Tommy Larsen and Eva Denmark. Tools Design's products are often characterized as very simple and innovative. To date, Henrik Holbæk and his partner have received more than 40 Danish and international awards, including two Focus in Silver awards in "Focus Way of Life", the 2002 Baden-Württemberg International Design Award, and have represented Danish design at several international exhibitions.

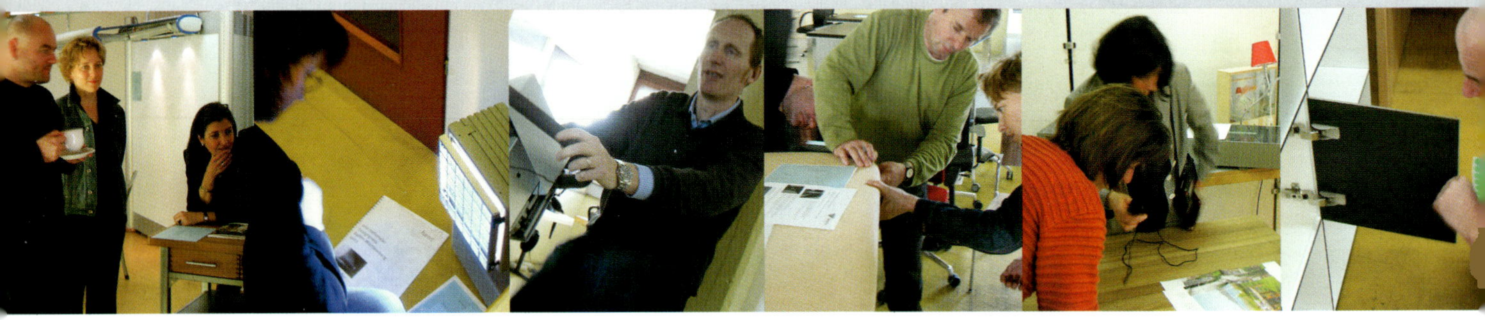

REGINE RITZ
Designerin, RitzSchultes Design in Stegen
·
Designer, RitzSchultes Design in Stegen

Vor dem Einstieg in das Produktdesign befasste sich Regine Ritz mit verschiedenen Studien- und Ausbildungsrichtungen, darunter Naturheilverfahren, Kunstgeschichte und Indologie. Danach absolvierte sie eine Ausbildung zur Damenschneiderin. Ab 1990 studierte sie Produktgestaltung an der Hochschule für Gestaltung Offenbach. Nach ihrem Studienabschluss und einem USA-Aufenthalt arbeitete sie 1997 zunächst freiberuflich für IDEO in London und war anschließend bis 2001 bei Siemens Design & Messe (heute: designafairs) in München angestellt. 2001 machte sie sich selbstständig und gründete zusammen mit Herbert H. Schultes das Designbüro RitzSchultes Design in Stegen. Ein Schwerpunkt ihrer Tätigkeiten liegt in den Bereichen Messe, Ausstellung und Event-Architektur. Dazu zählte unter anderem die gestalterische Projektleitung des EXPO-2000-Auftritts der Vereinten Arabischen Emirate.
·
Before turning to product design, Regine Ritz studied various subjects, including Natural Healing, the History of Art and Indology. Following that she completed an apprenticeship as a dressmaker. From 1990, she studied Product Design at the Offenbach Hochschule für Gestaltung. After graduation and a period of stay in the USA, she worked first of all on a freelance basis for IDEO in London, and was then employed by Siemens Design & Messe (now designafairs) in Munich until 2001. In 2001, she set up on her own, founding the RitzSchultes Design bureau in Stegen together with Herbert H. Schultes. Her work has focussed among other things on the trade fair, exhibition and event-architecture field. This has included work as the project manager responsible for the design of the United Arab Emirates' stand at EXPO 2000.

XÈNIA VILADÀS
Design-Managerin und Direktorin der Fakultät Design Management am Design College Elisava in Barcelona
·
Design manager and director of the Design Management faculty at the Elisava Design College in Barcelona

Nach dem Abschluss ihres Wirtschaftsstudiums in Barcelona 1981 war Xènia Viladàs bis 1989 verantwortlich für die internationale Ausrichtung und Werbung des spanischen Textilverbandes, der Asociación Industrial Textil de Proceso Algodonero AITPA. Später leitete sie das Design Center in Barcelona, war als Wirtschaftsberaterin tätig und machte den MBA in Design Management an der University of Westminster in London. Heute arbeitet sie als freiberufliche Design-Managerin und leitet als Direktorin die Fakultät Design Management am Design College Elisava, das zur Universität Pompeu Fabra gehört. Zahlreiche Veröffentlichungen zu Design-Management.

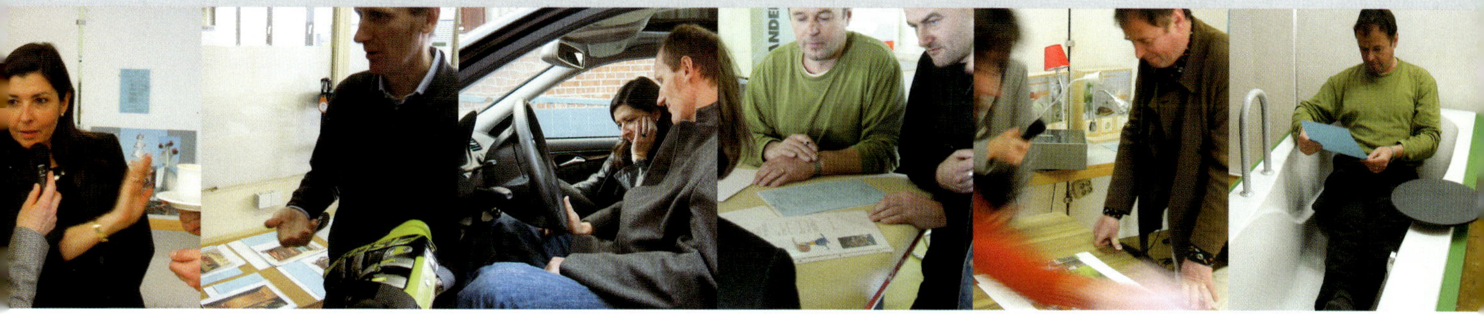

After graduating in Economics in Barcelona in 1981, Xènia Viladàs worked until 1989 for Asociación Industrial Textil de Proceso Algodonero AITPA, the Spanish textiles association, where she was responsible for international strategy and advertising. Later she was director of the Barcelona Design Centre, worked as a business consultant and took an MBA in Design Management at the University of Westminster in London. Today she works as a freelance design manager and is director of the Design Management faculty at the Elisava Design College in Barcelona, which is part of Pompeu Fabra University. Many publications on Design Management.

PROF. SIGMAR WILLNAUER
Professor für Produktgestaltung
Hochschule für Gestaltung in Schwäbisch Gmünd
Chefdesigner, Shelter and Roam in Mannheim
.
Professor of Product Design
Hochschule für Gestaltung in Schwäbisch Gmünd
Chief designer, Shelter and Roam, Mannheim

Sigmar Willnauer studierte Industrial Design in Braunschweig und Long Beach (Los Angeles). Danach arbeitete er in den USA, zunächst vier Jahre bei frogdesign, anschließend vier Jahre im Designbüro That in San Francisco. Er gestaltete Produkte für Apple Macintosh, für Goods in Berkeley entwarf er Möbel- und Leuchtenkollektionen. 1997 wurde er als Professor für Produktgestaltung an die Hochschule für Gestaltung in Schwäbisch Gmünd berufen. Der Schwerpunkt liegt hier auf der Entwicklung von Produkten und Produktsystemen unter Einbeziehung der soziologischen und ökologischen Kontexte. Noch heute ist Sigmar Willnauer durch internationale Lehrtätigkeiten mit den USA eng verbunden. So ist er Adjunkt Professor am Californian College of Arts and Crafts sowie Gastdozent an der Cranbrook Academy und Stanford University.

.

Sigmar Willnauer studied Industrial Design in Braunschweig and Long Beach (Los Angeles). Following his studies he worked in the USA, first for four years at frogdesign, and then for a further four years in the That design bureau in San Francisco. He designed products for Apple Macintosh, and designed furniture and luminaire collections for Goods in Berkeley. In 1997, he was called to the Chair of Product Design at the Hochschule für Gestaltung in Schwäbisch Gmünd. Here, his work has concentrated on the development of products and product systems, with reference to their sociological and ecological contexts. Sigmar Willnauer's international lecturing work means that he still has close contacts with the USA. He is adjunct professor at the Californian College of Arts and Crafts and visiting lecturer at Cranbrook Academy and Stanford University.

MIA SEEGER PREIS 2003	1. PREIS
MIA SEEGER PRIZE 2003	1ST PRIZE

DESIGN . DESIGNER
Ralf Baumunk
Fachhochschule Hannover

Erderstraße 33
D-30451 Hannover
Mobil +49.178.5 64 40 90
master@raballo.de

Joachim Möllmann
Fachhochschule Hannover

An der Strangriede 44
D-30167 Hannover
Mobil +49.177.5 43 68 27
j_moellmann@hotmail.com

Wasserabrasiv-System . Water jet-cutting system

Erst entwickeln Werkstofftechnologen ein Verfahren, Knochengewebe mit Wasserstrahl zu schneiden. Dann entwerfen Designer in einer Machbarkeitsstudie ein Modell, das zeigt, wie die Anwendung im Operationssaal aussehen kann. Herzstück des medizinischen Gerätesystems zur Vorbereitung der Knochen für die Kniegelenkimplantation ist ein Roboterarm, der ein Wasserstrahl-Schneidewerkzeug führt. Sein Sockelgehäuse enthält CPU, Steuerungselektronik, ein Dosiersystem für Schleifmittel und ein ausfahrbares Absaugmodul.
Die weiteren Module: auf dem Operationstisch die Cuttingbox (Knie fixieren, Wasser auffangen), daneben das mobile zentrale Steuerpult und im Hintergrund die stationäre Hochdruckpumpe.
Die medizinischen Vorteile liegen auf der Hand: Die Wasserstrahltechnik schont das Knochengewebe und schneidet viel präziser als die Hand, so dass Implantat und Knochen genau aufeinander passen, schneller zusammenwachsen und die Verbindung länger hält.

Materials application engineers have developed a method which uses a water jet to cut bone tissue. In a feasibility study, designers have now come up with a model that shows how its application might look in the operating theatre. The heart of this medical appliance to prepare bone for implantation in the knee joint is a robot arm with a water jet-cutting tool. The housing around its base contains a CPU, electronic controls, a system for dosage of abrasives and a retractable suction module. The other modules are a cutting box on the operating table (for immobilizing the knee and collecting water), a mobile central console and, in the background, a stationary high-pressure pump.
The medical benefits are obvious: the water-jet technique is less harmful to bone tissue and cuts more accurately than if bone is cut manually, with the result that the implant and the bone fit perfectly and grow together more quickly, and the joint lasts longer.

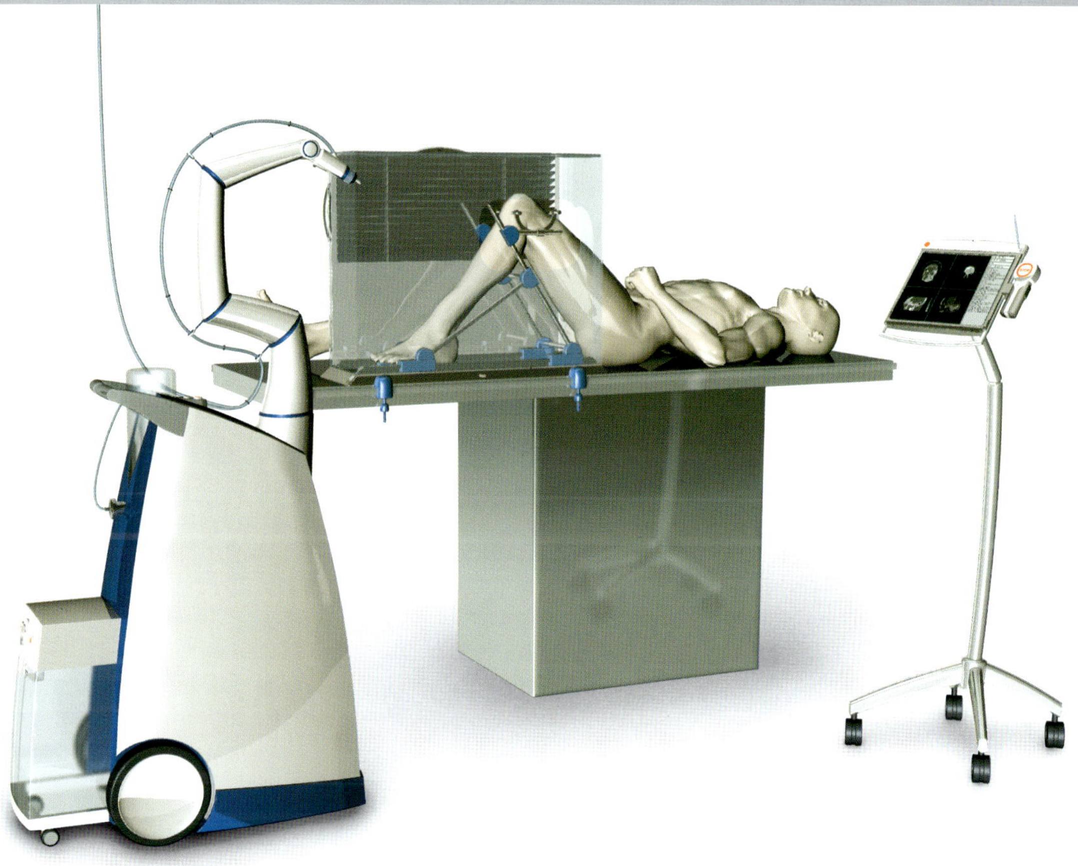

Jury . Judges panel

Ohne sich an ähnlichen oder vorangehenden Produkten orientieren zu können, haben Ralf Baumunk und Joachim Möllmann ein neuartiges Gerätesystem konzipiert und gestaltet. Eingehend haben sie die medizinischen Abläufe und Anwendungsbedingungen studiert. Entsprechend sinnvoll sind die Komponenten gegliedert. Deren einzelne Funktion, aber auch deren Zusammenwirken, ist genau bedacht. Bedienung, Reinigung, Sterilisation, Pflege und Wartung sind sorgsam berücksichtigt. Am Schluss der Studie kann stehen: Machbarkeit durch Gestaltung glaubhaft gemacht! Glaubwürdig auch, weil sie offen gebliebene Probleme nicht verschweigt.

.

Without being able to follow the lead set by predecessor or similar products, Ralf Baumunk and Joachim Möllmann have devised and designed an innovative system of appliances. They have studied medical processes and conditions of application in detail, with the result that the components are arranged in a logical way. Close thought has been given both to their individual functions and to the way they interact with other components. Operation, cleaning, sterilization, care and maintenance have been given careful consideration. At the end of the study, it can be said that design has been used to make feasibility credible. This credibility also comes from the fact that unsolved problems are not swept under the carpet.

MIA SEEGER PREIS 2003	2. PREIS
MIA SEEGER PRIZE 2003	2ND PRIZE

DESIGN . DESIGNER
Reno Büchner
Burg Giebichenstein Hochschule
für Kunst und Design Halle

Kurt-Eisner-Straße 7
D-06110 Halle/Saale
Telefon +49 . 3 45 . 8 06 28 20
reno_b@gmx.de

Hilfsmittel für Sehgeschädigte . Aids for the blind

Blinde haben das wichtigste Orientierungsorgan verloren. Der hier vorgestellte Katalog von technischen Hilfen kommt also nur recht. Darin finden sich unter anderem Konzepte für einen Blindenstock mit Scanreader für strichcodierte Wegmarkierungen, für einen zum Brailleschrift- und Reliefdrucker umgebauten Tintenstrahldrucker oder für einen Bodenbelag, der über Noppen oder Lamellen eine Richtung anzeigt. Zum einen sind es stationäre Informationssysteme, die haptisch gelesen werden, zum andern tragbare Geräte, die mittels Computer- bzw. Sensortechnologien visuelle Informationen in taktile oder akustische umsetzen. Herausragendes Beispiel dafür ist ein „Relief Touch Screen Navigator" (RTSN). Er rechnet satellitenübermittelte Navigationsdaten, Stadtplan- und allgemeine Bilddaten in ein Relief um, auf dem sich Blinde tastend orientieren können. Der Kartenausschnitt wird über die Randtasten, das Menü über „eingeblendete" taktile Icons und Brailleschrift gesteuert.

Blind people have lost their most important perceptual organ. This list of technical aids is very welcome, therefore. It includes the idea of a "white stick" with a scanner for bar-coded footpath markings, of a modified ink-jet printer that prints Braille and relief images, or of a floor covering whose pimpled or ribbed surface guides in the right direction. On the one hand, these are stationary information systems that are read by touch and, on the other, portable devices that use computer or sensor technology to translate visual information into tactile or acoustic information. One excellent example of this is a "Relief Touch Screen Navigator" (RTSN). It converts satellite-transmitted navigation data, town plan data and general image data into a relief which allows blind people to get their bearings. The map section is controlled via the side buttons, while the menu is controlled via "pop-up" tactile icons and Braille characters.

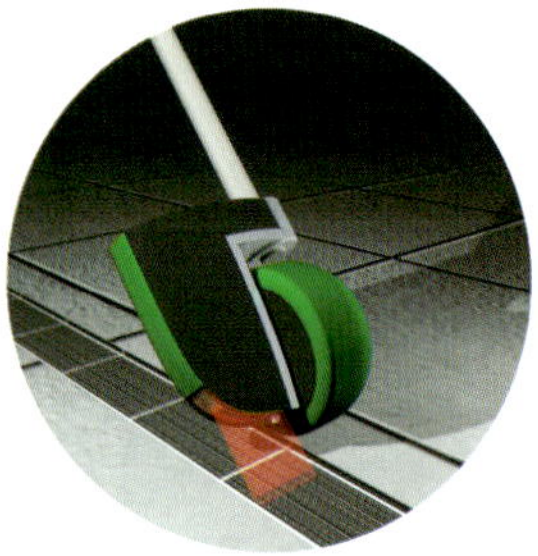

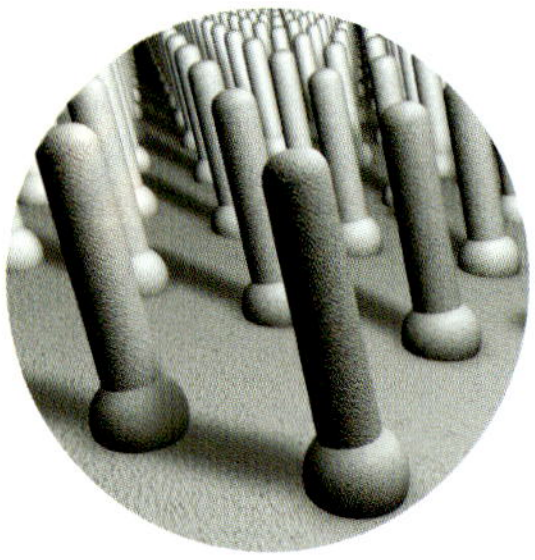

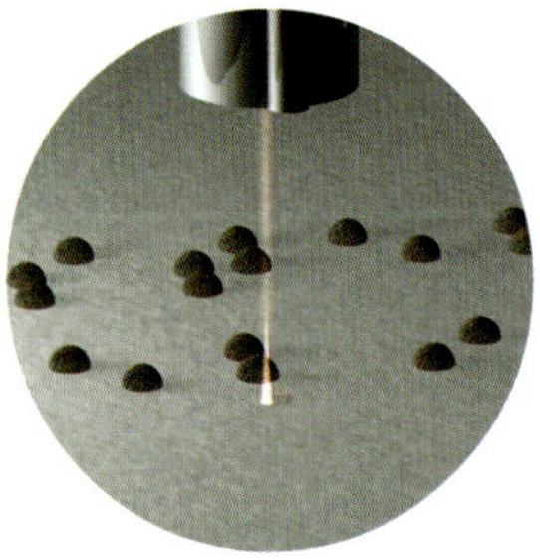

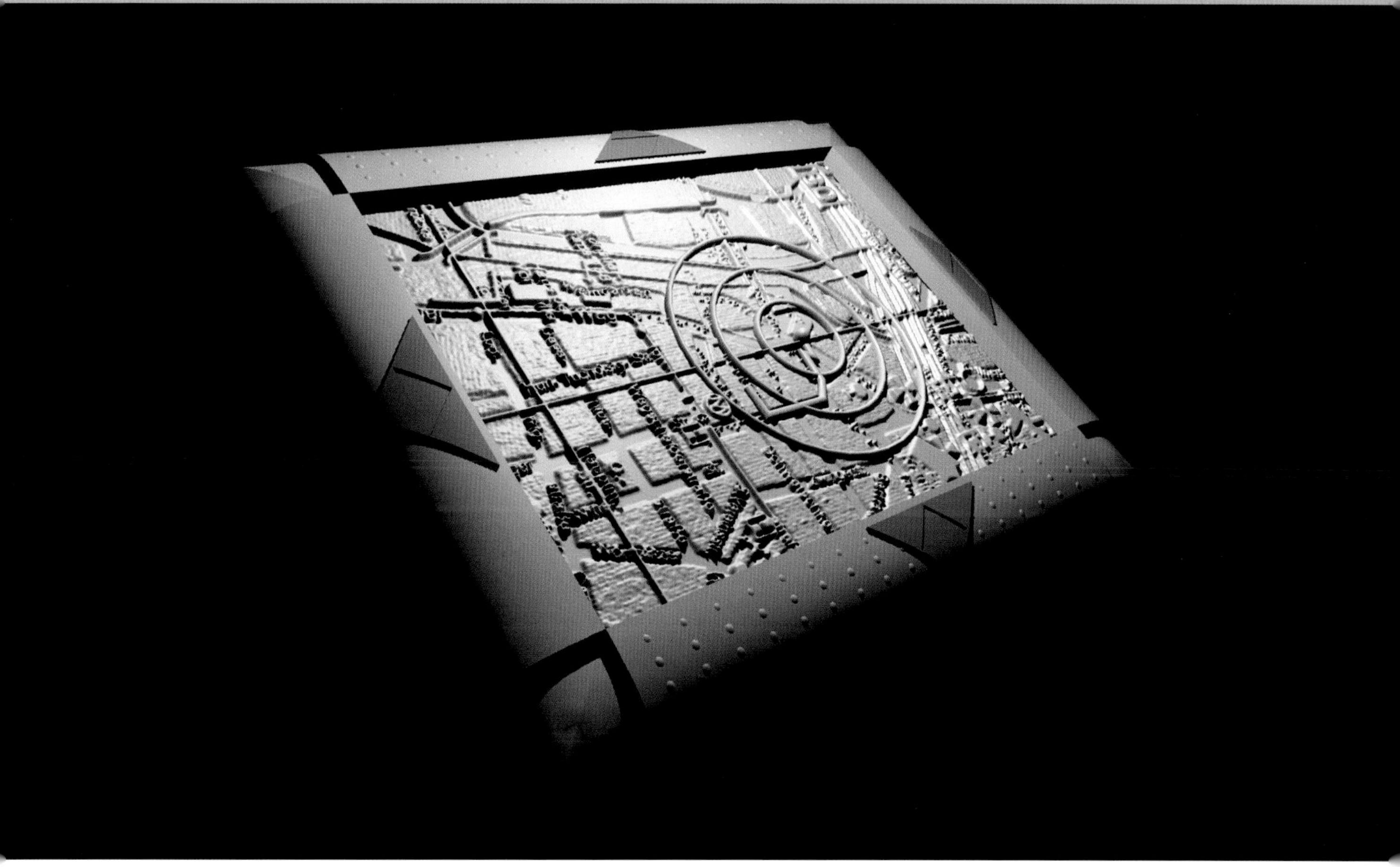

Jury . Judges panel

Reno Büchner breitet eine Fülle von Ideen aus. Nicht nur in den neuesten Kommunikationstechnologien, auch in neuen oder bekannten Materialien findet er Lösungen für das Transponieren visueller Informationen. Dabei bedenkt er immer, welche Reizbelegung den nicht-visuellen Sinnen zugemutet werden kann. Der Entwurf des RTSN ragt schon deswegen aus den übrigen heraus, weil er, freilich unter Annahme eines hochauflösenden Relief-Schirms, ein überraschend weites Feld von Anwendungen eröffnet. Die detailliert ausgearbeiteten Lösungen sind auch für den Normalsichtigen schön anzuschauen.

.

Reno Büchner has come up with a wealth of ideas. Not only in the latest communication technologies does he find ways of transposing visual information, but also in new or already known materials. He never forgets to consider how many stimuli the non-visual senses can be expected to tolerate. The design of the RTSN stands head and shoulders above the rest because it opens up a surprisingly wide field of applications – always assuming, of course, that a high-resolution relief touchscreen is possible. For those of us with normal sight, too, his detailed designs are a joy to behold.

MIA SEEGER PREIS 2003 | 3. PREIS

MIA SEEGER PRIZE 2003 | 3RD PRIZE

DESIGN . DESIGNER

Claudia Haßfurther
Muthesius-Hochschule Kiel

Michelsenstraße 25
D-24114 Kiel
Telefon +49 . 4 31 . 6 61 27 83
stern@muthesius.de

Bikester : unterwegs mit Kindern . on the move with children

Eigentlich gibt es kein alltagstaugliches Vehikel, mit dem auch behinderte Kinder per Fahrrad mitgenommen werden können. Das zeigt eine ausführliche Analyse. Dem abzuhelfen, hat Claudia Haßfurther eine Kabine entworfen, die einachsig und mit einem Fahrradrahmen verbunden zum Lastrad oder zweiachsig zum Handwagen wird. Der Dreh dabei: ein einfacher Verschiebemechanismus erlaubt die Umstellung von einer auf zwei Achsen und umgekehrt. Weitere wichtige Merkmale sind versenkbares Verdeck, vielfach verstellbarer Reha-Sitz, zwei Staufächer, robuster Profilrahmen. Das Gefährt ist so stabil und reparaturfreundlich ausgelegt, dass die Krankenkasse es als Alltagshilfe anerkennen kann. Natürlich können auch gesunde Kinder mitfahren.

.

There is in fact no practical everyday vehicle which allows handicapped children to be transported by bicycle. This is shown by a detailed analysis. To remedy this situation, Claudia Hassfurther has designed a cabin which, on a single axle and hooked up to a bicycle frame, becomes a bicycle trailer or, on a double axle, becomes a handcart. What is unusual is that a simple sliding mechanism allows it to be converted from one to two axles and back again. Other outstanding features include a retractable hood, an adjustable seat suitable for disabled occupants, two storage areas and a sturdy sectional frame. The vehicle's design makes it sturdy and easy to repair; so much so that it can be recognized by health insurance companies as an everyday aid. Of course, healthy children can also ride in it.

Jury . Judges panel

Die Arbeit beginnt mit einer ungewöhnlich sorgfältigen, umfassenden und tiefgehenden Voruntersuchung zum Transport von Kindern mit dem Fahrrad. Die Ergebnisse münden in den Entwurf eines Fahrzeugs, das die analysierten Anforderungen bis hinein in viele praktische, technisch-funktionale Details umsetzt. Vielfältige Nutzung, Sicherheit und bestmögliche Sitz-Ergonomie sind besonders bedacht. Ein derartiges Gerät kann wahrlich zu einem „normalen" Leben mit (behinderten) Kindern beitragen.

.

The dissertation begins with an unusually meticulous, comprehensive and in-depth preliminary study on transporting children by bicycle. Its results lead to the design of a vehicle which puts into practice the requirements it has analyzed, right down to many practical, technically functional details. Versatile use, safety and the best possible seating posture have been given special consideration. A vehicle such as this can indeed contribute to a "normal" life with (handicapped) children.

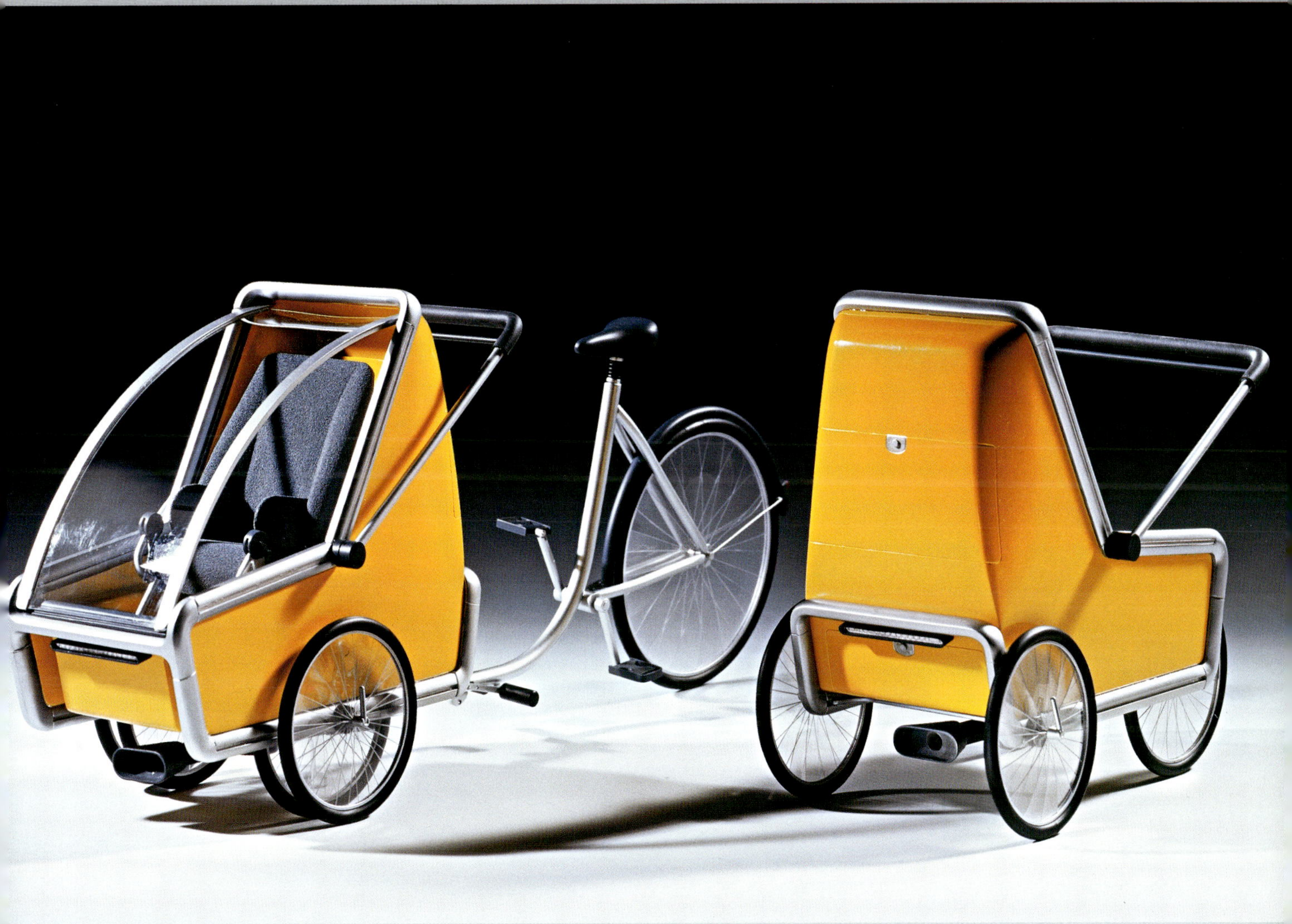

MIA SEEGER PREIS 2003 | ANERKENNUNG

MIA SEEGER PRIZE 2003 | HIGHLY COMMENDED

DESIGN . DESIGNER
Ansgar Howey
Technische Universität Dresden

Bäckergasse 9
D-86150 Augsburg
Telefon +49 . 8 21 . 2 09 73 59
ansgar@howey.de

Grünfutter-Vollerntemaschine . Harvester for soilage

Erst ganzjährig verfütterte Silage (d.i. konserviertes Gras) garantiert gleich bleibend hohe Milchqualität. Eine Grünfutter-Vollerntemaschine könnte wetterunabhängig hochwertige Silage in einem einzigen Arbeitsgang produzieren: Gras schneiden, reinigen, per Mikrowelle trocknen, zu Ballen pressen, wickeln und verpacken. Für dieses patentierte Prinzip liegt jetzt ein technisches Konzept mit einer Designstudie vor. Bewusst ist die Anmutung eines gefräßigen, sonst aber gutmütigen Insekts gewählt, dessen Leib die erforderlichen technischen Komponenten und Aggregate umschließt.

.

Consistently high milk quality can only be guaranteed if silage (i.e. partly fermented grass) is available as feed all year. A harvester for soilage could produce high-quality silage in a single pass, independently of the weather: cutting grass, cleaning, drying by microwave, pressing into bales, binding and packing. A technical concept and a design study have now been presented for this patented principle. It consciously alludes to a voracious but otherwise benign insect, whose abdomen houses the necessary technical components and units.

Jury . Judges panel

Eine Grünfutter-Vollerntemaschine käme einer landwirtschaftlichen Revolution gleich, wie sie vor Jahrzehnten der Mähdrescher mit sich brachte. Das setzt der Entwurf auch in einer ausdrucksstarken Gestalt um. Die Studie zeichnet in enger Verknüpfung von Technik und Design eine Richtung vor, in der die Entwicklung bis zur Marktreife vorangetrieben werden könnte.

.

A harvester for soilage would be the equivalent of an agricultural revolution, like that brought about decades ago by the combine harvester. The design lends impressive form to the idea. Closely linking technology and design, the study shows a possible way forward for making this development ready for the market.

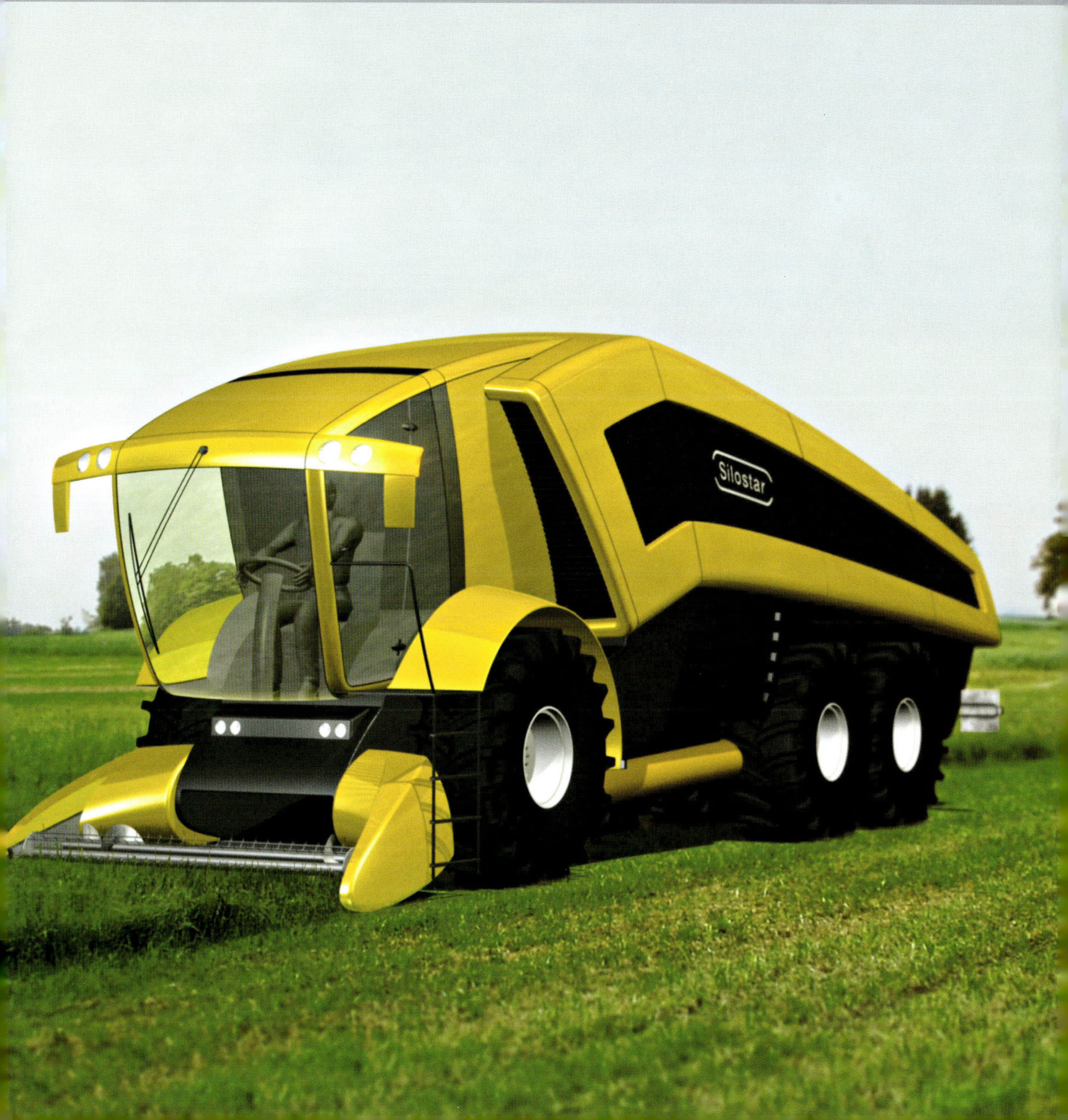
Silostar

MIA SEEGER PREIS 2003 | ANERKENNUNG

MIA SEEGER PRIZE 2003 | HIGHLY COMMENDED

DESIGN . DESIGNER
Cathrin Königsmann
Muthesius-Hochschule Kiel

Hardenbergstraße 23
D-24118 Kiel
Telefon +49 . 4 31 . 67 33 95
cathrin-k@gmx.de

Kiss : ein Kindersitzsystem . a child seat system

Ein Kindersitz, der als Kinderwagen-Aufsatz und im Auto Dienst tun und dabei dem Kind vom 9. Monat bis zum 4. Lebensjahr ergonomisch richtig Platz bieten soll. Das leistet eine Art gelenkig verbundener Profilrahmen, an dem Sitz- und Rückenpolster, Kopf-, Arm- und Fußstützen verstellbar angebracht sind. So ist der Sitz genau an die Körpergröße anpassbar, und verschlissene Teile können ausgetauscht werden.

.

A child seat designed as a pushchair attachment or for use in the car, intended to give ergonomically correct support for children aged between nine months and four years. This is achieved by a kind of articulated sectional frame, to which the adjustable seat and back cushion and head, arm and foot rests are fixed. This allows the seat to be adjusted accurately to body size, and worn parts can be replaced.

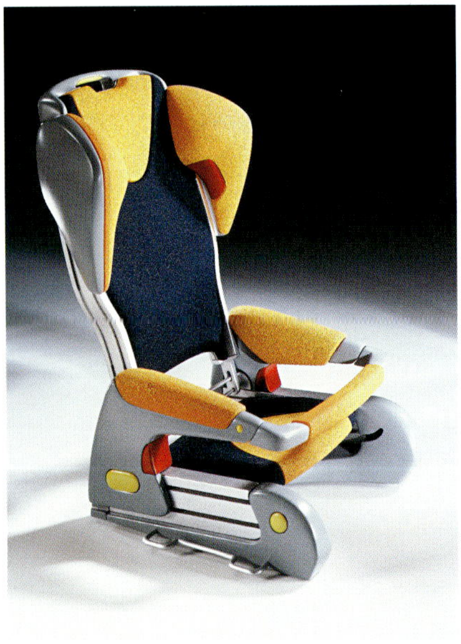

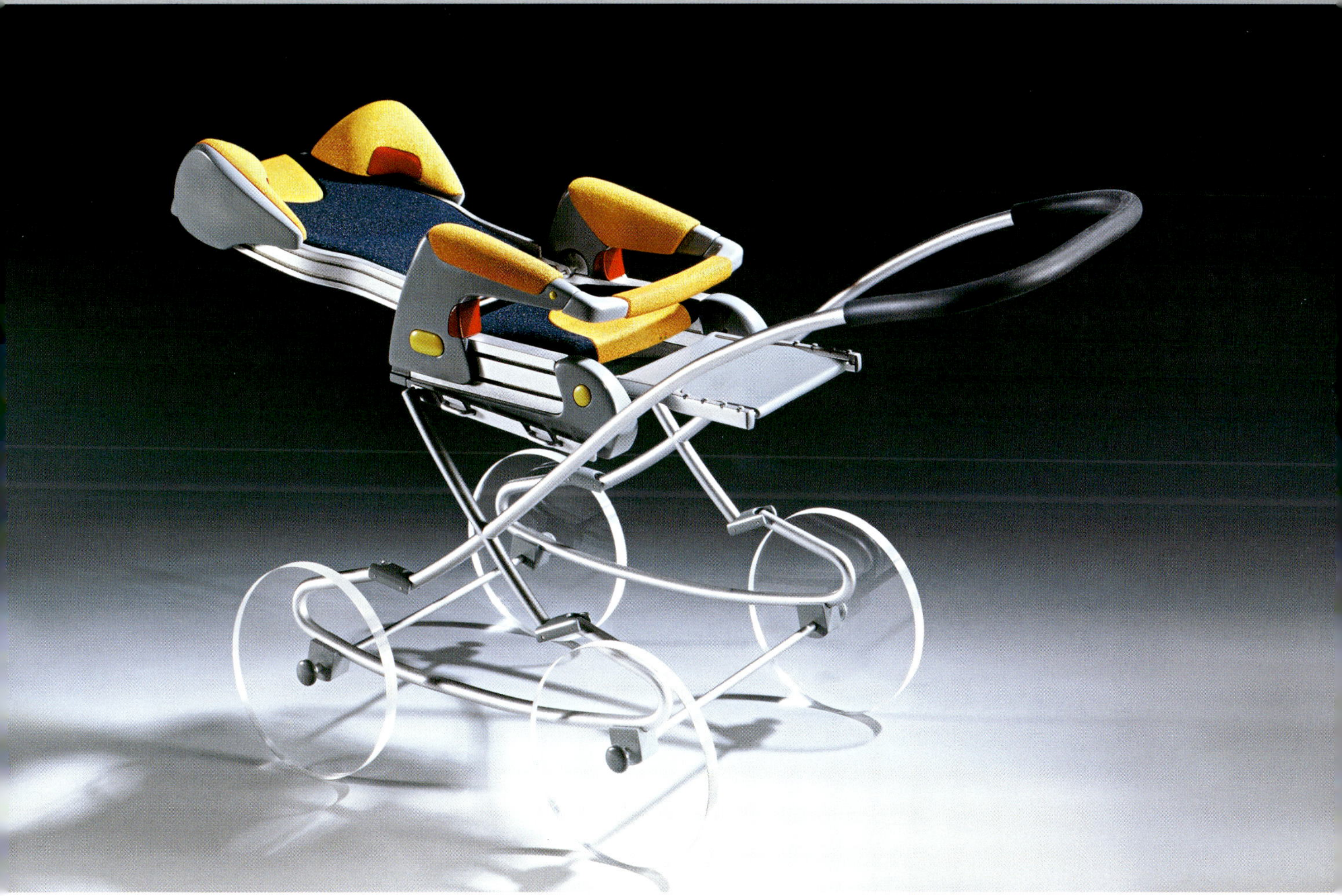

Jury . Judges panel

Autositz und Sportwagenaufsatz sind ohne Kompromisse in einem einzigen Produkt vereinbar. Persönliche Erfahrung und die Ergebnisse einer Elternbefragung fließen in den Entwurf ein. Konsequent ist die Dualität zwischen technisch-funktionalem Tragwerk und körpernahen Anbauelementen in Materialwahl, Formensprache und Farbgebung fortgeführt. Langlebigkeit und Doppelnutzen schonen den Geldbeutel der Eltern.

.

Car seat and pushchair attachment, combined without compromise in one single product. The design has been influenced by personal experience and findings from a poll of parents. The dichotomy between the technically functional supporting framework and the add-on elements with which the body has contact has found consistent expression in the choice of materials, formal expression and colour scheme. Longevity and dual use mean that the product puts less strain on family budgets.

MIA SEEGER PREIS 2003	ANERKENNUNG
MIA SEEGER PRIZE 2003	HIGHLY COMMENDED

DESIGN . DESIGNER
Andreas Lehnhoff
Hochschule für Gestaltung FH
Schwäbisch Gmünd

Stettinerstraße 29
D-61348 Bad Homburg v.d.H.
Mobil +49.179.135 42 79
lehnhoff-a@hfg-gmuend.de

Pressure Suit

Wohldosierte Druckluft kann helfen, den belasteten Rücken zu unterstützen. Ein flacher, aus röhrenförmigen Luftkammern gebildeter pneumatischer Muskel, mit Gurten an Schulter und Beinen befestigt, zieht sich dann zusammen. Gespeist wird das System aus einem Druckspeicher. Ein im Rücken eingearbeiteter Muskelsensor steuert das Ventil und löst bei hoher Belastung Vibrationsalarm aus.

Timely doses of compressed air can help support a back under strain. A flat, pneumatic muscle, made of tubular air chambers, and fixed by belts to the shoulders and legs, contracts when compressed air is injected. The system is fed from an accumulator. A muscle sensor integrated in the back section controls the valve and triggers a vibration alarm when there is excessive strain.

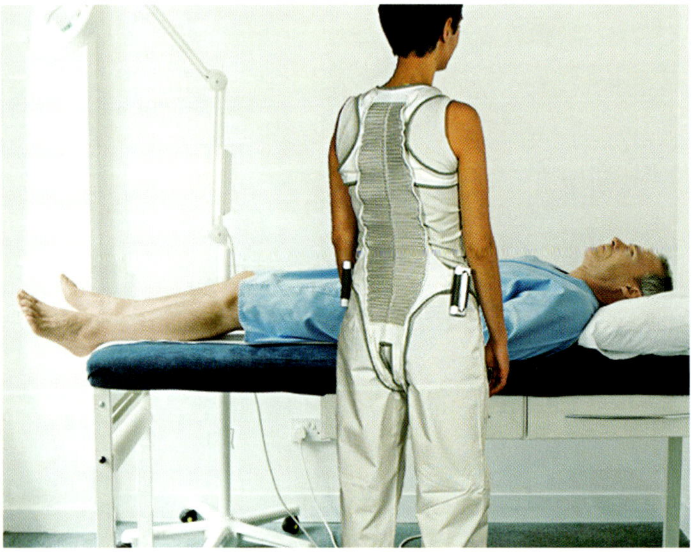

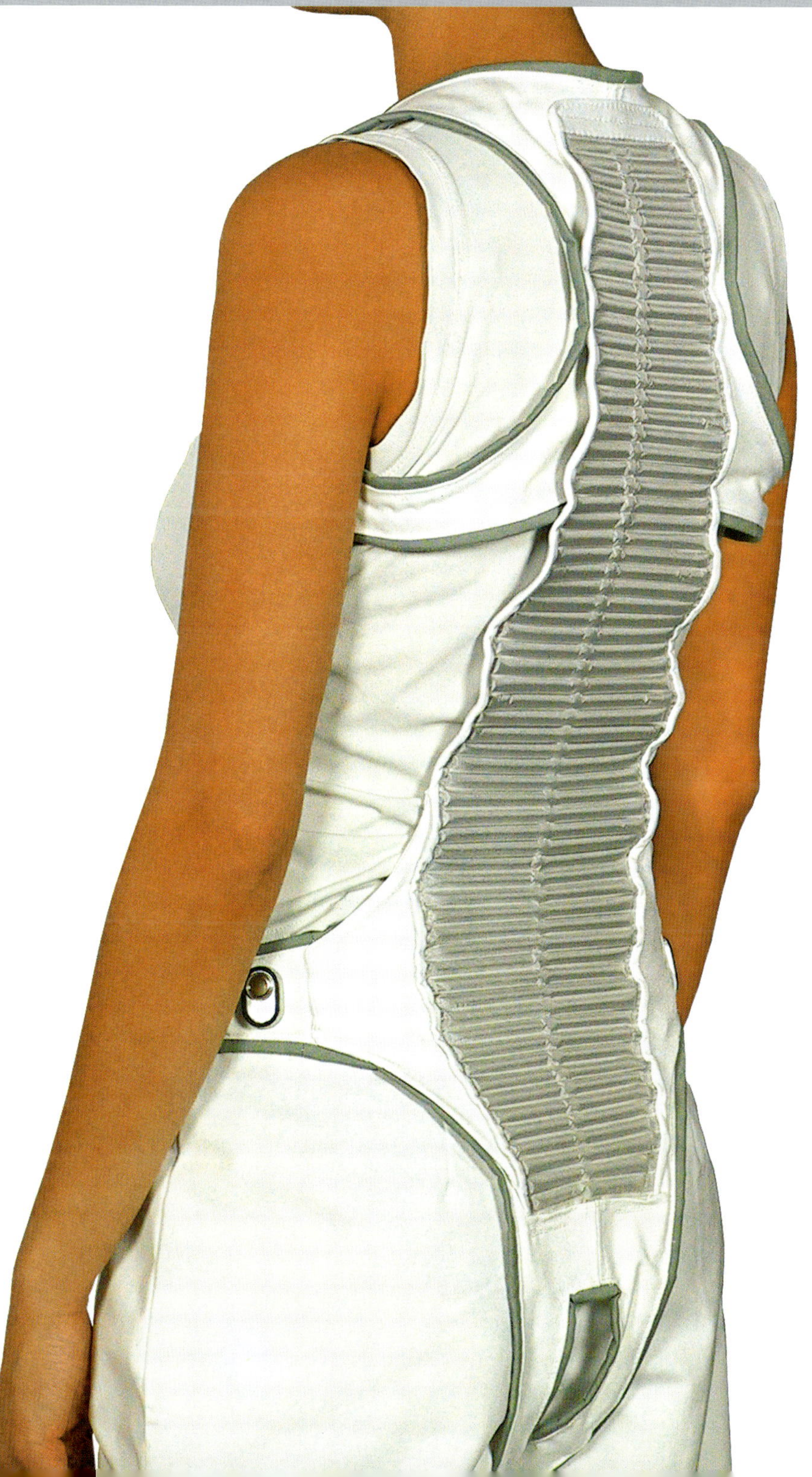

Jury . Judges panel

Das Konzept der „Smart Clothes“ findet hier in der Verbindung von Pneumatik und Sensortechnik eine sehr nützliche Anwendung. „Pressure Suit“ ist auch von der Anmutung her eine hochwertige und professionelle Ergänzung zur Arbeitskleidung, die es sinnvollerweise in zwei verschiedenen Versionen für Frauen und Männer geben soll.

.

This combination of pneumatics and sensor technology is a very useful application of the “smart clothes” idea. The impression the “pressure suit” gives is also one of a high-quality, professional addition to working clothing. Sensibly, it is designed to be available in two different versions, for men and women.

MIA SEEGER PREIS 2003	ANERKENNUNG
MIA SEEGER PRIZE 2003	HIGHLY COMMENDED

DESIGN . DESIGNER
Katrin Saier
Fachhochschule Darmstadt

Hochstraße 68
D-64285 Darmstadt
Telefon +49.61 51.49 45 59
Mobil +49.1 77.7 46 91 01
ks@id-1.de

Klein und mobil . Small and mobile

Kindersportwagen, fast wörtlich genommen. Gedacht ist er für sportive, mobilitätsfreudige Eltern, die ihr Kind oft im Auto und gern zum Joggen oder Inlineskaten mitnehmen. Eigentümlich für seine Struktur ist die zentrale Achse, die den Sitzrahmen und die Radstreben trägt und zugleich Hauptklappachse ist. Der Radstand ist variabel, kurz für die Stadt, lang für den Ausflug. Höhenverstellbar ist auch der Halte- bzw. Schiebebügel; in der Mitte die Bremse, die auf die Hinterräder wirkt.

A nippy little pushchair. It has been devised for sporty parents who love to be on the move, who often take their child with them in the car or like to take their child with them when they go jogging or inline skating. What is unusual about its structure is the central axle which supports the seating frame and wheel suspension, yet is also a collapsible axle. The wheelbase can be varied – short for the town, long for excursions. The height of the handlebar or push bar can also be adjusted; the rear-wheel brake is mounted in the middle.

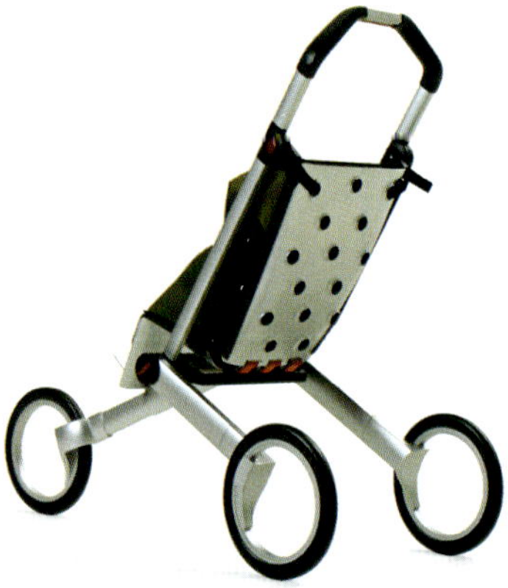

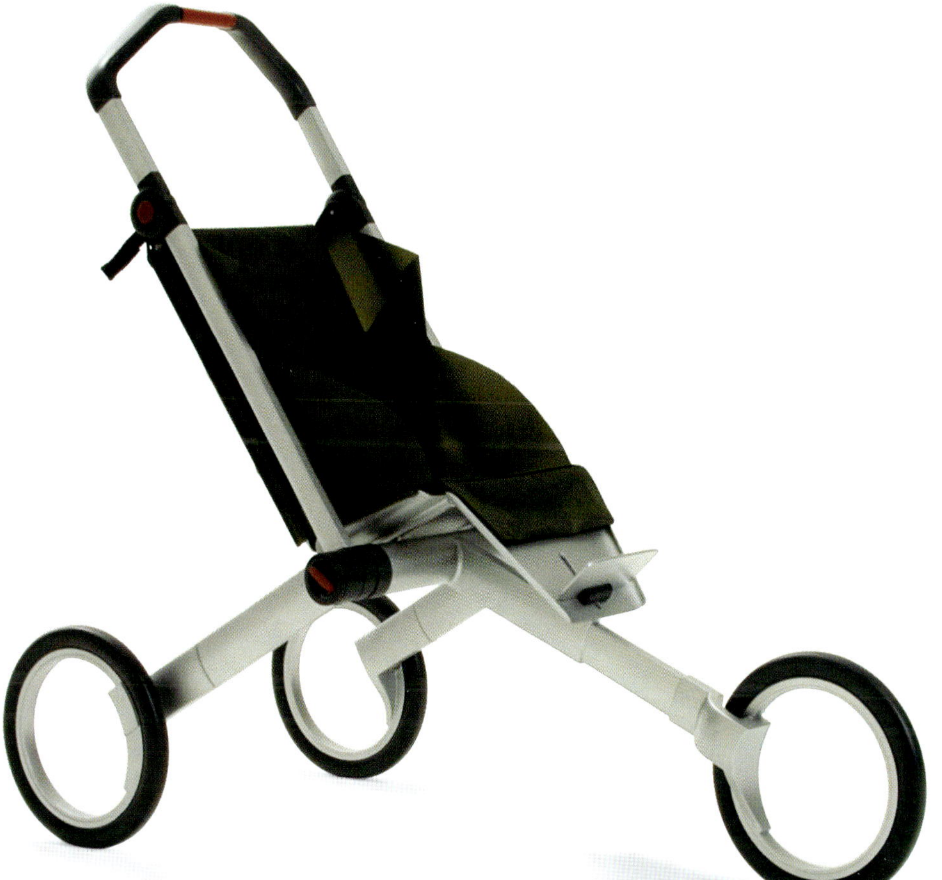

Jury . Judges panel

Dass es „mobil" ist, sieht man dem Fahrzeug unmittelbar an: reduzierte Struktur, glattflächiges Fahrwerk, integrierte Verstellelemente, nabenlose Räder. Da es sich beim Zusammenfalten, dank der abklappbaren Räder, ungewöhnlich „klein" machen kann, steigert das die Mobilität abermals.

.

The vehicle's "mobile" quality is immediately apparent: its reduced structure, the smooth surface of its chassis, the integrated adjusting elements and hubless wheels. The fact that it can make itself unusually small when it is folded up, thanks to the fold-down wheels, is a further boost to mobility.

MIA SEEGER STIFTUNG

MIA SEEGER FOUNDATION

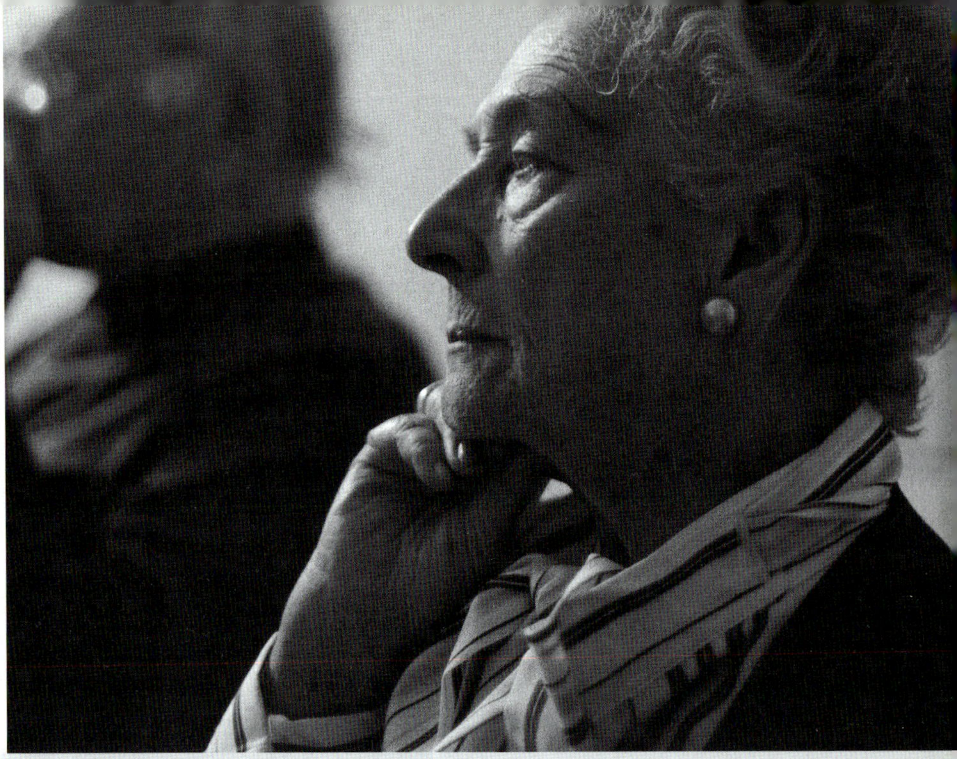

MIA SEEGER (1903 – 1991)

Ausführliche Informationen zu Stiftung, Preis oder Stipendium im Internet www.mia-seeger.de oder bei der Geschäftsstelle.

.

For more details about the Foundation, the Prize or the Scholarship, visit www.mia-seeger.de or contact the Foundation office.

Mia Seeger Stiftung
c/o Design Center Stuttgart
Willi-Bleicher-Straße 19
D-70174 Stuttgart
Telefon +49.711.1 23 26 84
oder +49.711.1 23 26 38
Fax +49.711.1 23 25 77
eMail wolfgang.berger@lgabw.de

Mia Seeger war die „Grande Dame" des Design. Vor dem Zweiten Weltkrieg wirkte sie an großen Ausstellungsprojekten des Deutschen Werkbundes mit, z.B. 1927 an der Weißenhofsiedlung in Stuttgart. Nach dem Krieg führte sie als Präsidialmitglied zwölf Jahre lang den Rat für Formgebung in Darmstadt. 1986 rief sie die nach ihr benannte Stiftung ins Leben, deren Zweck die umfassende Förderung des Design als Teil von Wirtschaft und Kultur ist. Namhafte Sponsoren aus der Wirtschaft haben sich ihren Zielen angeschlossen. Vorsitzender des Stiftungsvorstands ist Dr. Friedrich Bullinger, Präsident des Landesgewerbeamts Baden-Württemberg, Vorsitzende des Beirats Prof. Karin Kirsch, Stuttgart.
Mit der Absicht, besonders den Nachwuchs im Design zu fördern und ihn dabei zur Auseinandersetzung mit sozialen Fragen herauszufordern, schreibt die Stiftung jährlich den Mia Seeger Preis bundesweit aus, zusätzlich den Mia Seeger Förderpreis zur Unterstützung von Projekten.

.

Mia Seeger was the "grand old lady" of design. Before the Second World War, she was involved in major exhibition projects of the Deutscher Werkbund, such as the Weissenhof complex in Stuttgart in 1927. After the war, she sat for twelve years on the managing committee of the Rat für Formgebung in Darmstadt. In 1986, she established the foundation that bears her name, whose purpose is the comprehensive promotion of design as a part of economic and cultural life. Prominent sponsors from the business world have lent their support to her aims.
The chair of the foundation's management board is Dr. Friedrich Bullinger, President of the Baden-Württemberg Office of Trade and Industry, while Prof. Karin Kirsch, Stuttgart, chairs its advisory board.
With the aim of promoting young designers in particular, and of providing them with the additional challenge of coming to terms with social problems, the Foundation invites entrants from all over Germany to compete for the annual Mia Seeger Award. There is also a Mia Seeger Scholarship to support projects.

IMPRESSUM . PUBLISHING DETAILS
MIA SEEGER PREIS 2003
Herausgeber . Published by _ Mia Seeger Stiftung
Redaktion . Editor _ Wolfgang Berger
Grafikdesign . Graphic design _ Silke Nalbach
Ausstellungsgestaltung . Exhibition design _ Angelika Ertl
Plakat . Poster _ Uwe Lohrer
Fotos . Photographs _ Preisträger und Ausgezeichnete . prize-winners and commended entrants
Stuttgart 2003

ANHANG

AEG HAUSGERÄTE GMBH
Muggenhofer Straße 135
D-90429 Nürnberg
Telefon +49 . 9 11 . 32 30
Fax +49 . 9 11 . 3 23 17 70
info@aeg-hausgeraete.de
www.aeg-hausgeraete.de

AEG Industrial Design Center
Heinz-Peter Huxhold
Seite . Page 66

AUDI AG
D-85045 Ingolstadt
Telefon +49 . 8 41 . 8 90
Fax +49 . 8 41 . 8 94 04 94
www.audi.de

Audi Design Team
Seite . Page 60

BAMED AG FT.
H-9795 Vaskeresztes
Telefon +36 . 9 . 4 35 10 07
Fax +36 . 9 . 4 35 10 14
Seite . Page 38

RALF **BAUMUNK**
Erderstraße 33
D-30451 Hannover
Mobil +49 . 178 . 5 64 40 90
master@raballo.de
Seite . Page 194

BELUGA
TAUCHSPORT GMBH
Bremer Straße 13
D-27383 Scheeßel
Telefon +49 . 42 63 . 20 95
Fax +49 . 42 63 . 34 85
beluga-tauchsport@t-online.de
www. beluga-medicus.de
Seite . Page 32

SIEGFRIED **BENSINGER**
Meyersche Weg 37
D-21244 Buchholz i.d.N.
Telefon +49 . 41 87 . 64 12
Fax +49 . 41 87 . 66 97
Seite . Page 102, 106

MARKUS **BOGE**
Ederstraße 33
D-30451 Hannover
Telefon +49 . 5 11 . 4 50 14 55
mail@markusboge.de
www.markusboge.de
Seite . Page 96

ROBERT **BOSCH**
HAUSGERÄTE GMBH
Hochstraße 17
D-81669 München
Telefon +49 . 89 . 45 90 01
Fax +49 . 89 . 45 90 23 47
Seite . Page 70

ROBERT **BOSCH**
HAUSGERÄTE GMBH
Hans-Reinhart Janssen
Robert-Bosch-Straße
D-89537 Giengen
Telefon +49 . 73 22 . 92 25 52
Fax +49 . 73 22 . 92 25 12
hans-reinhart.janssen@bshg.com
Seite . Page 70

BREDEL.DE SIGN
Thomas Bredel
Böblinger Straße 32
D-70178 Stuttgart
Telefon +49 . 7 11 . 6 07 06 51
Fax +49 . 7 11 . 6 07 06 56
info@bredel.de
www.bredel.de
Seite . Page 104

BREE COLLECTION
GMBH & CO. KG
Gerberstraße 3
D-30916 Isernhagen
Telefon +49 . 51 36 . 8 97 60
Fax +49 . 51 36 . 8 97 62 29
bree.collection@bree.de
www.bree.com

Bree Designteam
Seite . Page 52

BRITAX RÖMER
KINDERSICHERHEIT GMBH
Blaubeurer Straße 71
D-89077 Ulm
Telefon +49 . 7 31 . 9 34 50
Fax +49 . 7 31 . 9 34 52 10
info@britax-roemer.de
www.britax-roemer.de
Seite . Page 46

BRÜHL & SIPPOLD GMBH
Alter Bad Stebener Weg 1
D-95138 Bad Steben
Telefon +49 . 92 88 . 95 50
Fax +49 . 92 88 . 9 55 99
info@bruehl.com
www.bruehl.com
Seite . Page 102, 106

RENO **BÜCHNER**
Kurt-Eisner-Straße 7
D-06110 Halle/Saale
Telefon +49 . 3 45 . 8 06 28 20
reno_b@gmx.de
Seite . Page 196

BUSSE DESIGN ULM GMBH
Nersinger Straße 18
D-89275 Nersingen
Telefon +49 . 73 08 . 81 80
Fax +49 . 73 08 . 8 18 65
bdg@busse-design-ulm.de
www.busse-design-ulm.de
Seite . Page 46

CARPET CONCEPT OBJEKT TEPPICHBODEN GMBH
Bunzlauer Straße 7
D-33719 Bielefeld
Telefon +49 . 5 21 . 92 45 90
Fax +49 . 5 21 . 9 24 59 20
info@carpet-concept.de
www.carpet-concept.de
Seite . Page 94

CODE.2.DESIGN
Michael Schmidt
Ludwigstraße 59
D-70176 Stuttgart
Telefon +49 . 7 11 . 50 53 13 50
Fax +49 . 7 11 . 50 53 13 55
ms@code2design.de
www.code2design.de
Seite . Page 118

CONDUX GMBH
Stammheimer Straße 22
D-70435 Stuttgart
Telefon +49 . 7 11 . 23 07 29 20
Fax +49 . 7 11 . 23 07 29 76
www.condux-gmbh.com
Seite . Page 34

D+S SANITÄRPRODUKTE GMBH
Industriestraße
D-69198 Schriesheim
Telefon +49 . 62 03 . 10 20
Fax +49 . 62 03 . 6 56 54
Seite . Page 118

JEAN-MARC **DA COSTA**
c/o Serien Raumleuchten GmbH
Hainhäuser Straße 3–7
D-63110 Rodgau
Telefon +49 . 61 06 . 6 90 90
Fax +49 . 61 06 . 69 09 22
serien@serien.com
www.serien.com
Seite . Page 156, 158

DESIGNPRAXIS DIENER
Irisweg 3
D-89079 Ulm
Telefon +49 . 73 05 . 9 64 80
Fax +49 . 73 05 . 96 48 30
design.diener@t-online.de
Seite . Page 76

DIMA FREIZEITMÖBEL GMBH
Dieselstraße 6
D-73278 Schlierbach
Telefon +49 . 70 21 . 7 27 60
Fax +49 . 70 21 . 72 76 40
info@fischer-moebel.de
www.fischer-moebel.de
Seite . Page 54

DUSCHOLUX GMBH
Industriestraße
D-69198 Schriesheim
Telefon +49 . 62 03 . 60 66 40
Fax +49 . 62 03 . 6 56 54
info@duscholux.de
www.duscholux.de
Seite . Page 118

MATTHIAS **EBERT**
Bebelstraße 18 A
D-70176 Stuttgart
Telefon +49 . 7 11 . 6 01 24 47
Fax +49 . 7 11 . 6 01 24 48
mebertdesign@t-online.de
www.mebertdesign.de
Seite . Page 30

EBM
Max-Emanuel-Straße 2
D-94036 Passau
Telefon +49 . 8 51 . 50 47 90
Fax +49 . 8 51 . 5 04 79 33
technik@ebmdesign.de
www.ebmdesign.de
Seite . Page 168

EICHER SIEBDRUCK
Willi-Rüsch-Straße 19
D-71394 Kernen
Telefon +49 . 71 51 . 4 66 60
Fax +49 . 71 51 . 4 74 10
Seite . Page 144

HUBERT **EILERS**
Altes Gutshaus
D-14974 Gröben
Telefon +49 . 33 78 . 87 41 10
Fax +49 . 33 78 . 87 41 19
eilers.architekten@t-online.de
Seite . Page 132

THOMAS **ELLERSTORFER**
Siechen 8
D-93413 Cham
Telefon +49 . 99 71 . 80 10 63
Fax +49 . 99 71 . 80 10 64
duo-design.ellerstorfer@t-online.de
www.duo-design.de
Seite . Page 178

EOOS DESIGN GMBH
Theresiengasse 11–15
A-1180 Wien
Telefon +43 . 1 . 4 05 39 87
Fax +43 . 1 . 4 05 39 87 80
design@eoos.com
Seite . Page 108, 112

FACTOR PRODUCT GMBH
Comeniusstraße 1 Rgb.
D-81667 München
Telefon +49 . 89 . 48 92 78 10
Fax +49 . 89 . 48 92 78 11
info@factor-product.com
www.factor-product.com
Seite . Page 56

FESTOOL GMBH
Wertstraße 20
D-73240 Wendlingen
Telefon +49 . 70 24 . 80 40
Fax +49 . 70 24 . 80 46 00
info@tts-festool.com
www.festool.com
Seite . Page 182

FISCHER GMBH
Fischerstraße 8
A-4910 Ried/Innkreis
Telefon +43 . 77 52 . 90 90
Fax +43 . 77 52 . 90 92 79
www.fischer-ski.com
Seite . Page 56

FISCHER MÖBEL GMBH
Dieselstraße 6
D-73278 Schlierbach
Telefon +49 . 70 21 . 7 27 60
Fax +49 . 70 21 . 72 76 40
info@fischer-moebel.de
www.fischer-moebel.de
Seite . Page 54

FISCHER UND LÖFFLER DEUTSCHLAND GMBH
Hirschberger Ring 10
D-94315 Straubing
Telefon +49 . 94 21 . 7 89 93 30
Fax +49 . 94 21 . 7 89 93 31
Seite . Page 56

PATRICK **FREY**
Leinaustraße 6
D-30451 Hannover
Telefon +49 . 5 11 . 2 10 94 17
mail@patrick-frey.de
www.patrick-frey.de
Seite . Page 96

TOBIAS **GRAU** GMBH
Siemensstraße 35 b
D-25462 Rellingen
Telefon +49 . 41 01 . 37 00
Fax +49 . 41 01 . 3 70 10 00
info@tobias-grau.com
www.tobias-grau.com

Werksdesign Inhouse design
Tobias Grau
Seite . Page 162, 164

GREINER GMBH
Riedbachstraße 5
D-74385 Pleidelsheim
Telefon +49 . 71 44 . 8 11 20
Fax +49 . 71 44 . 81 12 99
info@greiner-gmbh.de
www.greiner-gmbh.de
Seite . Page 30

CLAUDIA HASSFURTHER
Michelsenstraße 25
D-24114 Kiel
Telefon +49 . 4 31 . 6 61 27 83
stern@muthesius.de
Seite . Page 198

HELIT
Innovative Büroprodukte GmbH
Osemundstraße 23–25
D-58566 Kierspe
Telefon +49 . 23 59 . 90 50
Fax +49 . 23 59 . 90 52 20
info@helit.de
www.helit.de
Seite . Page 130

ROBERT HERDER GMBH & CO. KG WINDMÜHLENMESSER
Ellerstraße 16
D-42697 Solingen
Telefon +49 . 2 12 . 26 70 50
Fax +49 . 2 12 . 7 53 27
info@windmuehlenmesser.de
www.windmuehlenmesser.de

Werksdesign Inhouse design
Giselheid Herder-Scholz
Seite . Page 78

HESS FORM + LICHT GMBH
Schlachthausstraße 19–19/3
D-78050 Villingen-Schwenningen
Telefon +49 . 77 21 . 92 00
Fax +49 . 77 21 . 92 02 50
hess@hess-form-licht.de
www.hess-form-licht.de
Seite . Page 160

HOLLISTER'S MOTORCYCLES
Inh. Volker Sichler
Zimmerner Straße 48
D-78658 Zimmern-Horgen
Telefon +49 . 7 41 . 3 22 68
Fax +49 . 7 41 . 3 41 44
volker.sichler@hollisters.de

Werksdesign Inhouse design
Volker Sichler
Seite . Page 58

KATJA HORST
Amalienstraße 15 Rgb.
D-80333 München
Telefon +49 . 89 . 28 75 51 31
Fax +49 . 89 . 28 75 53 77
mail@katjahorst.de
Seite . Page 50

ANSGAR HOWEY
Bäckergasse 9
D-86150 Augsburg
Telefon +49 . 8 21 . 2 09 73 59
ansgar@howey.de
Seite . Page 200

HUMANSCALE
11 East 26th Street, 8th Floor
New York, NY 10010, USA
Telefon +1 . 2 12 . 7 25 47 49
Fax +1 . 2 12 . 7 25 75 45
info@humanscale.com
www.humanscale.com

Werksdesign Inhouse design
Manuel Saez
Seite . Page 148

ALEXANDER HUNGER
Art & Wohndesign
Brucknerstraße 9
D-88250 Weingarten
Telefon +49 . 7 51 . 5 69 95 16
Fax +49 . 7 51 . 5 69 95 17
hungersart@aol.com
www.alexforma.com

Werksdesign Inhouse design
Alexander Hunger
Seite . Page 86

I.D. POOL
Singerstraße 6
A-1010 Wien
Telefon +43 . 1 . 5 12 58 41
id.pooldesign@chello.at
Seite . Page 38

INTERLÜBKE
Gebr. Lübke GmbH & Co. KG
Ringstraße 145
D-33378 Rheda-Wiedenbrück
Telefon +49 . 52 42 . 1 22 32
Fax +49 . 52 42 . 1 23 11
info@interluebke.de
www.interluebke.de
Seite . Page 100

KAAD
Karl-Axel Andersson Design
Hyregatan 7
S-21121 Malmö
Schweden
Telefon +46 . 40 . 97 15 60
Fax +46 . 40 . 23 14 34

Werksdesign Inhouse design
Karl-Axel Andersson
Seite . Page 142

KAHLA/THÜRINGEN PORZELLAN GMBH
Christian-Eckardt-Straße 38
D-07768 Kahla

Hersteller Manufacturer
Telefon +49 . 3 64 24 . 7 92 80
Fax +49 . 3 64 24 . 7 92 11

Vertrieb Distributor
Telefon +49 . 3 64 24 . 7 91 17
Fax +49 . 3 64 24 . 7 92 05
info@kahlaporzellan.com
www.kahlaporzellan.com

Werksdesign Inhouse design
Cornelia Müller
Telefon +49 . 3 64 24 . 7 92 63
cm@kahlaporzellan.com
Barbara Schmidt
Telefon +49 . 3 64 24 . 7 91 55
bs@kahlaporzellan.com
Seite . Page 72, 74

W. F. **KAISER** UND CO. GMBH
Werner-von-Siemens-Straße 28
D-65582 Diez/Lahn
Telefon +49 . 64 32 . 91 50

Hersteller Manufacturer
Fax +49 . 64 32 . 91 51 99

Vertrieb Distributor
Fax +49 . 64 32 . 91 51 39
info@kaiser-backformen.de
Seite . Page 80

KEV KONSTRUKTION, ENTWICKLUNG UND VERTRIEB EDGAR + ANDREAS RENZ GBR
Jahnstraße 10
D-75417 Mühlacker
Mobil +49 . 175 . 4 13 87 51
andreas.renz@vg-ing.de

Werksdesign Inhouse design
Andreas Renz
Seite . Page 82

WALTER **KNOLL** AG & CO. KG
Bahnhofstraße 25
D-71083 Herrenberg
Telefon +49 70 32 . 20 80
Fax +49 70 32 . 20 82 54
info@walterknoll.de
www.walterknoll.de
Seite . Page 108, 110, 112, 128

KÖNIG + NEURATH AG
Industriestraße 1–3
D-61184 Karben
Telefon +49 . 60 39 . 48 30
Fax +49 . 60 39 . 48 32 14
info@koenig-neurath.de
www.koenig-neurath.de
Seite . Page 136

CATHRIN **KÖNIGSMANN**
Hardenbergstraße 23
D-24118 Kiel
Telefon +49 . 4 31 . 67 33 95
cathrin-k@gmx.de
Seite . Page 202

LANCASTER GROUP/ JIL SANDER COLOUR
Fort Malakoff Park
Rheinstraße 4E
D-55116 Mainz
Telefon +49 . 61 31 . 30 61 12
Fax +49 . 61 31 . 30 62 79
Seite . Page 138

ANDREAS **LEHNHOFF**
Stettinerstraße 29
D-61348 Bad Homburg v. d. H.
Mobil +49 . 179 . 135 42 79
lehnhoff-a@hfg-gmuend.de
Seite . Page 204

LINIE K.
Friedrichstraße 2–4
D-74613 Ohrnberg
Telefon +49 . 79 48 . 94 04 44
Fax +49 . 79 48 . 94 07 77
info@linie-k.de
www.linie-k.de
Seite . Page 138

LOEWE AG
Industriestraße 11
D-96317 Kronach
Telefon +49 . 92 61 . 9 90
Fax +49 . 92 61 . 9 54 11
loewe@loewe.de
www.loewe.de
Seite . Page 172

MAM BABYARTIKEL GMBH
Rudolf-Diesel-Straße 6
D-27383 Scheeßel
Telefon +49 . 42 63 . 9 31 70
Fax +49 . 42 63 . 93 17 20
Seite . Page 38

MAPA GMBH
Industriestraße 21–25
D-27404 Zeven
Telefon +49 . 42 81 . 7 30
Fax +49 . 42 81 . 7 32 41
mapazeven@mapa.de
www.mapa.de
Seite . Page 40

WOLFGANG C. R. **MEZGER**
Salacher Straße 88
D-73054 Eislingen
Telefon +49 . 71 61 . 8 99 98
Fax +49 . 71 61 . 8 99 88
design-mezger@t-online.de
Seite . Page 128

MIELE & CIE. GMBH & CO.
Carl-Miele-Straße 29
D-33332 Gütersloh
Telefon +49 . 52 41 . 8 90
Fax +49 . 52 41 . 89 20 90
info@miele.de
www.miele.de

Werksdesign Inhouse design
Telefon +49 . 52 41 . 89 41 41
Fax +49 . 52 41 . 89 41 40
Seite . Page 64

JOACHIM **MÖLLMANN**
An der Strangriede 44
D-30167 Hannover
Mobil +49 . 1 77 . 5 43 68 27
j_moellmann@hotmail.com
Seite . Page 194

NILS HOLGER **MOORMANN**
Möbel-Produktions- und Handels GmbH
Festhalle
D-83229 Aschau im Chiemgau
Telefon +49 . 80 52 . 9 04 50
Fax +49 . 80 52 . 90 45 45
info@moormann.de
www.moormann.de
Seite . Page 96, 98

NEXTSPACE
Claus Zimmermann und Roland Knieg
Friedrich-Ebert-Straße 100
D-34119 Kassel
www.nextspace.de
email@nextspace.de
Seite . Page 154

KERSTIN **NOLTE**
Hasenhöhe 17a
D-22587 Hamburg
Telefon +49 . 40 . 87 00 97 27
Fax +49 . 40 . 87 00 97 27
kerstin.nolte@hamburg.de
Seite . Page 84

NPK INDUSTRIAL DESIGN
Noordeinde 2d
NL-2311 CD Leiden
Telefon +31 . 71 . 5 14 13 41
Fax +31 . 71 . 5 13 04 10
npk@npk.nl
www.npk.nl
Seite . Page 64

FLOYD **PAXTON**
c/o D & W Designstudio
Oeder Weg 123
D-60318 Frankfurt
Seite . Page 152

PEARSONLLOYD
42–46 New Road
London E1 2AX
Telefon +44 . 20 . 73 77 05 60
Fax +44 . 20 . 73 77 05 50
mail@pearsonlloyd.co.uk
Seite . Page 110

KLAUS **PFEIFER**
Architekturbüro a:HP
Ahastraße 9
D-64285 Darmstadt
Telefon +49 . 61 51 . 31 55 05
Fax +49 . 61 51 . 31 17 62
kontakt@a-hp.de
www.a-hp.de
Seite . Page 44

PHOENIX PRODUCT DESIGN
Kölner Straße 16
D-70376 Stuttgart
Telefon +49 . 7 11 . 9 55 97 60
Fax +49 . 7 11 . 55 93 92
info@phoenixdesign.de
www.phoenixdesign.de
Seite . Page 172

PLANZWEI
Kai Ertel & Matthias Furch
Liststraße 24
D-70180 Stuttgart
Telefon +49 . 7 11 . 6 49 01 19
Fax +49 . 7 11 . 71 19 12
info@planzwei.net
Seite . Page 98

PORSCHE DESIGN GMBH
Ferdinand A. Porsche
Flugplatzstraße 29
A-5700 Zell am See
Telefon +43 . 65 42 . 57 22 70
Fax +43 . 65 42 . 57 22 72
info@porsche-design.at
www.porsche-design.at
Seite . Page 130

POWER JACK OEG
Steyrergasse 74
A-8010 Graz
Telefon +43 . 3 16 . 8 35 83 97
Fax +43 . 3 16 . 8 35 83 94
office@powerjack.at
www.powerjack.at
Seite . Page 176

KURT **RANGER** DESIGN
Stuttgarter Straße 77
D-70469 Stuttgart
Telefon +49 . 7 11 . 9 93 16 30
Fax +49 . 7 11 . 99 31 63 33
contact@ranger-design.com
www.ranger-design.com
Seite . Page 54

REFORM DESIGN
Industriestraße 25
D-70565 Stuttgart
Telefon +49 . 7 11 . 7 82 95 35 23
Fax +49 . 7 11 . 7 82 95 35 25
mathias.wagner@reform-design.de
www.reform-design.de
Seite . Page 34

REISENTHEL ACCESSOIRES
Benzstraße 3
D-82178 Puchheim
Telefon +49 . 89 . 8 00 64 40
Fax +49 . 89 . 80 60 35
info@reisenthel.de
www.reisenthel.de
Seite . Page 50

KATRIN **SAIER**
Hochstraße 68
D-64285 Darmstadt
Telefon +49 . 61 51 . 49 45 59
Mobil +49 . 1 77 . 7 46 91 01
ks@id-1.de
Seite . Page 206

ROY **SCHÄFER**
Weinbergweg 11
D-75181 Pforzheim
Telefon +49 . 72 31 . 56 53 44
Fax +49 . 72 31 . 56 52 94
Seite . Page 170

ALEXANDER **SCHENK**
Produktgestaltung & Designberatung
Grafenbergweg 21
D-73614 Schorndorf
Telefon +49 . 71 81 . 7 24 16
Fax +49 . 71 81 . 7 24 16
info@design-produkte.de
www.design-produkte.de
Seite . Page 170

PROF. WULF **SCHNEIDER** & PARTNER
Schellbergstraße 62
D-70188 Stuttgart
Telefon +49 . 7 11 . 2 86 49 00
Fax +49 . 7 11 . 2 62 10 07
Seite . Page 100

SEDUS STOLL AG
Brückenstraße 15
D-79761 Waldshut
Telefon +49 . 77 51 . 8 42 47
Fax +49 . 77 51 . 8 42 45
info@sedus.de
www.sedus.de
Werksdesign Inhouse design
Michael Kläsener
Telefon +49 . 77 51 . 8 42 63
Fax +49 . 77 51 . 8 43 60
michaelklaesener@sedus.de
Seite . Page 126, 134

MATHIAS **SEILER**
Kirchstraße 17
D-79539 Lörrach
Mobil +49 . 1 70 . 7 94 34 49
seiler.design@t-online.de
Seite . Page 126

SERIEN
RAUMLEUCHTEN GMBH
Hainhäuser Straße 3–7
D-63110 Rodgau
Telefon +49 . 61 06 . 6 90 90
Fax +49 . 61 06 . 69 09 22
serien@serien.com
www.serien.com
Seite . Page 152, 154, 156, 158

SIEMENS ELECTROGERÄTE GMBH
Hochstraße 17
D-81669 München

Hersteller Manufacturer
Telefon +49 . 89 . 45 90 23 87
Fax +49 . 89 . 45 90 29 58

Vertrieb Distributor
Telefon +49 . 89 . 45 90 01
Fax +49 . 89 . 45 90 23 47
www.siemens.de/hausgeraete

Werksdesign Inhouse design
Gerd E. Wilsdorf
gerd.wilsdorf@bshg.com
Seite . Page 68

SOLARPOWER GMBH
Zum Plom 2
D-08529 Plauen
Telefon +49 . 37 41 . 4 15 40
Fax +49 . 37 41 . 41 54 54
info@solarpower-gmbh.de
www.solarpower-gmbh.de
Seite . Page 178

STAEDTLER DEUTSCHLAND VERTRIEBS-GMBH
Moosäckerstraße 3
D-90427 Nürnberg
Telefon +49 . 9 11 . 9 36 52 32
Fax +49 . 9 11 . 9 36 52 30
info@staedtler.de
www.staedtler.de
Seite . Page 146

STAEDTLER MARS GMBH & CO. KG
Moosäckerstraße 3
D-90427 Nürnberg
Telefon +49 . 9 11 . 9 36 50
Fax +49 . 9 11 . 9 36 54 00
info@staedtler.de
www.staedtler.de

Staedtler Design Team
Telefon +49 . 9 11 . 9 36 52 31
Fax +49 . 9 11 . 9 36 54 01
Seite . Page 146

STATION NORD
Auguststraße 28
D-10107 Berlin
Telefon +49 . 30 . 28 39 06 06
Fax +49 . 30 . 28 39 06 07
Seite . Page 142

STEITZ SECURA GMBH
Vorstadt 40
D-67292 Kirchheimbolanden
Telefon +49 . 63 52 . 40 02 26
Fax +49 . 63 52 . 40 02 50
info@steitzsecura.com
www.steitzsecura.de

Werksdesign Inhouse design
Margit Ahnert
Telefon +49 . 63 52 . 4 00 20
margit.ahnert@steitzsecura.com
Seite . Page 180

STEULER FLIESEN GMBH
Industriestraße 77
D-75417 Mühlacker
Telefon +49 . 70 41 . 80 12 05
Fax +49 . 70 41 . 80 12 16
info@steuler-fliesen.de
www.steuler-fliesen.de
Seite . Page 116

STUDIOWERK DESIGN
Forellenstraße 11
D-82266 Inning
Telefon +49 . 81 43 . 18 69
Fax +49 . 81 43 . 87 11
contact@studiowerk-design.de
www.studiowerk-design.de
Seite . Page 182

THOMAS **TRAUTH** DESIGN GMBH
Bahnhofstraße 20/1
D-73547 Lorch
Telefon +49 . 71 72 . 1 84 96 10
Fax +49 . 71 72 . 1 84 96 40
t.trauth@thomas-trauth-design.de
www.thomas-trauth-design.de
Seite . Page 168

TROIKA BÖLL GMBH & CO. KG
D-57629 Müschenbach
Telefon +49 . 26 62 . 9 51 10
Fax +49 . 26 62 . 95 11 95
troika@troika.de
www.troika.de
Seite . Page 170

TTS TOOLTECHNIC SYSTEMS
DEUTSCHLAND GMBH
Wertstraße 20
D-73240 Wendlingen
Telefon +49 . 70 24 . 80 40
Fax +49 . 70 24 . 80 46 00
Seite . Page 182

TWO DESIGN
Claudia de Bruyn, Achim Nagel
Am kleinen Rahm
D-40878 Ratingen
Telefon +49 . 21 02 . 88 97 12
Fax +49 . 21 02 . 88 97 14
Seite . Page 94

BÜRO **UEBELE**
VISUELLE KOMMUNIKATION
Michaela Burger, Andreas Uebele
und Diane Ziegler
Heusteigstraße 94 A
70180 Stuttgart
Telefon +49 . 7 11 . 3 41 70 20
Fax +49 . 7 11 . 34 17 02 30
uebele@uebele.com
www.uebele.com
Seite . Page 144

VAUDE SPORT
GMBH & CO. KG
Vaudestraße 2
D-88069 Tettnang
Telefon +49 . 75 42 . 5 30 60
Fax +49 . 75 42 . 53 06 60
www.vaude.de
Seite . Page 42

BURKHARD **VOGTHERR** &
JONATHAN PRESTWICH
4, Rue du Molkenrain
F-68100 Mulhouse
Telefon +33 . 3 . 89 65 31 55
Fax +33 . 3 . 89 54 21 36
vogtherr.design@wanadoo.fr
www.vogtherrdesign.com
Seite . Page 136

VS VEREINIGTE
SPEZIALMÖBELFABRIKEN
GMBH + CO
Hochhäuser Straße 8
D-97941 Tauberbischofsheim
Telefon +49 . 93 41 . 8 80
Fax +49 . 93 41 . 8 83 14
vs@vs-moebel.de
www.vs-moebel.de
Seite . Page 132

WOLF UDO **WAGNER**
Hanauer Landstraße 161–173
D-60314 Frankfurt/Main
Telefon +49 . 69 . 92 87 05 74
Fax +49 . 69 . 92 87 05 75
wagner@wagner-design.de
www.wagner-design.de
Seite . Page 114

WAGNER-DESIGN
Hanauer Landstraße 161–173
D-60314 Frankfurt/Main
Telefon +49 . 69 . 92 87 05 74
Fax +49 . 69 . 92 87 05 75
marketing@wagner-design.de
www.wagner-design.de
Seite . Page 114

WEBER + HÄTINGER
Metzinger Straße 36
D-72622 Nürtingen
Telefon +49 . 70 22 . 50 37 00
Fax +49 . 70 22 . 50 37 08
info@weber-haetinger.de
www.weber-haetinger.de
Seite . Page 104

WIEGE
Hauptstraße 81
D-31848 Bad Münder
Telefon +49 . 50 42 . 99 99 00
Fax +49 . 50 42 . 99 99 01
wiege@wiege.com
www.wiege.com
Seite . Page 122, 124

WIGL-DESIGN
Steyrergasse 74
A-8010 Graz
Telefon +43 . 3 16 . 8 35 83 91
Fax +43 . 3 16 . 8 35 83 94
orleb@wigl-design.at
www.wigl-design.at
Seite . Page 176

M. H. WILKENS & SÖHNE GMBH
An der Silberpräge
D-28309 Bremen
Telefon +49 . 4 21 . 4 10 30

Hersteller Manufacturer
Fax +49 . 4 21 . 4 10 31 30

Vertrieb Distributor
Fax +49 . 4 21 . 4 10 31 67
contact@wilkens-design.de
www.wilkens-design.de
Seite . Page 84

WILKHAHN
Wilkening + Hahne GmbH + Co
Fritz-Hahne-Straße 8
D-31848 Bad Münder
Telefon +49 . 50 42 . 99 91 18
Fax +49 . 50 42 . 99 92 30
info@wilkhahn.de
www.wilkhahn.de
Seite . Page 122, 124

KARSTEN WINKELS
Leibnizstraße 8A
D-44147 Dortmund
Telefon +49 . 2 31 . 7 28 29 98
Fax +49 . 2 31 . 72 82 99 82
karsten.winkels@bauwerkstatt.net
Seite . Page 160

WMF WÜRTTEMBERGISCHE METALLWARENFABRIK AG
Eberhardstraße
D-73312 Geislingen/Steige
Telefon +49 . 73 31 . 2 51
Fax +49 . 73 31 . 4 53 87
info@wmf.de
www.wmf.de
Seite . Page 76

YELLOW CIRCLE
Georgstraße 5a
D-50676 Köln
Telefon +49 . 2 21 . 9 21 37 80
Fax +49 . 2 21 . 9 21 37 81
info@yellowdesign.com
www.yellowdesign.com
Seite . Page 40

ZUMTOBEL STAFF DEUTSCHLAND VERTRIEBS GMBH
Grevenmarschstraße 74–78
D-32657 Lemgo
Telefon +49 . 52 61 . 21 20
Fax +49 . 52 61 . 2 12 90 00
Seite . Page 90, 92

ZUMTOBEL STAFF GMBH
Schweizer Straße 30
A-6850 Dornbirn
Telefon +43 . 55 72 . 39 00
Fax +43 . 55 72 . 2 28 26
info@zumtobelstaff.co.at
www.zumtobelstaff.com
Seite . Page 90, 92

Die fettgedruckten Seitenzahlen verweisen auf den Katalogteil (Abbildungen), die mager gedruckten Seitenzahlen führen zum Adress-Verzeichnis.

FOTONACHWEIS . PHOTOGRAPHS (ACKNOWLEDGEMENTS)

soweit von den Einreichern erwünscht
information provided by competitors only

Die Arge Lola
Stuttgart
Seite . Page 31, 161

ArtKontor
Jena
Seite . Page 73

Fotostudio Schweiger GmbH
Hohberg
Seite . Page 87

Willebrand Photographie
Köln
Seite . Page 137

Volker Schrank
Stuttgart
Seite . Page 23, 25, 26, 27, 28–29, 36–37, 48–49, 62–63, 88–89, 120–121, 140–141, 150–151, 166–167, 174–175, Cover

HERAUSGEBER . EDITOR
Design Center Stuttgart
des Landesgewerbeamts
Baden-Württemberg
Willi-Bleicher-Straße 19
D-70174 Stuttgart
Telefon +49 . 7 11 . 1 23 26 84

TEXT + REDAKTION . TEXT + EDITORIAL WORK
Andrea Scholtz M.A.
Stuttgart

ÜBERSETZUNG . TRANSLATION
Philip Mann B.A., Ph.D.
Marbach

FOTOS ZU DEN PREISTRÄGERN + COVER . PHOTOS FOR THE AWARD WINNERS + COVER
Volker Schrank
Stuttgart

GRAFISCHE GESTALTUNG . LAYOUT
nalbach typografik
Silke Nalbach, Sophia Haas
Stuttgart

OFFSET-REPRODUKTIONEN . OFFSET REPRODUCTIONS
Repromayer
Reutlingen

DRUCK . PRINTING
Rung-Druck
Göppingen

Dieses Buch wurde gedruckt auf 100% chlorfrei gebleichtem Papier und entspricht damit dem TCF-Standard.

This book has been printed on paper that is 100% free of chlorine bleach in conformity with TCF standard.

Bibliographische Information
Der Deutschen Bibliothek
Die Deutsche Bibliothek verzeichnet diese Publikation in der Deutschen Nationalbibliografie; detaillierte bibliografische Daten sind im Internet über http://dnb.ddb.de abrufbar.

Bibliographic information published by
Die Deutsche Bibliothek
Die Deutsche Bibliothek lists this publication in the Deutsche Nationalbibliografie; detailed bibliographic data is available in the Internet at http://dnb.ddb.de.

ISBN 3-89790-195-1

Made in Germany, 2003

Die Publikation erscheint anlässlich der Ausstellung „Focus Balance – Internationaler Designpreis Baden-Württemberg 2003 und Mia Seeger Preis 2003"
14. September 2003 bis 7. Dezember 2003

The publication is published on the occasion of the exhibition "Focus Balance – Baden-Württemberg International Design Award 2003 and Mia Seeger Prize 2003"
14 September 2003 to 7 December 2003

VERANSTALTER . ORGANIZER
Design Center Stuttgart
des Landesgewerbeamts
Baden-Württemberg
Willi-Bleicher-Straße 19
D-70174 Stuttgart
Telefon +49 . 7 11 . 1 23 26 84

PROJEKTLEITUNG . PROJECT HEAD
Sabine Lenk

ORGANISATION . ADMINISTRATION
Hildegard Hild
Michael Kern

AUSSTELLUNGSGESTALTUNG . EXHIBITION DESIGN
Angelika Ertl
Landesgewerbeamt Baden-Württemberg